PHYSICS AND CHEMISTRY OF
EARTH MATERIALS

CAMBRIDGE TOPICS IN MINERAL PHYSICS AND CHEMISTRY

Editors
Dr. Andrew Putnis
Dr. Robert C. Liebermann

Physics and chemistry of Earth materials

ALEXANDRA NAVROTSKY

Princeton University

CAMBRIDGE
UNIVERSITY PRESS

Published by the Press Syndicate of the University of Cambridge
The Pitt Building, Trumpington Street, Cambridge CB2 1RP
40 West 20th Street, New York, NY 10011-4211, USA
10 Stamford Road, Oakleigh, Melbourne 3166, Australia

First published 1994

Printed in the United States of America

Library of Congress Cataloging-in-Publication Data

Navrotsky, Alexandra.

Physics and chemistry of earth materials / Alexandra Navrotsky.

p.   cm. — (Cambridge topics in mineral physics and chemistry : 6)
Includes bibliographical references.
ISBN 0-521-35378-5. — ISBN 0-521-35894-9 (pbk.)
1. Matter. 2. Materials. 3. Minerals. 4. Geochemistry.
I. Title. II. Series.
QC171.2.N38   1994
549′.13—dc20   93-43135
                                                              CIP

A catalog record for this book is available from the British Library

ISBN 0-521-35378-5 hardback
ISBN 0-521-35894-9 paperback

Contents

Acknowledgments

Sheryl Wasylenko typed and typed and typed and withstood an author's moods. Marina Newton and Eugene Smelik helped draft figures. The following people read drafts of the manuscript, made useful suggestions and additions, helped with figures, and provided sanity checks, humor, and proofreading: Juliana Boerio-Goates, Pamela Burnley, Liang Chai, Ron Cohen, Jerry Gibbs, Tony Hess, Raymond Jeanloz, Michael Hochella, Rebecca Lange, Vincent Lamberti, Robert Liebermann, Maureen McCarthy, Elena Petrovicova, Eugene Smelik, and Igor Tarina. The participants in GEO 501 in Spring 1993 read a draft of this book and caught mistakes and obscurities in it and in my lectures. My mother and my dogs put up with me and missed many walks because I was too busy. I thank all my colleagues who kept assuring me that the book was important and kept hounding me about when it would be ready. I hope they are not disappointed.

Illustration credits

Figure 2.2d From H. L. James, *Posts and rugs, the story of Navajo rugs and their homes* (p. 57). Copyright © 1976 by H. L. James. Reprinted by permission.

Figure 2.5 From *International tables for crystallography* (pp. 687–8) edited by Theo Hahn. Copyright © 1987 by D. Reidel Publishing Company. Reprinted by permission of Kluwer Academic Publishers.

Figure 2.11 Reprinted from Linus Pauling, *The nature of the chemical bond,* third edition (p. 410). Copyright © 1960 by Cornell University. Used by permission of the publisher, Cornell University Press.

Figure 2.39c From B. G. Hyde, T. J. White, M. O'Keeffe, and A. W. S. Johnson, Structures related to those of spinel and the β-phase, and a possible mechanism for the transformation olivine↔spinel, *Zeitschrift fur Kristallographie 160* (p. 55). Copyright © 1982 by R. Oldenbourg Verlag GmbH. Reprinted by permission.

Figure 2.39d From H. Horiuchi, M. Akaogi, and H. Sawamoto, Crystal structure studies on spinel-related phases, spinelloids: Implications to olivine–spinel phase transformation and systematics, *Advances in Earth and Planetary Science 12* (p. 398). Copyright © 1982 by Kluwer Academic Publishers. Reprinted by permission of Kluwer Academic Publishers.

Tables 2.13 and 2.14 From *Silicate crystal chemistry* by Dana T. Griffen (pp. 155, 205, 207). Copyright © 1992 by Oxford University Press, Inc. Reprinted by permission.
Table 2.16 From E. Eslinger and D. Pevear, *Clay minerals for petroleum geologists and engineers,* SEPM Short Course Notes No. 22 (p. 2–12). Copyright © 1988 by the Society for Sedimentary Geology. Reprinted by permission.
Figure 3.4 Reprinted from G. E. Bacon, *X-ray and neutron diffraction,* copyright © 1966, p. 70, with kind permission from Pergamon Press, Ltd, Headington Hill Hall, Oxford OX3 OBW, UK.
Figure 3.5a,b From D. R. Stewart and T. L. Wright, Al/Si order and symmetry of natural alkali feldspars and relationship of strained parameters to bulk composition cell, *Bulletin de la Société Française de Mineralogie et de Cristallographie 97* (p. 363). Copyright © 1974 by E. Schweiserbard'sche Verlagsbuchhandlung. Reprinted by permission.
Figure 3.6a From J. J. Papike, M. Ross, and J. M. Clark, *Crystal-chemical characterization of clinoamphiboles based on five new structure refinements,* Special Paper 2 (p. 128, fig. 4), copyright © 1969 by the Mineralogical Society of America. Reprinted by permission.
Figure 3.6b From F. C. Hawthorne and R. Grundy, The crystal chemistry of the amphiboles I: Refinement of the crystal structure of ferrotschermakite, *Mineralogical Magazine 39* (p. 44). Copyright © 1973 by The Mineralogical Society. Reprinted by permission.
Figure 3.7a From Y. Matsui and Y. Syono, Unit cell parameters of some synthetic olivine group solid solutions, *Geochemical Journal 2 (p. 56).* Copyright © 1968 by the Geochemical Society of Japan. Reprinted by permission.
Figure 3.11b From E. A. Smelik and D. R. Veblen, A transmission and analytical electron microscope study of exsolution microstructures and mechanisms in the orthoamphiboles and anthophyllite and gedrite, *American Mineralogist 78* (p. 522, fig. 16a), copyright © 1993 by the Mineralogical Society of America. Reprinted by permission.
Figure 3.12a,b From A. M. Hofmeister, Single crystal absorption and reflection infrared spectroscopy of forsterite and fayalite, *Physics and Chemistry of Minerals 14* (pp. 501, 502). Copyright © 1987 by Springer-Verlag. Reprinted by permission.
Figure 3.12c From K. Iishi, Lattice dynamics of forsterite, *American Mineralogist 63* (p. 1199, fig. 1), copyright © 1978 by the Mineralogical Society of America. Reprinted by permission.
Figure 3.13 and 3.25 From H. Rager, S. Hosoya, and G. Weiser, Electron paramagnetic resonance and polarized optical absorption spectra of Ni^{2+} in synthetic forsterite, *Physics and Chemistry of Minerals 15* (p. 386). Copyright © 1988 by Springer-Verlag. Reprinted by permission.
Figure 3.14 From R. G. Burns, D. A. Nolet, M. Parkin, C. A. McCammon, and K. B. Schwartz, Mixed-valence minerals of iron and titanium: Correlations of structural, Mössbauer and electronic spectral data, in D. B. Brown, ed., *Mixed-valence compounds: Theory and application in chemistry, physics, biology.* Copyright © 1980 by D. Reidel Publishing Company. Reprinted by permission of Kluwer Academic Publishers.
Figure 3.17 and 3.18 From G. E. Brown, G. Calas, G. A. Waychunas, and J. Petiau, X-ray absorption spectroscopy and its applications in mineralogy and geochemistry, *Reviews in Mineralogy 18* (p. 434, fig. 1-1; p. 440, fig. 2-2), copyright © 1988 by the Mineralogical Society of America. Reprinted by permission.
Figure 3.19a,b From G. A. Waychunas, M. J. Apted, and G. E. Brown, Jr., X-ray K-edge absorption spectra of Fe minerals and model compounds: Near-edge structure, *Physics and Chemistry of Minerals 10* (pp. 4, 5). Copyright © 1983 by Springer-Verlag. Reprinted by permission.

Figure 3.19c From G. Calas, G. E. Brown, Jr., G. A. Waychunas, and J. Petiau, X-ray absorption spectroscopic studies of silicate glasses and minerals, *Physics and Chemistry of Minerals 15* (pp. 20, 25). Copyright © 1987 by Springer-Verlag. Reprinted by permission.

Figure 3.21 Reprinted from *Geochimica et Cosmochimica, 52,* M. E. Brandriss and J. F. Stebbins, Effects of temperature on the structure of silicate liquids: ^{29}Si NMR P results, p. 529, copyright © 1988, with kind permission from Pergamon Press Ltd, Headington Hill Hall, Oxford OX3 OBW, UK.

Figure 3.22 From B. L. Phillips, D. A. Howell, R. J. Kirkpatrick, and T. Gasparik, Investigation of cation order in $MgSiO_3$-rich garnet using ^{29}Si and ^{27}Al MAS NMR spectroscopy, *American Mineralogist 77* (p. 706, fig. 1), copyright © 1992 by the Mineralogical Society of America. Reprinted by permission.

Figure 3.24 From R. J. Kirkpatrick, MAS NMR spectroscopy of minerals and glass, *Reviews in Mineralogy 18* (p. 360, fig. 12), copyright © 1988 by the Mineralogical Society of America. Reprinted by permission.

Figure 3.27a From E. Murad, Magnetic ordering in andradite, *American Mineralogist 69* (p. 723, fig. 1), copyright © 1984 by the Mineralogical Society of America. Reprinted by permission.

Figure 3.27b From L. Zhang and S. S. Hafner, Gamma resonance of ^{57}Fe in grunerite at high pressures, *American Mineralogist 77* (p. 475, fig. 1), copyright © 1992 by the Mineralogical Society of America. Reprinted by permission.

Figure 3.27c From M. D. Dyar, C. L. Perry, C. R. Rebbert, B. L. Dutrow, M. J. Holdaway, and H. M. Lang, Mössbauer spectroscopy of synthetic and naturally occurring staurolite, *American Mineralogist 76* (p. 36, fig. 2), copyright © 1991 by the Mineralogical Society of America. Reprinted by permission.

Table 3.4 From R. J. Kirkpatrick, MAS NMR spectroscopy of minerals and glass, *Reviews in Mineralogy 18* (p. 357, table 1), copyright © 1988 by the Mineralogical Society of America. Reprinted by permission.

Figures 4.4 and 4.6 From J. R. Holloway and B. J. Wood, *Simulating the Earth experimental geochemistry* (pp. 26, 30). Copyright © 1988 by Chapman and Hall, Inc. Reprinted by permission.

Figures 5.6 and 5.7 From *Electronic structure to the properties of solids* by Harrison. Copyright © 1980 by W. H. Freeman and Company. Reprinted with permission.

Figure 5.11a, b From D. G. Isaak, R. E. Cohen, M. J. Mehl, and D. J. Singh, Phase stability of wustite at high pressure from first principles LAPW calculations, *Physical Review B47,* copyright © 1993 by The American Physical Society.

Figure 5.12a From P. F. McMillan and A. M. Hofmeister, Infrared and raman spectroscopy, *Reviews in Mineralogy 18* (p. 104, fig. 3a–e), copyright © 1988 by the Mineralogical Society of America. Reprinted by permission.

Figure 5.12b Reprinted from *Solid-State Communications 63,* S. Ghose, J. M. Hastings, L. M. Corliss, K. R. Rao, S. L. Chaplot, and N. Choudbury, Study of phonon dispersion relations in forsterite, Mg_2SiO_4 by inelastic neutron scattering, p. 1047, copyright © 1987, with kind permission from Pergamon Press, Ltd, Headington Hill Hall, Oxford OX3 OBW, UK.

Figure 5.12c From K. R. Rao, S. L. Chaplot, N. Choudbury, S. Ghose, and D. L. Price, Phonon density of states and specific heat of forsterite, Mg_2SiO_4 *Science 236* (p. 64), copyright © 1987 by the American Association for the Advancement of Science. Reprinted by permission.

Figures 5.15 and 5.17 From R. S. Berry, S. A. Rice, and J. Ross, *Physical chemistry* (pp. 176, 297), copyright © 1980 by John Wiley & Sons, Inc. Reprinted by permission of John Wiley & Sons, Inc.

Figures 5.21 and 5.44 From G. V. Gibbs, E. P. Meagher, M. D. Newton, and D. K.

Swanson in O'Keeffe and Navrotsky, eds., *Structure and bonding in crystals,* vol. 1, copyright © 1981 by Academic Press. Reprinted by permission.

Figure 5.25a,b From K. L. Geisinger and G. V. Gibbs, SiSSi and SiOSi bonds in molecules and solids: A comparison, *Physics and Chemistry of Minerals 7* (pp. 205, 207). Copyright © 1981 by Springer-Verlag. Reprinted by permission.

Figure 5.28a From X. Liu and C. T. Prewitt, High-temperature X-ray diffraction study of Co_3O_4: Transition from normal to disordered spinel, *Physics and Chemistry of Minerals 17* (p. 170). Copyright © 1990 by Springer-Verlag. Reprinted by permission.

Figure 5.37 From G. V. Gibbs, M. A. Spackman, and M. B. Boisen, Jr., Bonded and promolecule radii for molecules and crystals, *American Mineralogist 77* (p. 742, fig. 1), copyright © 1992 by the Mineralogical Society of America. Reprinted by permission.

Figure 5.38a,b From M. A. Spackman, R. J. Hill, and G. V. Gibbs, Exploration of structure and bonding in stishovite with fourier and pseudoatom refinement methods using single crystal and powder X-ray diffraction data, *Physics and Chemistry of Minerals 14* (p. 142). Copyright © 1987 by Springer-Verlag. Reprinted by permission.

Figure 5.38c From R. E. Cohen, Bonding and elasticity of stishovite SiO_2 at high pressure: Linearized augmented plane wave calculations, *American Mineralogist 76* (p. 740, fig. 6a—d), copyright © 1991 by the Mineralogical Society of America. Reprinted by permission.

Figure 5.41 From W. B. Pearson, *The crystal chemistry and physics of metals and alloys* (p. 247), copyright © 1972 by John Wiley & Sons, Inc. Reprinted by permission of John Wiley & Sons, Inc.

Figure 5.43a From A. Zunger in O'Keeffe and Navrotsky, eds., *Structure and bonding in crystals,* vol. 1, copyright © 1981 by Academic Press. Reprinted by permission.

Figure 5.43b Reprinted with permission from J. K. Burdett, New ways to look at solids, *Accounts of Chemical Research 15* (p. 39). Copyright © 1982 American Chemical Society.

Figure 5.45a,b From R. M. Hazen, A useful fiction: Polyhedral modeling of mineral properties, *American Journal of Science 288A* (p. 254), copyright © 1988. Reprinted by permission of *American Journal of Science.*

Figure 5.46 From R. M. Hazen and C. T. Prewitt, Effect of temperature and pressure on interatomic distances in oxygen-based minerals, *American Mineralogist 62* (p. 311, Fig. 1), copyright © 1977 by the Mineralogical Society of America.

Tables 5.1 and 5.7 From *Theoretical geochemistry: Applications of quantum mechanics in the earth and mineral sciences* by J. A. Tossell and D. J. Vaughan (pp. 11, 454–7). Copyright © 1992 by Oxford University Press, Inc. Reprinted by permission.

Figure 6.8 From M. M. Elcombe, Some aspects of the lattice dynamics of quartz, *Proc. Phys. Soc. 91* (p. 954, Fig. 3). Copyright © 1967 by IOP Publishing Limited. Reprinted by permission.

Figure 6.9 From N. L. Ross in *Stability of minerals* (pp. 150—1). Copyright © 1993 by Chapman and Hall, Inc. Reprinted by permission.

Figure 6.13 From Arnulf Muan and E. F. Osborn, *Phase equilibria among oxides in steelmaking* (Fig. 13), © 1965 by Addison-Wesley Publishing Company, Inc. Reprinted by permission.

Figure 6.24 From Y. Fei, H.-K. Mao, and B. O. Mysen, *Journal of Geophysical Research 96* (p. 2167), 1991, copyright by the American Geophysical Union.

Figure 7.2 From D. R. Gaskell, *Introduction to metallurgical thermodynamics,* 2d ed. (p. 387). Copyright © 1981 by McGraw-Hill, Inc. Reproduced with permission of McGraw-Hill, Inc.

Figure 7.15a From B. P. Burton, Theoretical analysis of cation ordering in binary rhombohedral carbonate systems, *American Mineralogist 72* (p. 331, fig. 2, fig. 3), copyright © 1987 by the Mineralogical Society of America. Reprinted by permission.

Figure 8.1c From Y. Waseda, *The structure of non-crystalline materials, liquids and amorphous solids* (p. 134). Copyright © 1980 by McGraw-Hill, Inc. Reproduced with permission of McGraw-Hill, Inc.

Figure 8.2 From D. L. Griscom, Borate glass structure, *Materials Science Research 12* (p. 45). Copyright © 1977 by Plenum Publishing Corporation. Reprinted by permission.

Figure 8.9 From C. A. Angell, Strong and fragile liquids, in Ngai and Wright, eds., *Relaxation in complex systems,* copyright © 1985 National Technical Information Service. Reprinted by permission.

Figure 8.11 From M. A. McMillan and J. R. Holloway, Water solubility in aluminosilicate melts, *Contribution Mineralogy and Petrology 97 (pp. 325, 326),* copyright © 1987 by Springer-Verlag. Reprinted by permission.

Figures 8.13 and 8.14 From C. Meade, R. J. Hemley, and H. K. Mao, High pressure X-ray diffraction of SiO_2 glass, *Physical Review Letters 69* (p. 1389), copyright © 1992 by The American Physical Society.

Figure 8.16 From A. Putnis, *Introduction to mineral sciences* (p. 335). Copyright © 1992 Cambridge University Press. Reprinted with the permission of Cambridge University Press.

1

Introduction

Minerals, naturally occurring solid-state compounds, form the firm earth under our feet, the hard rocks that make up continents, the ores of commercial importance, and the inaccessible depths of the earth and other planets. Mineralogy and crystallography have long been linked both esthetically, because many minerals occur as beautiful crystals, and intellectually, because the periodic arrangement of atoms in a crystal provides the fundamental explanation for all mineral properties. Indeed, elucidating crystal structures has been the main business of mineralogy for fifty years, from the first X-ray diffraction experiments around 1910 to the automation of diffraction techniques that came with the first small computers in the late sixties. This technological development spelled both the end and the beginning of an era. The end came in the sense that solving or refining a crystal structure no longer represents a laborious task in itself worthy of an advanced degree. The beginning was, and is, the opportunity to use crystallography as a tool, often in a comparative sense, to study many related structures and how they vary with imposed constraints of pressure, temperature, and composition. X-ray crystallography can now be used in combination with other emerging structural techniques such as neutron diffraction and electron microscopy, with spectroscopic studies, and with work on phase equilibria and thermochemistry. Crystallography and many spectroscopic tools have been rejuvenated by access to much higher intensity sources of radiation, such as those provided by synchrotrons. New synthetic methods, both at atmospheric and high pressure, provide the means to synthesize materials, grow crystals, and map out their regions of stability. Analytical techniques enable one to determine with increasing sensitivity and spatial resolution the composition and homogeneity (or lack thereof) of a natural crystal or its synthetic counterpart with ever-increasing speed, sensitivity, and accuracy.

While instrumentation has blossomed, solid-state theory has advanced to the point of being able to deal with crystal structures much more complex than the simple metals and III–V semiconductors that early solid-state physics largely addressed. Band theory and lattice dynamics taken from solid-state physics vie with ideas of chemical bonding and molecular orbital theory taken from the physical chemist's toolbox. These ideas challenge the traditional mineralogical faith in the ionic model with its arsenal of powerful empirical systematics, including ionic radii and Pauling's rules. Each of these approaches is facilitated, indeed made possible in the modern context, by increasing computational power.

Earth science as a whole has undergone several major revolutions in the last twenty-five years. Plate tectonics provides a set of unifying descriptions of geologic processes. Though the manifestations of these processes are now quite well documented, their underlying causes and mechanisms probably lie deep in the Earth's interior, tied to as yet poorly understood phenomena of convection and energy balance in the lower mantle and core. These regions are inaccessible to the geologist's rock hammer, however, their conditions can be simulated both in the laboratory and by computer. Such studies have provided many new, and often surprising, insights about the behavior of rocks and minerals under extremes of pressure and temperature.

Emerging from classical approaches to metallurgy, ceramics, and polymers, modern materials science seeks to understand the links among structure, properties, and processes in a wide range of potentially technologically useful solids. Its concern with materials at high temperature, in terms of structure-property relations, and of defects and their influence on mechanical strength and electrical and magnetic behavior is directly relevant to earth materials. Glass science and the study of natural silicate melts, that is, magmas, share common concerns. Catalysis and surface chemistry share common phenomena with studies of the formation and dissolution of fine-grained low-temperature minerals.

Under these influences, mineralogy has been reborn as a broader and more sophisticated science with, especially in the United States, the name *mineral physics* or *physics and chemistry of earth materials (PACEM)*. In the most general sense, this field addresses the following five questions: (1) What is the structure of a given mineral or related synthetic material, including fine details of order–disorder, heterogeneity, or other microscopic complications? (2) Why does the material have its observed structure in terms of interactions at the molecular level and the effect of pressure and temperature on closely balanced forces? (3) How are the physical properties of a mineral related to its structure and to pressure and temperature? (4) How do minerals react, both in terms of rates and mechanisms of processes such as dissolution-precipitation, melting-crystallization, grain growth and removal of porosity, and flow under either brittle or ductile regimes? (5) How can this detailed knowledge of the microscopic basis of mineral behavior be applied to understand geologic process on both a global and a local scale?

You as a student of mineral physics are challenged as are students in any other interdisciplinary and rapidly evolving field. You must become conversant with ideas from chemistry, physics, and materials science while keeping the geological context firmly in mind. For the student entering the field from geology, the necessary mathematics, physics, and chemistry need to be mastered. The geophysicist must learn mineralogy, petrology, and the lore of rocks. The chemist, physicist, or engineer must learn geological and mineralogical jargon and become attuned to the vast time and distance scales and immense variations of temperature and pressure that planets experience.

This book emerges from a one-semester course I taught in two different contexts: in the Chemistry Department at Arizona State University as a course in solid-state chemistry and in the Department of Geological and Geophysical Sciences and the Princeton Materials Institute at Princeton University as a course in the chemistry and physics of minerals and materials. In both places the class has been a mixture of geologists, chemists, physicists, and engineers. The course is an introduction to mineral physics and solid-state chemistry aimed at seniors and first-year graduate students. I seek to review crystal chemistry in a context somewhat more general than that of the rock-forming minerals, to summarize quantitative concepts of chemical bonding, to discuss physical properties, and to relate microscopic structural features to macroscopic thermodynamic behavior. At the same time I hope to acquaint the student, albeit in a superficial fashion, with the tantalizing variety of modern structural and spectroscopic tools. I conclude with discussions of high-pressure phase transitions, of amorphous materials (melts and glasses), and of solid-state reactions. To achieve this breadth, I have chosen to take an approach more phenomenological than rigorous. My hope in my lectures and in this book is that this introductory survey, by identifying a set of fascinating problems, phenomena, and approaches, will spur you to specialize further in some subset of the field.

The students in my class, and you, the reader, will generally have been exposed to at least one year each of college chemistry and physics and to mathematics through calculus. The geologists will have had an introduction to mineralogy and phase equilibria. The chemists and physicists probably have learned a little about condensed matter in physical chemistry or solid-state physics. At best, these courses will have given you some fundamental concepts and understanding. At worst, they will have conferred immunity to the subjects covered, but left you some books you may wish to reread or use as references. In any case, a subject is learned in stages, by going over the same material with increasing levels of understanding. Truly, the basic questions remain the same as you progress, but the answers become deeper, more interconnected, more satisfying.

Specifically, the book is organized as follows. Chapter 2 is concerned with crystal chemistry. A brief review of symmetry, crystallographic nomenclature, and some useful equations sets the stage for a discussion of crystal structures. Starting with concepts of close packing and substitution of atoms, the crystal structures of the elements and simple binary and ternary compounds are described and compared. The systematics of silicate structures are reviewed, and various ways of depicting and describing them are presented. Useful geometric concepts, such as bond lengths, radii, and coordination polyhedra, are introduced. Ideas related to sublattices, superlattices, and order–disorder are presented. The chapter ends with a discussion of relations among structures.

Chapter 3 deals with experimental means of studying structure. Diffraction techniques using X rays, neutrons, and electrons, as well as electron microscopy are summarized. Spectroscopy (ultraviolet, optical, infrared, and Raman methods, as well as the newer X-ray absorption and photoelectron spectroscopies) are discussed. Solid-state nuclear magnetic resonance is introduced, as is electron paramagnetic resonance. In each case the emphasis is both on the fundamental principles and on which methods are especially suited to what sorts of problems and materials.

Chapter 4 is a survey of means of studying thermodynamic properties of minerals. Phase equilibria, high-pressure methods, and calorimetry are discussed. Various approaches to pressure-volume-temperature relations are described.

The fifth chapter deals with chemical bonding. First, concepts from solid-state physics are introduced to describe both electrons and lattice vibrations in solids in terms of constraints imposed by the periodicity of the crystal. Band theory and the distinctions among metals, insulators, semiconductors, and superconductors are discussed. The role of defects and impurities, leading to activated transport processes, is considered. The metal-insulator transition is discussed. Possible implications of these phenomena for earth processes are emphasized. Advances in computational capabilities are outlined. Second, bonding is looked at from a local point of view in terms of small groups of atoms and their molecular orbitals. This approach is particularly useful for structures containing linked silicate tetrahedra, and the insights and implications of these cluster calculations are considered. This leads naturally to a discussion of covalency and directional bonding. A local bonding approach is also useful for systems containing transition metals, and ensuing ideas of crystal field and ligand field theory are enumerated. Third, the ionic approach to crystal chemistry is assessed, with emphasis on modern appraisals of Pauling's rules, of ionic radii, and of lattice energy calculations. The three approaches are compared, and their differences are found to be far more superficial than one might first imagine. The chapter closes with a discussion of comparative crystal chemistry, of how and why specific bonding arrangements evolve over a wide range of conditions.

Chapter 6 looks into mineral thermodynamics, in particular, linking structure on the microscopic scale with macroscopic thermodynamic properties. The behavior of heat capacity is discussed in some detail. The relation of lattice vibrations to entropy is demonstrated, using both simple and sophisticated models. Systematics in heats and free energies of formation of binary and ternary compounds are presented. Phase transitions of geophysical importance are discussed, and the general systematics of phase transition energetics is stressed. These are linked to mantle mineralogy.

Chapter 7 deals with the intertwined topics of solid solution formation and order–disorder. Enough thermodynamic formalism is presented to give the reader some sense of the similarities and differences among different approaches and to realize the importance of "making the punishment fit the crime" when choosing a model to apply to experimental data. These concepts are illustrated by application to spinels, carbonates, and feldspars. Chapters 6 and 7 go into somewhat more detail than does my one-semester course, and they borrow from a separate course I have taught on thermodynamics of geological and ceramic materials.

Chapter 8 considers amorphous materials, particularly silicate glasses and melts. What is known about their structures on a local, mid-range, and long-range scale? How are melts and glasses related and how do their relaxation processes differ? How do silicate systems compare to inorganic, metallic, and polymeric melts and glasses? How does one describe the thermodynamics of melting, vitrification, and the glass transition, and how are these related to concepts of structure? What is the role of water and carbon dioxide in silicate melts and glasses? How does melt structure change with pressure?

The final chapter explores the interface between mineral physics and materials science. The synthesis and tailoring of new materials for specific applications increasingly demand phases of great structural complexity, such as zeolites for catalysis and perovskite-related multicomponent phases for superconductivity. The mineral physicist's crystal chemical insight and experimental ability to control pressure, temperature, and chemical potential complements the materials scientist's emphasis on structure-property relations. Problems of mineral resources, of the technological uses of natural earth materials, of the management of nuclear and chemical waste – all present challenges and opportunities to mineral physics.

The very important subjects of solid-state reactions and transport properties are given little attention. They could (and do) form books and courses in their own right. To keep this volume within bounds, and because I never quite do justice to these subjects in class, either, I have chosen to exclude kinetic aspects of the solid state from systematic discussion.

I should warn you of some of my biases, which I am sure I have brought to this book. I am a thermodynamicist and solid-state chemist with a point of view that comes from training in physical chemistry. I have worked extensively on high-pressure phase transitions and mantle mineralogy, on solid solution thermodynamics and order–disorder, and on properties of silicate melts and glasses. I have a passion for spinels, perovskites, and dogs, especially German shepherds.

Because this book aims at an overview of a large and diverse subject, it can not be exhaustive in the details. The references at the end of each chapter are texts, monographs, and review articles selected to explore various topics in more depth. They are grouped as General References and Bibliography (books and reviews) and Specific References. Not all the former are cited specifically in the text, but they form a basic reading list for the next level of detail in each area. The Reviews in Mineralogy series published by the Mineralogical Society of America and Short Course Handbook series published by the Mineralogical Association of Canada provide excellent summaries of selected topics and are the logical next step for the student needing more information. After that, review articles and the scientific literature offer endless detail and examples. It is a characteristic of modern science that you can not become an expert in all fields relevant to your interests. Yet at the same time you must be conversant with their general substance and with what they can do for you. If this book exposes you to a set of concepts, tools, and phenomena, motivates you to delve deeper, and provides a useful set of road maps, signposts, and references, I will consider my purpose fulfilled.

2

Crystal chemistry

2.1 Crystals: real and idealized

Physically, a *crystal* is an object of finite size containing various impurities and imperfections but adhering to an approximately regular arrangement of fundamental units (atoms, molecules, or ions). These units undergo vibrations about their equilibrium positions, the magnitude of which increases with increasing temperature. Mathematically, a *lattice* is an infinite space-filling construct adhering to certain operations that relate equivalent points. These operations are described in terms of symmetry elements about a point (rotations and reflections) that define a *point group* and symmetry operations involving movement along a vector (translations, glides, and screws) that, together with the point symmetry, define a *space group*. Because of its almost perfect periodicity, a real crystal can have associated with it a lattice that describes its symmetry or, more precisely, describes what the symmetry of a perfect crystal would be. The actual arrangement of atoms in the crystal, subject to the symmetry constraints of the lattice, is the *crystal structure*. A given space group or lattice can accommodate many different configurations of points corresponding to different atomic arrangements and therefore to distinct crystal structures. Both the idealized perfect crystal structure and deviations from that perfection (defects, disorder, interfaces) are of immense fundamental and practical importance.

2.2 Symmetry, space groups, and lattices

Symmetry about a point can be described in terms of rotations and mirror planes. An n-fold rotation is one of $360/n$ degrees (n an integer), where n such rotations are required to bring the object back to its original position. However, if that n-fold rotation is a symmetry element of the object in question, each of the intermediate positions is indistinguishable from the original; thus a total of n equivalent positions are generated. Some rotational symmetries are illus-

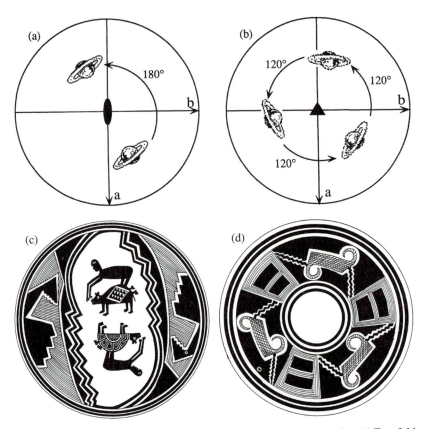

Figure 2.1. Stereographic projections of two simple rotational symmetries. (a) Two-fold
axis normal to the plane of the page. (b) Three-fold axis normal to the plane of the page.
(c) two-fold axis, (d) three-fold axis. Examples of these symmetries (or approximations
thereto) from Mimbres (prehistoric native American) pattern designs (From Giammattei and Reichert 1975).

trated in Figure 2.1, both in terms of molecules and by some artistic designs
that approximately have those elements.

Rotations preserve right- or left-handed objects, but reflections or mirrors
transform left- to right-handed (compare Fig. 2.1a, a two-fold axis, or diad,
perpendicular to the plane of the paper to Fig. 2.2a, a mirror plane perpendicular to the paper). A mirror located at $z = 0$ transforms a point (x,y,z) to $(x,y,-z)$.
A center of symmetry decrees that any point (x,y,z) is equivalent to one
at $(-x,-y,-z)$. An inversion axis combines an n-fold rotation with a mirror
plane perpendicular to that axis. These symmetries are further illustrated in
Figure 2.2.

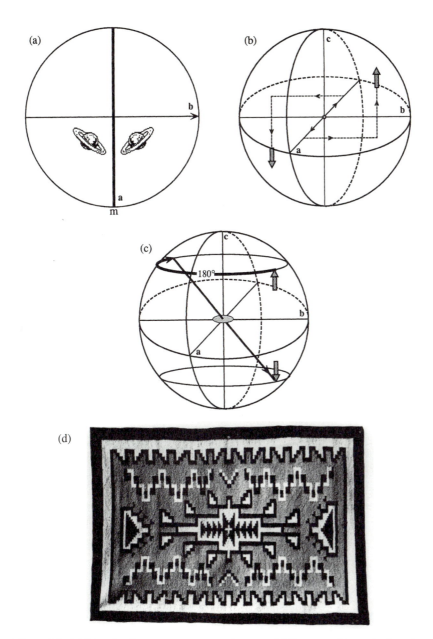

Figure 2.2. Additional point group symmetry elements. (a) Stereographic projection of a vertical mirror plane normal to the page. (b) Center of symmetry. This symmetry operation relates equivalent points by inversion through the center. (c) Two-fold inversion axes. This operation combines a two-fold rotation with an inversion through the central point along that axis. (d) This Two Gray Hills Navajo rug design has two mirror planes and two two-fold axes perpendicular to the plane (From James 1976).

Table 2.1. *Some common point groups*

Point group symbol				Example of
Hermann-Maughin	Schoenflies	Name	Symmetry elements	coordination polyhedron
$\bar{4}3m$	T_d	Regular tetrahedron	$3\bar{A}_4, 4A_3, 6m$	SiO_4
$4/m\ \bar{3}\ 2/m$	O_h	Regular octahedron	$3A_4, 4\bar{A}_3, 6A_2, 9m$	MgO_6 in periclase
$4/m\ 2/m\ 2/m$	$D4_h$	Distorted octahedron (one long axis)	$i, 1A_4, 4A_2, 5m$	MnO_6 in Mn_2O_3
4	C_4	Square planar	$1A_4$	CuO_4 in CuO
$4m\ \bar{3}\ 2/m$	O_h	Cubic	$3A_4, 4\bar{A}_3, 6A_2, 9m$	CaF_8 in fluorite

Note: Regular octahedron and cube (hexahedron) have same point group.

The set of all symmetry operations about a point generates a point group. Because the presence of certain symmetry elements automatically implies other ones, the naming of point and space groups is not a simple enumeration of all the elements of symmetry. Two systems of point group nomenclature have emerged, the Schoenflies notation common to physical chemistry and molecular symmetry and the Hermann-Maughin or international system used by crystallographers. The reader is referred to crystallography texts listed at the end of this chapter (Boisen and Gibbs 1985; Burns and Glazer 1978) and to the *International Tables for Crystallography* (Hahn 1983) for a full description. Table 2.1 lists some common point groups in both notations, their symmetry elements, and some molecular species having those symmetries.

Symmetry in three dimensions combines point symmetry with lateral periodic movement. The simplest such is translation by some repeat distance **a** (see Fig. 2.3a). More complex motifs are generated by a glide plane, where translation is accompanied by a mirror reflection (see Fig. 2.3b). A screw axis combines translation with an *n*-fold rotation (see Fig. 2.3c). Both left-turning and right-turning screws can be generated if other features of the lattice make the space group enantiomorphic (left- and right-handed).

The study of crystallography is made immensely easier, more economical, and more elegant by the use of vector notation and group theory. This is exploited to good advantage in the text by Boisen and Gibbs (1985).

The unit cell, a volume whose corners are equivalent points in the lattice, is defined in three dimensions by vectors, **a**, **b**, and **c**. In scalar notation these correspond to three unit cell distances or parameters, a, b, and c, and the angles between them, α, β, and, γ, where α is the angle between b and c, and so on.

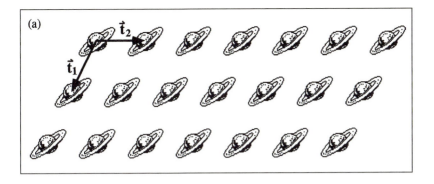

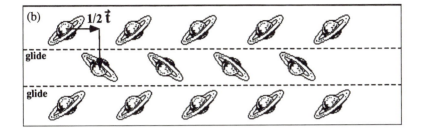

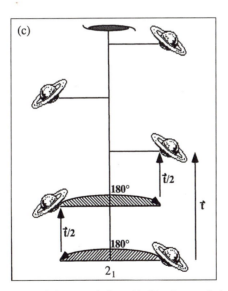

Figure 2.3. Symmetry involving translation. (a) Simple translational symmetry. This two-dimensional pattern is generated by repeating the motif in two directions along the vectors t_1 and t_2. (b) Horizontal glide planes normal to the page (dashed lines). Equivalent points are related by a translation of $t/2$ parallel to the glide plane followed by reflection across the glide plane. (c) Two-fold screw axes (2_1) in the plane of the page. Equivalent points are related by a translation of $t/2$ parallel to the axis followed by a 180° rotation about the axis.

The seven crystal systems are listed in Table 2.2. The cubic system requires only one lattice parameter, a, to describe the unit cell; all three axes are equal and at right angles. One can imagine a deviation from cubic symmetry occurring that makes one dimension longer or shorter, leaving two equivalent and one unique direction and resulting in tetragonal symmetry. Physically such a deviation or distortion may occur when two ions order, as, for example, during the ordering of Mg and Ti on octahedral sites in the spinel Mg_2TiO_4, or when magnetic ordering occurs, as in NiO. The orthorhombic system, with three unequal axes at right angles, represents a further lowering of symmetry. Though these systems represent distinct symmetries, in practice it is sometimes hard to detect a small distortion, for example, to tell whether a given structure is tetragonal or slightly orthorhombic. Furthermore, in some cases a change to a lower symmetry can occur gradually (see later discussions of phase transitions), and the ability to tell just where a transition starts may depend on the sensitivity of the instrument and the homogeneity of the sample under study. Thus distinctions of symmetry that are clear-cut in the abstract often become disputed in real solids.

There are four crystal systems whose axes are not all at right angles: hexagonal, trigonal, monoclinic, and triclinic. The hexagonal system has two equal axes, a and b, at an angle of 120° and one different axis, c, perpendicular to the a-b plane. The points in the a-b plane define a hexagonal array and one can easily draw a hexagonal unit cell with lattice points of the corners at the cell (see Fig. 2.4a). Such a cell is called primitive (P) and contains one lattice point per cell. In the trigonal subsystem of the hexagonal system, a rhombohedral unit cell (R) is often chosen (see Fig. 2.4b). This cell has equal edges and lattice points at each corner. The interaxial angle is given as α_R.

The monoclinic system has three unequal edges, two angles of 90°, and one different angle. The difference between orthorhombic and monoclinic, and the transition from one to the other, which involves a small angular distortion, is often hard to pinpoint in practice. The system of lowest symmetry, triclinic, has a unit cell with no equal edges and no equal angles.

A primitive cell (P) has one repeating point or motif: A centered cell has additional lattice points related by translational symmetry. Centering may occur in the interior of the cell at the unique point where the body diagonals cross (body centered, or I), on all of the exterior faces (face centered, or F), or on one set of faces (C, B, or A centered). The combination of crystal system and centering permissible with that symmetry gives rise to fourteen Bravais lattices (see Table 2.3.).

By combining the 32 point groups with allowable translational symmetries

Table 2.2. *The seven crystal systems*

Symmetry elements	System	Lengths and angles in unit cell
1(E) or $\bar{1}$(i)	Triclinic	$a \neq b \neq c$ $\alpha \neq \beta \neq \gamma$
2(C_2) or $\bar{2}$(σ)	Monoclinic	$a \neq b \neq c$ $\alpha = \beta = 90° \neq \gamma$ (first setting) $\alpha = \gamma = 90° \neq \beta$ (second setting)
Two 2-fold or $\bar{2}$ axes	Orthorhombic	$a \neq b \neq c$ $\alpha = \beta = \gamma = 90°$
4(C_4) or $\bar{4}$($S_4{}^3$)	Tetragonal	$a = b \neq c$ $\alpha = \beta = \gamma = 90°$
Four 3-fold or $\bar{3}$ axes	Cubic	$a = b = c$ $\alpha = \beta = \gamma = 90°$
6(C_6) or $\bar{6}$($S_3{}^5$)	Hexagonal	$a = b \neq c$ $\alpha = \beta = 90°; \gamma = 120°$
3(C_3) or $\bar{3}$($S_6{}^5$)	Trigonal (Rhombohedral)	same as hexagonal $(a = b = c; \alpha = \beta = \gamma)$

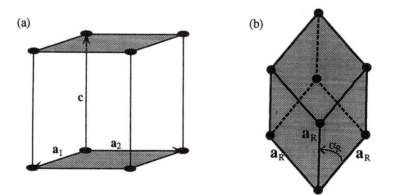

Figure 2.4. (a) Primitive (P) hexagonal unit cell with lattice points only at the corners. The angle γ between a_1 and a_2 is 120°. (b) Rhombohedral unit cell (R) of the trigonal subsystem of the hexagonal system. The cell edges are all equal in length and the interaxial angle is given as α_R.

Table 2.3. *The fourteen Bravais lattices*

System	Centering	Total number of space groups
Cubic	Primitive (P)	15
	Body centered (I)	10
	Face centered (F)	11
Tetragonal	Primitive (P)	49
	Body centered (I)	19
Orthorhombic	Primitive (P)	30
	Body centered (I)	9
	Face centered (F)	5
	One face centered (A or C)	15
Monoclinic	Primitive (P)	8
	One face centered (B)	5
Hexagonal	Primitive (P)	27
Trigonal	Primitive (P)	18
	Rhombohedral (R)	7
Triclinic	Primitive (P)	2
		230

(centering, glide planes, screw axes), all of the 230 space groups can be generated. The reader is referred to crystallography texts listed at the end of this chapter for the derivation and the enumeration of these space groups as well as for a full description of the notation used to describe symmetry elements. The *International Tables for Crystallography*, volume one, is excellent in this regard (Hahn 1983).

2.3 Some useful equations and concepts

It is useful, especially for diffraction (see Chapter 3), to consider planes within the unit cell: These often contain specific atoms that interact with X rays or neutrons. A plane is defined by its Miller indices (*hkl*), such that a given plane cuts the unit cell axes at a/h, b/k, c/l. Translational symmetry generates a family of $\{hkl\}$ planes once a set of indices is given. The distance between adjacent planes, d-spacing or d_{hkl}, is easily calculated for cubic systems as

$$\frac{1}{d_{hkl}^2} = \frac{1}{a^2}(h^2 + k^2 + l^2) \tag{2.1}$$

For the orthorhombic system, it is

$$\frac{1}{d_{hkl}^2} = \frac{h^2}{a^2} + \frac{k^2}{b^2} + \frac{l^2}{c^2} \tag{2.2}$$

For lower symmetry cases, the proper trigonometric or vector relations are used. The general expression is

$$\frac{1}{d_{hkl}^2} = \left(\frac{abc}{V^2}\right)\left(\sum\frac{h^2}{a^2}\sin^2\alpha + 2\sum\frac{hk}{ab}(\cos\alpha\,\cos\beta - \cos\gamma)\right) \quad (2.3)$$

where V is the volume of the unit cell and each summation is three terms, obtained by the simultaneous permutations $a \to b \to c$, $h \to k \to l$, $\alpha \to \beta \to \gamma$.

The volume of the unit cell for the general case is

$$V = abc(1 - \cos^2\alpha - \cos^2\beta - \cos^2\gamma + 2\cos\alpha\,\cos\beta\,\cos\gamma)^{1/2} \quad (2.4)$$

For monoclinic crystals, it is (with $\beta \neq 90°$)

$$V = abc\,\sin\beta \quad (2.5)$$

For the cubic case,

$$V = a^3 \quad (2.6)$$

Unit cell edges are usually given in ångstroms, Å (or in nanometers, nm). To convert volume in cubic ångstroms to molar volume (in cm³/mol) use

$$V(\text{cm}^3/\text{mol}) = V(\text{Å}^3) \times (10^{-8})^3 \times \frac{6.024 \times 10^{23}}{Z} \quad (2.6a)$$

or

$$V(\text{cm}^3/\text{mol}) = V(\text{nm}^3) \times (10^{-9})^3 \times \frac{6.024 \times 10^{23}}{Z} \quad (2.6b)$$

where 6.023×10^{23} is Avogadro's number, the number of atoms or molecules in a mole, Z is the number of formula units per unit cell, and 10^{-24} (or 10^{-27}) is the conversion from Å³ (or nm³) to cm³.

The actual atomic positions in a unit cell are described by fractional coordinates (x, y, z), which are decimal fractions of the unit cell axes a, b, c, respectively. A general position, with site symmetry one, is a point invariant only to the identity operator but to no other symmetry operator of the space group. A high-symmetry space group has a large number of general positions that are crystallographically equivalent; a low symmetry space group has few. In addition, especially for cases of high symmetry, there are special positions within the unit cell. Commonly, several distinct sets of such positions exist, each having additional symmetry arising from being located on specific symmetry elements, for example, on a mirror plane or at an inversion center. When one constructs a structure consistent with a given space group, placing an atom at one position immediately implies that the same atom (or a statistically equivalent mixture for disordered systems) is placed at all other symmetry-related positions in the unit cell. Since the ratios of atoms in a given material are usually given by a chemical formula, the preceding considerations often constrain the number of formula units in the unit cell.

A lower symmetry space group may be related to one of higher symmetry simply by removal of one or more symmetry elements. Then the lower symmetry group is considered a subgroup of the higher. Physically, the symmetry lost in going from the higher to the lower space group may result from the tilting of polyhedra, the ordering of species with different bond lengths, or other atomic displacements. These can be related by phase transitions that occur as temperature, pressure, or composition are varied. A high symmetry structure, such as the cubic perovskite (space group Pm3m) may have many lower symmetry derivatives (e.g., P4mm, C2mm, R$\bar{3}$m, F4/mmc). (See Table 2.19.)

If one starts with the higher symmetry structure and orders atoms within it, a structure of lower symmetry but with a larger unit cell often results. This larger cell is referred to as a supercell of the smaller one, the new ordered structure as a superstructure, and the extra allowable diffraction peaks as superstructure reflections. The ordering of the calcite structure (R$\bar{3}$c) to form the dolomite structure (R$\bar{3}$) is an example (refer to Fig. 2.22).

When a structure contains several different kinds of atoms (e.g., large and small cations and anions), each generally occupies positions of distinct geometry. Often each set of sites can be thought of as a distinct space-filling arrangement with its own symmetry and other properties. Somewhat loosely, such a group of sites is called a sublattice, and the crystal structure can often be described as consisting of several interpenetrating sublattices.

2.4 An illustration of crystallographic concepts: space group number 227, Fd3m and the spinel structure

Figure 2.5 shows a page from the *International Tables for Crystallography* (Hahn 1983) for space group Fd3m. It lists the numbers and point symmetries of general and special positions and the conditions limiting possible diffraction peaks. Because of the high symmetry, there are 192 symmetry equivalent general positions. In addition, there are eight different sets of special positions, with multiplicity ranging from 8 to 96. If the 8*a* positions are filled, the diamond structure (see Sec. 2.5.2) results. Each atom is tetrahedrally coordinated, with the point group 43m (T$_d$) of a regular tetrahedron. The same space group

Generators selected (1); $t(1,0,0)$; $t(0,1,0)$; $t(0,0,1)$; $t(0,\tfrac12,\tfrac12)$; $t(\tfrac12,0,\tfrac12)$; $t(\tfrac12,\tfrac12,0)$; (2); (3); (5); (13); (25)

Positions

Multiplicity, Wyckoff letter, Site symmetry	Coordinates $(0,0,0)+$ $(\tfrac12,\tfrac12,0)+$ $(0,\tfrac12,\tfrac12)+$ $(\tfrac12,0,\tfrac12)+$			Reflection conditions h,k,l permutable

192 i 1

(1) x,y,z (2) $\bar{x},\bar{y}+\tfrac12,z+\tfrac12$ (3) $\bar{x}+\tfrac12,y,\bar{z}+\tfrac12$ (4) $x+\tfrac12,\bar{y}+\tfrac12,\bar{z}$
(5) z,x,y (6) $z+\tfrac12,\bar{x},\bar{y}+\tfrac12$ (7) $\bar{z},\bar{x}+\tfrac12,y+\tfrac12$ (8) $\bar{z}+\tfrac12,x+\tfrac12,\bar{y}$
(9) y,z,x (10) $\bar{y}+\tfrac12,z+\tfrac12,\bar{x}$ (11) $y+\tfrac12,\bar{z},\bar{x}+\tfrac12$ (12) $\bar{y},\bar{z}+\tfrac12,x+\tfrac12$
(13) $y+\tfrac34,x+\tfrac14,\bar{z}+\tfrac34$ (14) $\bar{y}+\tfrac14,\bar{x}+\tfrac14,\bar{z}+\tfrac14$ (15) $\bar{y}+\tfrac34,x+\tfrac14,z+\tfrac34$ (16) $y+\tfrac14,\bar{x}+\tfrac14,z+\tfrac14$
(17) $x+\tfrac34,z+\tfrac14,\bar{y}+\tfrac34$ (18) $\bar{x}+\tfrac14,\bar{z}+\tfrac14,\bar{y}+\tfrac14$ (19) $\bar{x}+\tfrac34,z+\tfrac14,y+\tfrac34$ (20) $x+\tfrac14,\bar{z}+\tfrac14,y+\tfrac14$
(21) $z+\tfrac34,y+\tfrac14,\bar{x}+\tfrac34$ (22) $\bar{z}+\tfrac14,\bar{y}+\tfrac14,\bar{x}+\tfrac14$ (23) $\bar{z}+\tfrac34,y+\tfrac14,x+\tfrac34$ (24) $z+\tfrac14,\bar{y}+\tfrac14,x+\tfrac14$
(25) $\bar{x}+\tfrac14,\bar{y}+\tfrac14,\bar{z}+\tfrac14$ (26) $x+\tfrac14,y+\tfrac34,\bar{z}+\tfrac34$ (27) $x+\tfrac34,\bar{y}+\tfrac14,z+\tfrac34$ (28) $\bar{x}+\tfrac34,y+\tfrac34,z+\tfrac14$
(29) $\bar{z}+\tfrac14,\bar{x}+\tfrac14,\bar{y}+\tfrac14$ (30) $\bar{z}+\tfrac34,x+\tfrac14,y+\tfrac34$ (31) $z+\tfrac14,x+\tfrac34,\bar{y}+\tfrac34$ (32) $z+\tfrac34,\bar{x}+\tfrac34,y+\tfrac14$
(33) $\bar{y}+\tfrac14,\bar{z}+\tfrac14,\bar{x}+\tfrac14$ (34) $y+\tfrac34,\bar{z}+\tfrac34,x+\tfrac14$ (35) $\bar{y}+\tfrac34,z+\tfrac14,x+\tfrac34$ (36) $y+\tfrac14,z+\tfrac34,\bar{x}+\tfrac34$
(37) $\bar{y}+\tfrac12,\bar{x},z+\tfrac12$ (38) y,x,z (39) $y+\tfrac12,\bar{x},\bar{z}+\tfrac12$ (40) $\bar{y},x,\bar{z}$
(41) $\bar{x}+\tfrac12,\bar{z},y+\tfrac12$ (42) x,z,y (43) $x+\tfrac12,\bar{z},\bar{y}+\tfrac12$ (44) $\bar{x},z,\bar{y}$
(45) $\bar{z}+\tfrac12,\bar{y},x+\tfrac12$ (46) z,y,x (47) $z+\tfrac12,\bar{y},\bar{x}+\tfrac12$ (48) $\bar{z},y,\bar{x}$

Reflection conditions

General:

hkl : $h+k=2n$ and
$\quad\quad h+l,k+l=2n$
$0kl$: $k+l=4n$ and
$\quad\quad k,l=2n$
hhl : $h+l=2n$
$h00$: $h=4n$

96 h ..2

$\tfrac18,y,\bar{y}+\tfrac14$ $\tfrac78,\bar{y}+\tfrac12,y+\tfrac34$ $\tfrac38,y+\tfrac12,\bar{y}+\tfrac34$ $\tfrac58,\bar{y},y+\tfrac14$
$\bar{y}+\tfrac14,\tfrac18,y$ $\bar{y}+\tfrac34,\tfrac78,\bar{y}+\tfrac12$ $y+\tfrac14,\tfrac38,y+\tfrac12$ $\bar{y},\tfrac58,y+\tfrac14$
$y,\bar{y}+\tfrac14,\tfrac18$ $\bar{y}+\tfrac12,y+\tfrac34,\tfrac78$ $y+\tfrac12,\bar{y}+\tfrac34,\tfrac38$ $\bar{y},y+\tfrac14,\tfrac58$
$\tfrac18,\bar{y}+\tfrac14,y$ $\tfrac78,y+\tfrac34,\bar{y}+\tfrac12$ $\tfrac38,\bar{y}+\tfrac34,y+\tfrac12$ $\tfrac58,y+\tfrac14,\bar{y}$
$y,\tfrac18,\bar{y}+\tfrac14$ $\bar{y}+\tfrac12,\tfrac78,y+\tfrac34$ $y+\tfrac12,\tfrac38,\bar{y}+\tfrac34$ $y+\tfrac14,\tfrac58,\bar{y}$
$\bar{y}+\tfrac14,y,\tfrac18$ $y+\tfrac34,\bar{y}+\tfrac12,\tfrac78$ $\bar{y}+\tfrac34,y+\tfrac12,\tfrac38$ $y+\tfrac14,\bar{y},\tfrac58$

Special: as above, plus

no extra conditions

Figure 2.5. *(pp. 17–19)* Description of space group Fd3̄m (Hahn 1983).

Generators selected (1); $t(1,0,0)$; $t(0,1,0)$; $t(0,0,1)$; $t(0,\tfrac12,\tfrac12)$; $t(\tfrac12,0,\tfrac12)$; (2); (3); (5); (13); (25)

Positions

Multiplicity, Wyckoff letter, Site symmetry	Coordinates $(0,0,0)+$	$(0,\tfrac12,\tfrac12)+$	$(\tfrac12,0,\tfrac12)+$	$(\tfrac12,\tfrac12,0)+$			Reflection conditions h,k,l permutable
96 g $..m$	x,x,z	$\bar{x},\bar{x}+\tfrac12,z+\tfrac12$	$\bar{x}+\tfrac12,x+\tfrac12,\bar{z}$	$x+\tfrac12,\bar{x},\bar{z}+\tfrac12$			no extra conditions
	z,x,x	$z+\tfrac12,\bar{x},\bar{x}+\tfrac12$	$\bar{z},\bar{x}+\tfrac12,x+\tfrac12$	$\bar{z}+\tfrac12,x+\tfrac12,\bar{x}$			
	x,z,x	$\bar{x}+\tfrac12,z+\tfrac12,\bar{x}$	$x+\tfrac12,\bar{z},\bar{x}+\tfrac12$	$\bar{x},\bar{z}+\tfrac12,x+\tfrac12$			
	$x+\tfrac34,x+\tfrac14,\bar{z}+\tfrac34$	$\bar{x}+\tfrac14,\bar{x}+\tfrac14,\bar{z}+\tfrac14$	$x+\tfrac14,\bar{x}+\tfrac34,z+\tfrac34$	$\bar{x}+\tfrac34,x+\tfrac34,z+\tfrac14$			
	$x+\tfrac34,z+\tfrac14,\bar{x}+\tfrac34$	$\bar{x}+\tfrac34,z+\tfrac34,x+\tfrac14$	$\bar{x}+\tfrac14,\bar{z}+\tfrac14,\bar{x}+\tfrac14$	$x+\tfrac14,\bar{z}+\tfrac34,x+\tfrac34$			
	$z+\tfrac34,x+\tfrac14,\bar{x}+\tfrac34$	$z+\tfrac14,\bar{x}+\tfrac34,x+\tfrac34$	$\bar{z}+\tfrac34,x+\tfrac34,x+\tfrac14$	$\bar{z}+\tfrac14,\bar{x}+\tfrac14,\bar{x}+\tfrac14$			
48 f $2.m\,m$	$x,0,0$	$\bar{x},\tfrac12,\tfrac12$	$0,x,0$	$\tfrac12,\bar{x},\tfrac12$	$0,0,x$	$\tfrac12,\tfrac12,\bar{x}$	hkl : $h=2n+1$
	$\tfrac34,x+\tfrac14,\tfrac34$	$\tfrac14,\bar{x}+\tfrac14,\tfrac14$	$x+\tfrac34,\tfrac14,\tfrac34$	$\bar{x}+\tfrac34,\tfrac34,\tfrac14$	$\tfrac34,\tfrac14,\bar{x}+\tfrac34$	$\tfrac14,\tfrac34,x+\tfrac34$	or $h+k+l=4n$
32 e $.3m$	x,x,x	$\bar{x},\bar{x}+\tfrac12,x+\tfrac12$					no extra conditions
	$\bar{x}+\tfrac12,x+\tfrac12,\bar{x}$	$x+\tfrac12,\bar{x},\bar{x}+\tfrac12$					
	$x+\tfrac34,x+\tfrac14,\bar{x}+\tfrac34$	$\bar{x}+\tfrac14,\bar{x}+\tfrac14,\bar{x}+\tfrac14$					
	$x+\tfrac14,\bar{x}+\tfrac34,x+\tfrac34$	$\bar{x}+\tfrac34,x+\tfrac34,x+\tfrac14$					
16 d $.\bar{3}m$	$\tfrac58,\tfrac58,\tfrac58$	$\tfrac38,\tfrac78,\tfrac18$	$\tfrac78,\tfrac18,\tfrac38$	$\tfrac18,\tfrac38,\tfrac78$			hkl : $h=2n+1$ or $h,k,l=4n+2$ or $h,k,l=4n$
16 c $.\bar{3}m$	$\tfrac18,\tfrac18,\tfrac18$	$\tfrac78,\tfrac38,\tfrac58$	$\tfrac38,\tfrac58,\tfrac78$	$\tfrac58,\tfrac78,\tfrac38$			
8 b $\bar{4}\,3m$	$\tfrac12,\tfrac12,\tfrac12,$	$\tfrac14,\tfrac34,\tfrac14$					hkl : $h=2n+1$ or $h+k+l=4n$
8 a $\bar{4}\,3m$	$0,0,0$	$\tfrac34,\tfrac14,\tfrac34$					

Symmetry of special projections

Along [001] $p\,4m\,m$
$\mathbf{a}'=\tfrac14(\mathbf{a}-\mathbf{b})$ $\mathbf{b}'=\tfrac14(\mathbf{a}+\mathbf{b})$
Origin at $0,0,z$

Along [111] $p\,6m\,m$
$\mathbf{a}'=\tfrac16(2\mathbf{a}-\mathbf{b}-\mathbf{c})$ $\mathbf{b}'=\tfrac16(-\mathbf{a}+2\mathbf{b}-\mathbf{c})$
Origin at x,x,x

Along [110] $c\,2m\,m$
$\mathbf{a}'=\tfrac12(-\mathbf{a}+\mathbf{b})$ $\mathbf{b}'=\mathbf{c}$
Origin at $x,x,\tfrac18$

ORIGIN CHOICE 1

Maximal non-isomorphic subgroups

I [3]$F\,4_1/d\,1\,2/m$ $(I\,4_1/a\,m\,d)$ (1; 2; 3; 4; 13; 14; 15; 16; 25; 26; 27; 28; 37; 38; 39; 40)+
   [3]$F\,4_1/d\,1\,2/m$ $(I\,4_1/a\,m\,d)$ (1; 2; 3; 4; 17; 18; 19; 20; 25; 26; 27; 28; 41; 42; 43; 44)+
   [3]$F\,4_1/d\,1\,2/m$ $(I\,4_1/a\,m\,d)$ (1; 2; 3; 4; 21; 22; 23; 24; 25; 26; 27; 28; 45; 46; 47; 48)+
   [4]$F\,1\,\bar{3}\,2/m$ $(R\,\bar{3}m)$ (1; 5; 9; 14; 19; 24; 25; 29; 33; 38; 43; 48)+
   [4]$F\,1\,\bar{3}\,2/m$ $(R\,\bar{3}m)$ (1; 6; 12; 13; 18; 24; 25; 30; 36; 37; 42; 48)+
   [4]$F\,1\,\bar{3}\,2/m$ $(R\,\bar{3}m)$ (1; 7; 10; 13; 19; 22; 25; 31; 34; 37; 43; 46)+
   [4]$F\,1\,\bar{3}\,2/m$ $(R\,\bar{3}m)$ (1; 8; 11; 14; 18; 22; 25; 32; 35; 38; 42; 46)+
   [2]$F\,d\,\bar{3}\,1$ $(F\,d\,\bar{3})$ (1; 2; 3; 4; 5; 6; 7; 8; 9; 10; 11; 12; 25; 26; 27; 28; 29; 30; 31; 32; 33; 34; 35; 36)+
   [2]$F\,4_1\,3\,2$ (1; 2; 3; 4; 5; 6; 7; 8; 9; 10; 11; 12; 13; 14; 15; 16; 17; 18; 19; 20; 21; 22; 23; 24)+
   [2]$F\,\bar{4}\,3m$ (1; 2; 3; 4; 5; 6; 7; 8; 9; 10; 11; 12; 37; 38; 39; 40; 41; 42; 43; 44; 45; 46; 47; 48)+

IIa   none

IIb   none

Maximal isomorphic subgroups of lowest index

IIc   [27]$F\,d\,\bar{3}\,m$ $(a' = 3a, b' = 3b, c' = 3c)$

Minimal non-isomorphic supergroups

I   none

II   [2]$P\,n\,\bar{3}m$ $(2a' = a, 2b' = b, 2c' = c)$

can accommodate a more complex structure, that of spinel (see Sec. 2.6.2). In this AB_2X_4 compound, A atoms are located in the tetrahedral $8a$ positions, B atoms in the (octahedral) $16c$ positions, and oxygens in the $32e$ positions. The unit cell contains thirty-two oxygens and, therefore, eight AB_2X_4 formula units. Several depictions of this structure are shown in Figure 2.6. The $8a$ and $16c$ positions have no free parameters, but the $32e$ positions have one free parameter, labeled x in the table but often referred to as u, the oxygen parameter. When u has a value of 0.375, the X atoms can be described as being in exact cubic close packing (see Sec. 2.5.1), and the A sites are regular tetrahedra and the B sites regular octahedra (see Fig. 2.6a). When this oxygen parameter deviates from its ideal value of 0.375, the space group symmetry does not change, but the AX and BX bond distances change. For $u > 0.375$, the tetrahedra become larger but keep their symmetry, and the octahedra become smaller

Figure 2.6. (*pp. 21–22*) The spinel structure AB_2O_4. The idealized cubic unit cell contains thirty-two oxygens in cubic close packing, eight A cations in tetrahedral coordination, and sixteen B cations in octahedral coordination. One way to show the cation distribution is to construct two smaller cubes as shown in (a). One of these, the AO_4 cube, represents an AO_4 tetrahedron. The other is a B_4O_4 cube. As shown in (b), these cubes are arranged in the octants of the larger cubic unit cell so that the B cations (half-filled circles) are in octahedral coordination. Note there are additional A cations (solid circles) at the unit cell corners and the face-centers. Projections of the structure on (001) are shown in (c) and (d). The atom z-coordinates are given in terms of eighths of the unit cell edge. The oxygens are shown in both projections as large open circles. Note that there are two layers of oxygens that overlie each other. In (c) the positions of the A cations are shown and the coordination tetrahedron for the A cation at (1/4, 1/4, 1/4) is drawn. This cation is coordinated by oxygens at z-coordinates of 1/8 and 3/8. The projection in (d) shows the half-filled B cations and the larger oxygens. One coordination octahedron is drawn for the B cation at (3/8, 3/8, 5/8) and is coordinated by six oxygens, four at a z-coordinate of 5/8, one at 3/8, and one at 7/8. Part (e) shows a polyhedral representation of one layer of the structure projected on (001). This layer contains B cations at $z = 5/8$ and A cations at $z = 3/4$ and 1/2. The z-coordinates for the oxygens are shown. The dashed line outlines the unit cell. There are strips of edge-sharing BO_6 octahedra parallel to the face-diagonal that are linked by AO_4 tetrahedra. In the next layer down, the octahedral strips are parallel to the other face diagonal. The structure may be thought of as an arrangement of these layers. (Parts (a), (b), and (e) after Putnis 1992.) (f) Packing model showing the contents of the unit cell. Tetrahedrally coordinated A cations are shown as small solid circles, octahedrally coordinated B cations as medium-sized solid circles, and anions as large open circles.

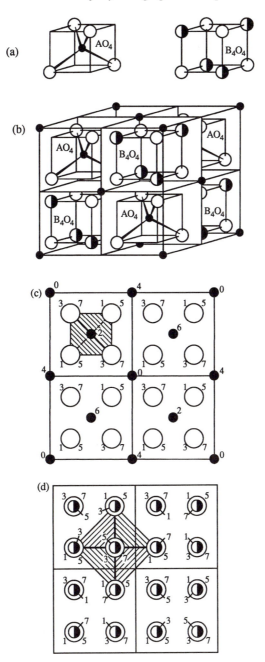

(e)

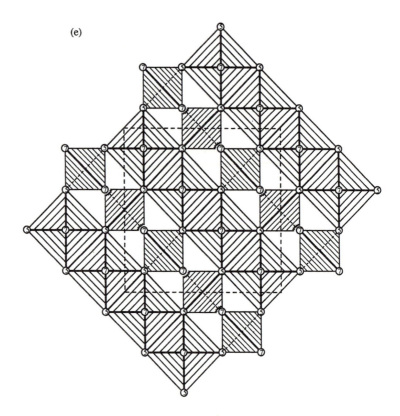

(f)

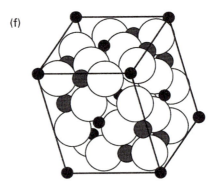

and angularly distorted. The relations between interatomic distances (r_{AX}, r_{BX}, r_{AA}, r_{BB}), unit cell parameter, a, and oxygen parameter, u, are:

$$r_{AX} = \sqrt{3}a(u - 0.25) \tag{2.7a}$$

$$= 0.217a \quad \text{for } u = 0.375 \tag{2.7b}$$

$$r_{BX} = a(0.625 - u) \tag{2.8a}$$

$$= 0.25a \quad \text{for } u = 0.375 \tag{2.8b}$$

$$r_{AA} = \frac{\sqrt{3}}{4} a = 0.433a \tag{2.9}$$

$$r_{BB} = \frac{\sqrt{2}}{4} a = 0.354a \tag{2.10}$$

Thus the octahedral bond lengths are greater than the tetrahedral (for $u <$ 0.387), but the next nearest neighbor octahedral cations are closer together than the next nearest neighbor tetrahedral cations because anion octahedra share edges whereas tetrahedra do not (see Fig. 2.6e). Because of the linkage of octahedra and tetrahedra with only one variable positional parameter, the octahedra must shrink as the tetrahedra expand or vice versa. Put differently, for a spinel with given A, B, and X atoms, optimizing the AX and BX bond lengths will determine not only the unit cell parameter but also the oxygen parameter. This becomes important in discussions of order–disorder (see following and Sec. 6.6).

The spinel structure is capable of several more complications. First, although the "normal" distribution of cations in AB_2O_4 is to place all A ions in tetrahedral (8a) sites and all B atoms in octahedral (16c) sites, an "inverse" distribution that places B atoms in 8a and a random mixture of A and B in 16c sites has also been observed. Furthermore, intermediate distributions, including the random $(A_{1/3} B_{2/3})_{8a} (A_{2/3} B_{4/3})_{16c} X_4$ or $(A_{1/3}B_{2/3}) [A_{2/3} B_{4/3}] X_4$, where ( ) denote tetrahedral and [ ] octahedral sites, are also possible. In general, the cation distribution can vary with temperature for a given composition without changing the space group but with small changes in a and u. To complicate matters more, when a sample held at high temperature is cooled rapidly (quenched) to room temperature, it is not always clear to what extent the high temperature cation distribution is preserved. Thus a detailed structural description of a given sample may be specific not just to that composition but to a given thermal history.

Superstructures form a different set of complications. When two different ions occupy the same sublattice (as in an inverse spinel), there may be a tendency for them to order on alternate sites or in a more complex pattern. The

Table 2.4. *Superstrucutres based on the spinel structure*

Type of ordering	Symmetry	Space group	Example
None	Cubic	Fd3m	$MgAl_2O_4$
1:1 cation order, octahedral	Tetragonal	$P4_122$	Mg_2TiO_4 $ZnCoGeO_4$, Low T Fe_3O_4
1:3 cation order, octahedral	Tetragonal	$P4_132$	$LiFe_5O_8$
1:5 cation and vacancy order, octahedral	Tetragonal	$P4_1$	γ-Fe_2O_3
1:1 cation order, tetrahedral	Tetragonal		$LiFeCr_5O_8$
Two types of cations plus vacancies, octahedral	Tetragonal	$P4_322$	$Ti_4Li_2Co_3O_{12}$
Magnetic	Cubic	F$\bar{4}$3m	$Co_3O_4 < 40K$

Table 2.5. *Some representative spinels, AB_2X_4*

Formula	a(Å)	Cation distribution, x = fraction of tetrahedral sites occupied by B ions
$MgAl_2O_4$	8.086	0 to ~ 0.3
$MnAl_2O_4$	8.241	0.3
$FeAl_2O_4$	8.10	0 to 0.1
$CoAl_2O_4$	8.105	0 to 0.1
$NiAl_2O_4$	8.046	0.7 to 0.9
$CuAl_2O_4$	8.086	0.4
$ZnAl_2O_4$	8.086	0 to 0.05
$MgFe_2O_4$	8.360	0.7 to 0.9
$MgCr_2O_4$	8.333	0
$MgGa_2O_4$	8.282	0.7 to 0.9
MgV_2O_4	8.413	0
Fe_3O_4	8.394	0.67 to 1.0
$FeCr_2O_4$	8.377	0
$FeCrS_4$	9.998	0
Mg_2SiO_4	8.082	0
Fe_2SiO_4	8.234	0
Co_2SiO_4	8.140	0
Ni_2SiO_4	8.044	0
Mg_2GeO_4	8.255	0
Mg_2TiO_4	8.440	1
Mg_2SnO_4	8.601	1
γ-Al_2O_3	7.905	Octahedral vacancies
Li_2NiF_4	8.310	1

same is true if one of the "ions" is in fact a vacancy or empty site, giving the spinel a chemical formula with a cation-anion ratio different from 3:4. Some examples of superstructures are given in Table 2.4.

The spinel structure is ubiquitous; several hundred compounds have been synthesized. Some representative members, their lattice parameters, and cation distributions are shown in Table 2.5.

The following main points gleaned from the preceding discussion are applicable to all crystalline materials: (1) Diverse structures (e.g., diamond and spinel) can belong to the same space group. (2) The space group and positional parameters allow one to calculate all atomic positions and interatomic distances in a structure. (3) Even a relatively simple high-symmetry structure, such as spinel with only one variable positional parameter, can show considerable complexity in terms of disorder, superstructures, and structural features that depend on thermal history.

2.5 Structure types: elements and binary compounds

This section is a summary of the common structure types in minerals and related inorganic compounds. Of necessity, this is a much abbreviated tour of crystal chemistry. In addition to providing descriptions of common structures, I wish to stress how structures are related and how, in certain cases, structures can be derived from each other both conceptually and mechanistically. For more detailed expositions, the reader is referred to a number of textbooks, monographs, and review papers (see General References).

2.5.1 Close packing, polytypes, and metal structures

If one envisions atoms as spheres, a structure may be thought of as a packing of spheres of different sizes. If all the spheres are of one type, then they can be arranged in space such that they are touching and occupy the smallest possible volume. There are two ways of doing this in a crystalline array, namely, hexagonal close packing (hcp) and cubic close packing (ccp).

Consider a layer of touching spheres, as in Figure 2.7a. Each sphere is surrounded by six others in a hexagonal array. A second layer fits over the first such that its sphere centers are over the interstices between the spheres in the first layer, as in Figure 2.7b. One can choose either of two equivalent sets of interstices for placing this second layer. The third layer can be placed in one of two inequivalent orientations. If it is placed directly over the first layer, the stacking sequence can be called AB, and if repeated indefinitely, ABABAB. . . . Hexagonal close packing (hcp) results, as illustrated in Figure

(a)

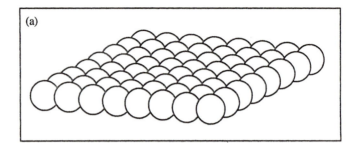

(b)

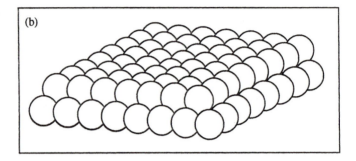

(c)

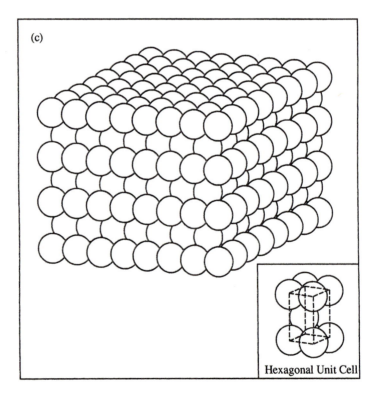

Hexagonal Unit Cell

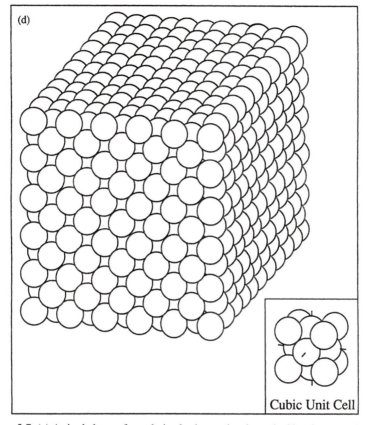

Cubic Unit Cell

Figure 2.7. (a) A single layer of equal-sized spheres closely packed in a hexagonal array. (b) A second layer of spheres placed over the interstices of the first layer. (c) When a third layer is placed exactly above the first layer, the stacking sequence is ABABAB. . . . This is hexagonal close packing (hcp) and results in a hexagonal unit cell. (d) If the third layer of spheres is placed in the other set of interstices, not above layer one, then the stacking sequence is ABCABCABC . . . and a face-centered cubic (fcc) unit cell results. This is cubic close packing (ccp).

2.7c. The c-direction is then different from directions in the a-b plane, and a hexagonal structure with the ideal c/a ratio of 1.634 results. The spheres occupy 74.6 percent of the available space. If the third layer is placed, not over the first, but with sphere centers in the other remaining set of interstices, the sequence ABC results, as shown in Figure 2.7d. Its repetition, ABCABCABC ..., develops cubic close packing (ccp) with the same packing efficiency as hcp. Several depictions of these basic arrangements are shown in Figure 2.8.

Other variants on these packing schemes are possible. A double hexagonal arrangement, ABACABACABAC ..., has been observed in cobalt (Pearson 1972). Any basic sequence can have mistakes, that is, wrong stacking sequences, in it. A random stacking sequence under the constraint that no two adjacent layers are identical is the limiting case of such disorder. Stacking sequences with long repeat distances, for example, one in which a single repeat is ABCABACABCABCAB are also possible, though they generally occur in more complex layered systems. Two structures are defined as being polytypes if they are identical in composition and in basic structure within a layer, but differ in the stacking sequence of layers in a given direction. This concept is important for materials as diverse as zinc sulfide and micas. The thermodynamic, structural, and mechanistic implications of stacking disorder will be discussed in later chapters.

In both fcc and hcp, each atom has twelve nearest neighbors (see Fig. 2.8). These form a *coordination sphere* and each atom is said to have a coordination number of twelve. In the ideal fcc and hcp structures, all twelve nearest neighbors are equidistant. In more complex structures, or in distortions of hcp with c/a different from its ideal value, not all neighbor distances (bond lengths) are equal. In severely distorted coordination polyhedra, the definition of a coordination number becomes less obvious; one can not clearly decide which atoms are too far away to be counted.

Another structure commonly encountered is the body-centered cubic (bcc), illustrated in Figure 2.9. Each atom is surrounded by a cube of others, giving a coordination number of eight. Considered as a packing of equal spheres, here the spheres touch on a body diagonal. The packing efficiency, 67 percent, is less than in hcp or fcc, so one might expect a bcc structure to be less dense than fcc or hcp for a given substance. However, atoms are not, in fact, hard spheres of fixed size, and, in general, as coordination number decreases, the remaining bonds become shorter and stronger (see Chapter 5). The result is that the fcc and bcc forms can be very similar in volume and energy, and a metal like iron shows complex polymorphism (occurrence of different structures) with changes in temperature and pressure (see Fig. 2.10). Such polymorphism is important both to metallurgy and to the structure of the Earth's solid inner core.

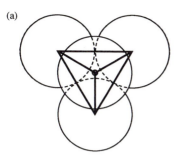

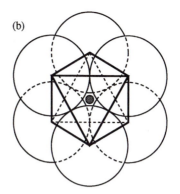

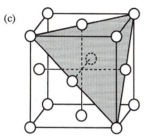

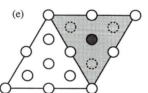

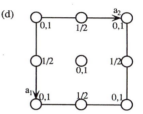

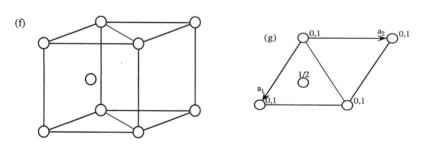

Figure 2.8. Cubic close packing and hexagonal close packing. (a), (b) Nature of interstitial sites that occur between two close-packed layers of equal sized spheres. A four-coordinated, or tetrahedral, site is shown in (a) and a six-coordinated, or octahedral, site is shown in (b). (c) A perspective view of cubic close packing (ccp). The close-packed layers are stacked along the cube diagonal. The (111) plane containing a close-packed layer is shaded. (d) A plan view on (001) of the face-centered cubic unit cell. (e) A projection onto (111) of a close-packed layer. The shaded plane is the same as in (c) with the stippled circle lying above the plane and the dotted circles below the plane. (f) A perspective view of hexagonal close packing (hcp) and (g) a plan view on (001).

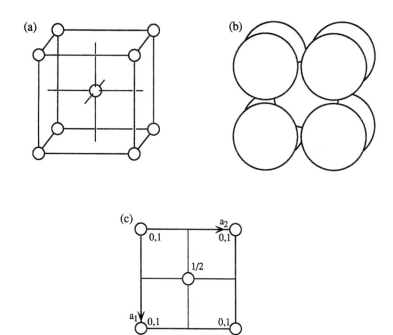

Figure 2.9. Body-centered cubic arrangement (bcc). (a) Perspective view showing atoms at each corner of the unit cell and one at the body center. (b) Packing model of bcc. (c) Projection of the bcc arrangement.

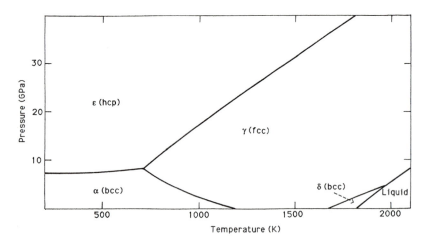

Figure 2.10. Polymorphism of iron as a function of temperature and pressure, α, $\alpha' =$ bcc, $\gamma =$ fcc, $\varepsilon =$ hcp (modified after a compilation by Presnall in preparation).

Element	Z	Structure	Smallest interatomic distances (coordination)
Li	3	A2	3.039 (8)
Be	4	A3	2.226(6), 2.286(6)
Na	11	A2	3.716(8)
Mg	12	A3	3.197(6), 3.209(6)
Al	13	A1	2.864(12)
Si	14	A4	2.353(4)
K	19	A2	4.544(8)
Ca	20	A1 / A3	3.947(12) / 3.940(6), 3.955(6)
Sc	21	A3 / A1	3.256(6), 3.309(6) / 3.212(12)
Ti	22	A3	2.896(6), 2.951(6)
V	23	A2	2.622(8)
Cr	24	A2	2.498(8)
Mn	25	A12, A13, A6, A1, A2	
Fe	26	A2	2.482(8)
Co	27	A1 / A2	2.506(12) / 2.501(6), 2.507(6)
Ni	28	A1	2.492(12)
Cu	29	A1	2.556(12)
Zn	30	A3	2.665(6), 2.913(6)
Ga	31	A11	2.442(1), 2.712(2), 2.742(2), 2.801(2)
Ge	32	A4	2.450(4)
Rb	37	A2	4.95(8)
Sr	38	A1	4.303(12)
Y	39	A3	3.551(6), 3.647(6)
Zr	40	A3	3.179(6), 3.231(6)
Nb	41	A2	2.858(8)
Mo	42	A2	2.725(8)
Tc	43	A3	2.703(6), 2.735(6)
Ru	44	A3	2.650(6), 2.706(6)
Rh	45	A1	2.690(12)
Pd	46	A1	2.751(12)
Ag	47	A1	2.889(12)
Cd	48	A3	2.979(6), 3.293(6)
In	49	A6	3.251(4), 3.373(8)
Sn	50	A4 / A5	2.810(4) / 3.022(4), 3.181(2)
Cs	55	A2	5.324(8)
Ba	56	A2	4.347(8)
La	57	A1 / A3	3.745(12) / 3.739(6), 3.770(6)
Hf	72	A3	3.127(6), 3.195(6)
Ta	73	A2	2.800(8)
W	74	A2	2.742(8)
Re	75	A3	2.741(6), 2.760(6)
Os	76	A3	2.675(6), 2.735(6)
Ir	77	A1	2.775(12)
Pt	78	A1	2.775(12)
Au	79	A1	2.884(12)
Hg	80	A10	3.000(6), 3.466(6) (-46 °C)
Tl	81	A3	3.408(6), 3.457(6)
Pb	82	A1	3.500(12)
Fr	87		
Ra	88		
Ac	89	A1	3.756(12)
Th	90	A1	3.595(12)
Pa	91	A6	3.212(8), 3.238(2)
U	92	α	2.77(2), 2.86(2), 3.28(4), 3.37(4)
Np	93	α	
Pu	94	A1	3.285(12)
Ce	58	A1 / A3	3.650(12) / 3.620(6), 3.652(6)
Pr	59	A1 / A3	3.649(12) / 3.640(6), 3.673(6)
Nd	60	A3	3.638(6), 3.658(6)
Pm	61		
Sm	62	A	3.587(6), 3.629(6)
Eu	63	A2	3.989(8)
Gd	64	A3	3.573(6), 3.636(6)
Tb	65	A3	3.525(6), 3.601(6)
Dy	66	A3	3.503(6), 3.590(6)
Ho	67	A3	3.486(6), 3.577(6)
Er	68	A3	3.468(6), 3.559(6)
Tm	69	A3	3.447(6), 3.538(6)
Yb	70	A1	3.880(12)
Lu	71	A3	3.453(6), 3.503(6)

Figure 2.11. The periodic table showing the structure types for the metallic elements. A1 = ccp, A2 = bcc, A3 = hcp, A4 = diamond, others are more complex structures. The numbers below the symbols are the smallest interatomic distances and the coordination number (in parentheses). (Pauling 1960, p. 410.)

Figure 2.11 shows the periodic table, with the structures of the elements indicated. Note that about two thirds of all elements are metallic at room temperature and atmospheric pressure and that almost all the metals have hcp, fcc, or bcc structures with occasional polymorphism.

2.5.2 Nonmetallic elements

The structures of the nonmetallic elements offer greater variety. Carbon has two common polymorphs stable under different pressure and temperature conditions. Diamond, a three-dimensional tetrahedral framework structure (see Fig. 2.12), belongs to space group Fd3m. It is a high-pressure form occurring in kimberlites and is denser and harder than the other polymorph, graphite. Graphite (see Fig. 2.13) is a layer structure with hexagonal arrays of carbon forming planes that slip easily past each other, which makes graphite a good lubricant. Because the planes are only loosely bonded to each other, graphite can easily absorb (intercalate) many different molecules between the carbon layers and dramatically expand its interlayer spacing. It shares this property of expandability and intercalation with a number of phyllosilicates, notably expanding clays. In diamond, the carbon forms four bonds to other carbons at the corners of a tetrahedron by using sp^3 hybridization of its orbitals, whereas in graphite, carbon is sp^2 hybridized and forms three bonds at 120° angles. The former is analogous to a carbon in an aliphatic hydrocarbon, the latter to carbon in an aromatic. Buckminsterfullerite (buckyballs) is a recently characterized (metastable?) form of carbon consisting of C_{60} molecular clusters.

Silicon and germanium of semiconductor fame crystallize in the diamond structure, whereas tin, below them in the periodic table, is dimorphic between fairly complex semimetallic and metallic structures. Lead is metallic. As one proceeds down a column of the periodic table, the gradual transition from nonmetallic to metallic behavior is a common chemical trend.

The halogens, all diatomic gases at room temperature, at low temperature form crystals based on packings of these diatomic molecules. This is also true of hydrogen, nitrogen, and oxygen. It has long been speculated that under immense pressures (in the 2 to 5 megabar range), attainable in the deep interiors of Jupiter and Saturn, hydrogen becomes a metal analogous to the alkalis. Achieving this transition is one of the goals of high-pressure research.

The heavier members of the oxygen and nitrogen groups crystallize in complex crystal structures. Sulfur and phosphorus show extensive polymorphism in both crystalline and liquid states. In the context of mineral physics, these structural changes, which reflect differences in the size distribution of chains and rings, have conceptual similarities to polymerization equilibria in silicate melts (see Chapter 8), though the analogy is by no means direct.

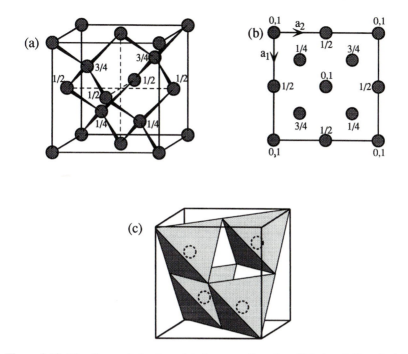

Figure 2.12. The diamond structure. (a) A perspective view. This is a ball-and-stick representation showing the C–C bonds. Each carbon atom is coordinated by four other C atoms. The z-coordinates of the atoms inside the unit cell are given. (b) A projection of the structure on (001) with z-coordinates given. (c) A polyhedral representation showing that each unit cell contains four corner-sharing tetrahedra with C atoms at their corners and at their centers.

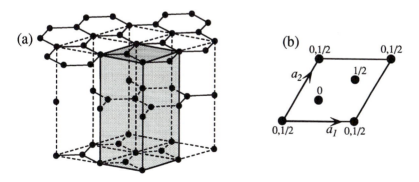

Figure 2.13. The structure of graphite. (a) The structure is composed of layers of hexagonal arrays of C atoms. The vertical dotted lines represent the weak van der Waals forces that join the layers. Four hexagonal unit cells are outlined and one of them shaded. (b) A (001) projection of the structure (after Smith 1982).

2.5.3 Binary compounds of type AX

Compounds of stoichiometry AX, where A is a cation and X is an anion, form a variety of structures. One can derive tetrahedrally coordinated structures from the parent diamond structure by substituting the nearest neighbors of a given atom with a different atom. This can be done in two inequivalent ways (see Fig. 2.14), leading to the cubic (sphalerite) and hexagonal (wurtzite) structure type. In both these structures, each A atom is tetrahedrally coordinated by four X atoms (nearest neighbor coordination is the same), but the arrangement of the next A atoms is different, namely, directly over the previous layer in the hexagonal sphalerite or zinc blend polytype but rotated 120° to it in the cubic wurtzite polytype. The two polytypes can also be thought of as derived from ccp and hcp packings of atoms by filling the tetrahedral interstices with the other atom or ion (see Fig. 2.14). By common convention, though not necessarily because of any unique physical reality, anions are thought of as being large and cations as small, so a structure derived from close packing of anions by cations filling the interstices is a useful conceptual construct. Dimorphism between wurtzite and sphalerite structures, and polytypes involving long and complex stacking sequences occur frequently in zinc sulfide. ZnS and CdS, many III–V semiconductors, and ZnO crystallize in these structures or in significantly distorted variants thereof (see Table 2.6). The transition from tetrahedral to octahedral coordination, as it depends on bonding properties or is induced by pressure, is discussed in Chapter 6.

The filling of octahedral interstices in close packing results in the sodium chloride (rocksalt) structure derived from ccp (see Fig. 2.15) and the nickel arsenide structure derived from hcp (see Fig. 2.16). In each, the cation is surrounded by six anions forming an octahedron. In the sodium chloride type, the anion is likewise octahedrally coordinated by cations, whereas in the nickel arsenide structure, even though the coordination number of the anion is also six, its coordination geometry is a trigonal prism. The common halides and oxides (NaF, NaCl, KCl, MgO, CaO, MnO, FeO, CoO, NiO) adopt this struc-

Figure 2.14. (*Facing page*) The structures of ZnS. (a) The sphalerite structure. The Zn atoms are shown as solid circles and the S atoms by open circles. This cubic structure is similar to the diamond structure with the Zn atoms each coordinated by four S atoms in tetrahedral coordination. One such tetrahedron is shaded. (b) An (001) projection. (c) The wurtzite structure. Each Zn atom is coordinated by four S atoms and vice versa. Both the Zn and S atoms are in hcp arrangements, giving rise to a hexagonal unit cell. (d) An (001) projection of the unit cell.

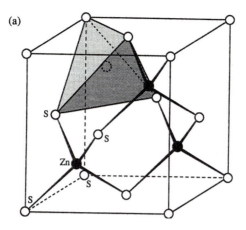

(a)

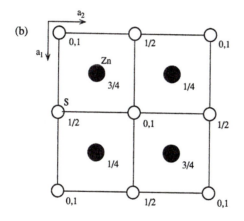

(b)

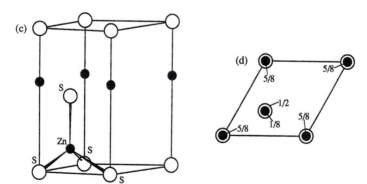

(c)

(d)

Table 2.6. *Phases crystallizing in the wurtzite and sphalerite structures*

Wurtzite	Sphalerite
ZnO[a]	ZnS
CuF	ZnSe
CuCl	ZnTe
AgI	CdS
BeO[a]	CdSe
CoO (metastable)	CdTe
	CoO (metastable)

[a] Heavily distorted

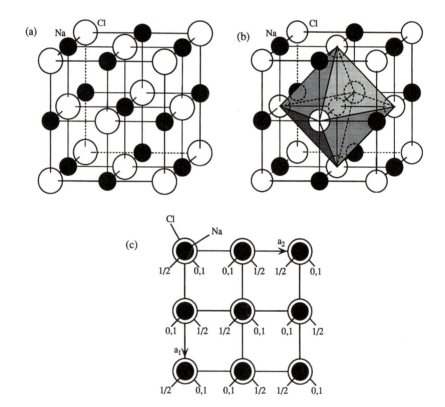

Figure 2.15. The NaCl, or rocksalt, structure. (a) Perspective drawing. The Cl atoms are shown as larger open circles and the Na atoms are solid. Each Na atom is coordinated by six Cl atoms and vice versa, all in regular octahedral coordination. One octahedron is shaded in (b). (c) A projection of the structure.

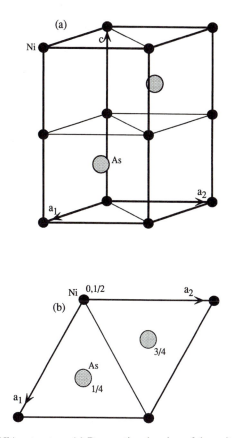

Figure 2.16. The NiAs structure. (a) Perspective drawing of the unit cell with Ni atoms (solid circles) at each corner and at z-coordinates of 1/2 along the cell edges. Two As atoms (shaded) are inside the unit cell, each coordinated by six Ni atoms that define a trigonal prism. The As atoms are in hcp. (b) A projection of the structure on (001).

ture (see Table 2.6). In the NiAs structures derived from hcp, the c-direction is unique, and changes in the c/a ratio, which make the structure deviate from ideal packing, offer a degree of freedom for distortions, possibly a factor in stabilizing those structures relative to the ones derived from ccp in which all three directions are equivalent. The NiAs structure is more common in sulfides and other chalcogenides than in oxides, for example, NiAs, NiS, CoS, MnSe, and many others (see Table 2.7).

The cesium chloride structure is derived from a bcc packing by substituting

Table 2.7. *Phases crystallizing in the rocksalt, nickel arsenide,*
and cesium chloride structures

Rocksalt	Nickel arsenide	Cesium chloride
CsCl (high T)	FeS	CsCl (low T)
RbCl	CoS	RbCl (high P)
KCl	NiS	NH$_4$Cl
NaCl	MnSe	RbBr
LiCl	FeSe	CsBr
CsF	CoSe	NH$_4$Br
RbF	MnTe	RbI
KF	FeTe	CsI
NaF	NiAs	NH$_4$I
LiF		BaO (high P)
BaO		SrO (high P)
SrO		CaO (high P)
CaO		
MgO		
MnO		
FeO		
CoO		
ZnO (high P)		
CdO		
MnS		
MgS		
CaS		
LiH		

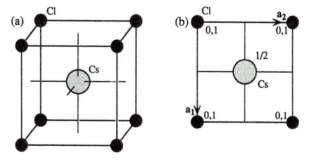

Figure 2.17. The CsCl structure. This structure is based on a bcc arrangement as shown in (a). The central Cl atom (shaded) is surround by eight Cs ions (solid circles) in cubic coordination. Each Cl ion is also coordinated by eight Cs atoms. (b) A projection of the structure.

two kinds of atoms in an ordered fashion such that each atom is coordinated by a cube of neighbors of the opposite type (see Fig. 2.17). It occurs in a number of metallic alloys, in CsCl and RbCl, and in other alkali halides and oxides under high pressure (see Table 2.7).

These five structure types (wurtzite, sphalerite, rocksalt, nickel arsenide, cesium chloride) and distortions thereof account for most of the nonmetallic binary inorganic compounds and for many alloy structures as well. Other AX structures, generally dictated by specific bonding requirements, include the tenorite structure virtually unique to CuO and having square planar coordination for copper, ordered layer structures derived from graphite for boron nitride and silicon carbide, and a structure based on packing of CO molecules for solid carbon monoxide.

For a fixed composition, the density generally increases in the order wurtzite, sphalerite, rocksalt, nickel arsenide, cesium chloride. Phase transitions among these structures can be triggered by both temperature and pressure and are discussed in Section 6.5.

2.5.4 Stoichiometry AX$_2$

The fluorite structure (Fig. 2.18) is related to the cesium chloride type by having the same anion array but missing the cation in every other cube. This arrangement requires the charge of the cation to be twice that of the anion. The coordination number of the cation is eight, as in the cesium chloride type, and that of the anion is four. Generally this structure forms for very ionic compounds with cations significantly larger than anions. Examples are the type-mineral fluorite itself, CaF_2; heavier alkaline earth fluorides, UO_2 and ThO_2; ZrO_2 at high temperature; and the "stabilized" cubic zirconias, ZrO_2 doped with altervalent substituents like Ca and Y (see Table 2.8). The latter materials show fast anionic conductivity (see Chapter 9) and are used as high-temperature electrolytes and, because their high density results in a high index of refraction, as diamond substitutes in the gem trade. At lower temperatures, pure ZrO_2 shows polymorphism among several distorted monoclinic and orthorhombic structures; additional polymorphs, including one with Zr in nine-fold coordination, appear at high pressure. The antifluorite structure (one in which the role of cations and anions is reversed) occurs in Na_2O. TiO_2 is the type-mineral for the rutile structure. Other examples are transition metal fluorides, the low-temperature polymorph of GeO_2, and the high-pressure

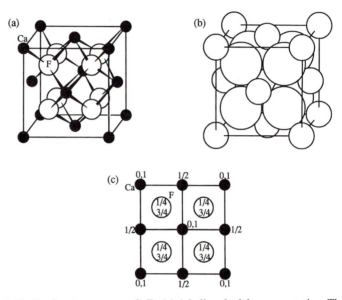

Figure 2.18. The flourite structure, CaF$_2$. (a) A ball-and-stick representation. The cubic unit cell has Ca atoms at the corners and at the face-centers, each of which are coordinated by eight F atoms (open circles). There are four F atoms inside the cell. (b) Packing model of the flourite structure. (c) Projection of the structure onto (001).

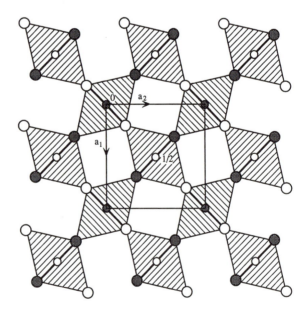

Table 2.8. *AX$_2$ structure types*

Quartz and related	Rutile	Fluorite	Layer structures
SiO$_2$	TiO$_2$	ZrO$_2$	MgCl$_2$
GeO$_2$	GeO$_2$ (low T)	HfO$_2$	MnCl$_2$
BeF$_2$	SiO$_2$ (high P)	CeO$_2$	FeCl$_2$
	SnO$_2$	PrO$_2$	CoCl$_2$
	PbO$_2$	UO$_2$	NiCl$_2$
Chain structures		NpO$_2$	ZnCl$_2$
PbCl$_2$	NbO$_2$	PuO$_2$	CdCl$_2$
CuCl$_2$	TaO$_2$	AmO$_2$	MgBr$_2$
CuBr$_2$	CrO$_2$	CaF$_2$	MnBr$_2$
SiS$_2$	MnO$_2$	SrF$_2$	FeBr$_2$
	WO$_2$	BaF$_2$	CoBr$_2$
Isolated molecules	IrO$_2$	CdF$_2$	NiBr$_2$
CO$_2$	MnF$_2$		ZnBr$_2$
HgCl$_2$	FeF$_2$		MgI$_2$
	CoF$_2$		MnI$_2$
	NiF$_2$		CoI$_2$
	ZnF$_2$		CoI$_2$
	MgF$_2$		NiI$_2$
	PdF$_2$		ZnI$_2$
	CaBr$_2$		TiS$_2$
			ZrS$_2$
			SrS$_2$
			TaS$_2$
			PtS$_2$
			MoS$_2$
			WS$_2$

polymorph of SiO$_2$, stishovite. This structure provides octahedral coordination for cations (see Fig. 2.19) and shows a fairly complex pattern of edge sharing of octahedra. Chlorides of the alkaline earth and transition metals crystallize in a variety of layer structures in which the cation is octahedrally coordinated, but anion layers above and below each cation layer (see Fig. 2.20) are only

Figure 2.19. (*Facing page*) The structure of rutile, TiO$_2$. The tetragonal structure consists of edge-sharing chains of TiO$_6$ octahedra running parallel to the *c*-axis. Octahedra in adjacent chains are joined at their corners. This polyhedral representation is a projection down the *c*-axis, parallel to the chains. The shaded symbols are atoms at a *z*-coordinate of 0, and the open circles are at *z* = 1/2. The unit cell is outlined.

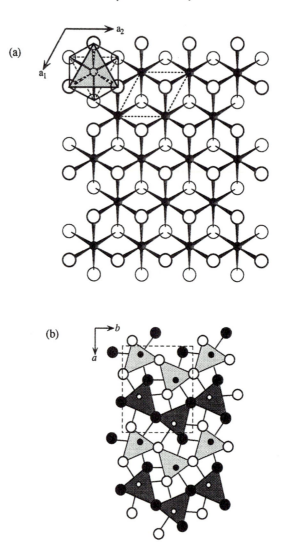

Figure 2.20. Layered AX_2 structures. (a) The basic layers are close packed arrays of anions, shown by open circles. The cations (solid circles) lie in the plane of the page. The anions above the page have darker outlines than those below the page. Each cation is coordinated by six anions. When the anion layers are stacked in an ABAB (hcp) arrangement, the CdI_2 structure results. When the layers are stacked in an ABCABC (ccp) sequence, then the $CdCl_2$ structure results. (b) The $PbCl_2$ structure projected down the *c*-axis. Each Pb (small circles) is coordinated by nine Cl's (larger circles), six of which lie at the corners of a trigonal prism and the other three just beyond the centers of the prism faces. (Data from Hyde and Andersson 1989.)

Table 2.9. *Silica polymorphs*

Polymorph	Space group	Molar volume (cm³/mol) at 298 K	Structural features	Enthalpy relative to α-quartz (kJ/mol)
α-quartz	P3$_2$21 or P3$_1$21^a	22.67	6 rings	0
β-quartz	P6$_2$22 or P6$_4$22^b		6 rings	0.3
Cristobalite	Fd3̄m	25.78	6 rings	2.8
Tridymite	P6$_3$mmc^c	26.53	6 rings	3.2
Glass (fused)	P1	27.27	Distribution of ring size, but 6 rings predominant	7.0
Coesite	C2/c	20.64	4 and 8 rings	2.9
Stishovite	P4$_2$mmm	14.01	Octahedral Si, rutile structure	51.9

^{a} Low-temperature form, space groups reflect chirality (handedness).
^{b} High-temperature form, related by nonquenchable displacive transition.
^{c} High-temperature form. Low-temperature forms related by displacive transitions are very complex.

loosely bonded to each other. Water is easily absorbed between these layers, leading to hygroscopicity and the use of, for example, magnesium chloride as a desiccant.

To a mineralogist, the most ubiquitous and important AX_2 structures are those in which each cation sits in the center of a tetrahedron of anions, and each tetrahedron is linked to four others in an infinite three-dimensional framework. The linkages of tetrahedra into rings result in fairly open structures with internal cavities and/or channels. The silica polymorphs (see Fig. 2.21 and Table 2.9) quartz, cristobalite, tridymite, the high-pressure phase coesite, silica glass, and synthetic zeolitic silicas with large pores all represent variations on the theme of tetrahedral architecture, variations that are numerous and close in energy. This richness reflects the strength and covalency of the Si–O bond and the stability of the Si–O–Si linkage over a large variation of intertetrahedral angles (see Chapter 5). Similar bonding factors allow for the variety of multi-component silicate structures (see following). A coordination number of two is found for silicon in its sulfide, SiS_2, a chain structure, and solid CO_2 represents a loose packing of linear O=C=O molecules.

(a)

(b)

(c)

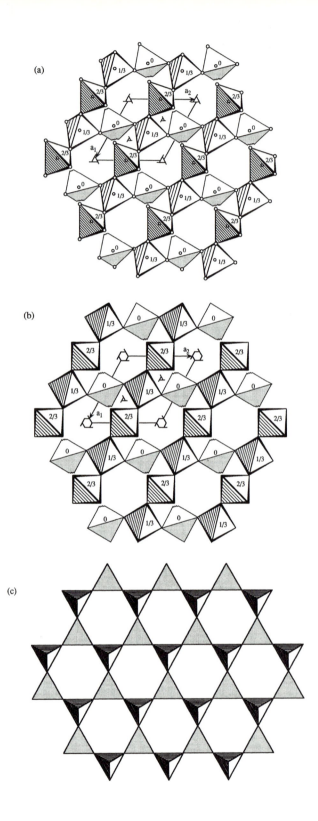

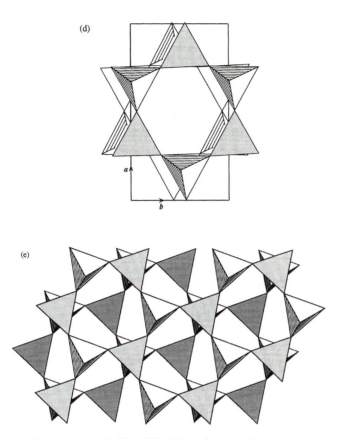

Figure 2.21. The structures of silica, SiO_2. These framework structures are composed of a three-dimensional network of edge-sharing SiO_4 tetrahedra. (a) The structure of α quartz (low-temperature). This polyhedral representation is a projection down the c-axis. At the centers of the large ditrigonal channels are 3_2 screw axes. The z-coordinates of the Si atoms are shown. (b) A similar projection of β quartz (high-temperature) down the c-axis. The channels now have a hexagonal shape with 6_2 screw axes at their centers. (c) Portion of the ideal hexagonal structure of high-temperature tridymite. It consists of six-membered rings of tetrahedra arranged in layers parallel to (001). Half the tetrahedra point up and the other half point down, thereby connecting to adjacent layers. These are stacked in an AB arrangement, resulting in a hexagonal unit cell. (d) Projection of the unit cell for high-tridymite down c. This structure represents a slight tilting of the tetrahedra joining adjacent layers, resulting in orthorhombic symmetry. (e) Projection of a portion of the high-cristobalite structure onto (111). This structure is also composed of six-membered rings of up- and down-pointing SiO_4 tetrahedra, but the layers are stacked in an ABC sequence. (f) Projection of the coesite structure onto (001). Coesite occurs at higher pressures and is composed of four-membered rings of tetrahedra that lie in layers parallel to (001). The rings form chains running parallel to b. These are connected by glide-related chains above and below. (Crystallographic data from Papike and Cameron 1976; Dollase 1967; Peacor 1973; and Zoltai and Buerger 1959.)

2.5.5 Other binary structures

The reader is referred to other texts (e.g., Wells 1984) for discussions of stoichiometry AX_3, A_2X, A_2X_5, and others. A_2X_3 oxides form the corundum structure, a filling of layers of octahedral interstices in an hcp oxygen framework (discussed further in Sec. 2.6.1 in the context of ternary ABX_3 compounds) when the cations are small. Several rare earth oxide structures, with seven-, eight- or nine-fold coordination, are formed for large cations. These oxides often show polymorphism.

2.6 Ternary compounds $A_aB_bX_n$

This section describes structures common to oxides, fluorides, and some silicates. Structures containing T–O–T linkages (T is a tetrahedral atom such as Si, Ge, Al, Ga) will be discussed in Section 2.7. Only a few frequently encountered structures are presented; for an encyclopedic view, the reader is referred to Wells (1984). It is useful to categorize these structures in terms of the relative sizes of the two cations as reflected in the AX and BX bond lengths. When the two bond lengths are similar, the two cations are likely to occupy similar coordination sites and order–disorder may occur. When the two cations are very different in size, ordered structures with very distinct coordination polyhedra for each are formed. If one bond (e.g., B–X) is much stronger than the other (e.g., A–X), the structure may contain discrete complex ions (e.g., BX_n). This transfer of anions to the bonding sphere of the more "powerful" cation may be regarded as an "acid–base reaction" and results in great thermodynamic stability of the ternary compound relative to its binary components (see Sec. 6.2).

2.6.1 Stoichiometry ABX₃

Starting with the case where the B cation bonds much more strongly to the anion than the A cation, one encounters the nitrates and carbonates, A^+NO_3 and $A^{2+}CO_3$. In the rhombohedral carbonates (calcite-related structures, see Fig. 2.22a), which contain relatively small cations, layers of CO_3 groups provide octahedral coordination for the cations that are between these roughly planar layers. Carbon or nitrogen is coordinated in the center of a triangle of

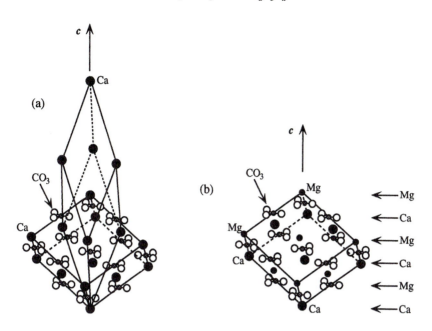

Figure 2.22. (a) The structure of calcite, $CaCO_3$. Ca atoms are shown as solid circles. A face-centered cleavage rhomb is shown that can be derived from the NaCl structure, with $CaCO_3$ groups replacing Na, and the cell compressed along the cube diagonal. The steep, true rhombohedral unit cell for calcite is also shown. (b) The dolomite structure. This structure is derived from the calcite structure by a regular alternation of Ca and Mg (smaller solid circles) in every other layer, doubling the c-axis.

oxygens. Examples are $CaCO_3$, $FeCO_3$, $MnCO_3$, and $NaNO_3$. Because the CO_3 or NO_3 group is almost planar and surrounded above and below by cations, the ease with which it can rotate in the plane defined by its three oxygens differs from the ease of rotation perpendicular to that plane. In addition, slight distortions in the positions of the carbonate or nitrate groups relative to the cation positions can occur. These possibilities lead to a number of complex phase transitions in carbonates and nitrates (see also Chapter 6). When two cations of somewhat different sizes, e.g., Ca and Mg, form a carbonate phase, an ordered dolomite structure, $CaMg(CO_3)_2$, can form. In it the basic calcite-like relation between cation layers and carbonate layers is maintained, but each cation layer contains either only Mg or only Ca, with the two types of layers alternating and leading to a doubling of the unit cell and an ordered superstructure (see Fig. 2.22b). This type of ordering optimizes both Mg–O and Ca–O bond lengths. Order–disorder transitions in dolomite-type structures are discussed

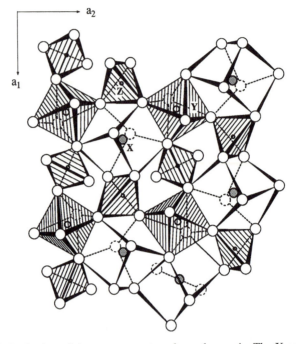

Figure 2.23. Projection of the garnet structure down the c-axis. The X cations are in irregular eight-fold coordination defined by the surrounding oxygens. The Y cations are in six-fold, or octahedral, coordination. The Z cations are in tetrahedral coordination. The YO_6 octahedra and ZO_4 tetrahedra are sharing corners to form a three-dimensional framework. (Crystallographic data from Novak and Gibbs 1971.)

in Chapter 6. For larger cations, orthorhombic carbonate structures with the cations in seven- to nine-fold coordination occur. Reeder (1983) presents a review of carbonate crystal chemistry.

When the B cation is in tetrahedral coordination, as for B $=$ Si or Ge, the metasilicate composition, $ASiO_3$, contains infinite silicate chains (see Section 2.7).

The garnet structure (see Fig. 2.23 and Table 2.10), contains isolated tetrahedra (commonly occupied by Si), octahedra (commonly occupied by Al), and cubic sites (commonly occupied by divalent cations). Though aluminosilicate garnets of the type $A_3B_2Si_3O_{12}$ (A $=$ Mg, Ca, Mn, Fe^{2+}; B $=$ Al, Cr, Fe^{3+}) are the most familiar to mineralogists, garnets containing all trivalent ions, e.g., $Y_3Al_5O_{12}$ (yttrium aluminum garnet, YAG) and its iron counterpart, $Y_3Fe_5O_{12}$ (YIG), are well known as electronic materials, and silicate garnets with no

Table 2.10. *ABX₃ structures*

Pyroxenoid	Pyroxene	Garnet	Ilmenite	Lithium niobate	Perovskite
$CaSiO_3$	$MgSiO_3$	$Mg_3Al_2Si_3O_{12}$	$FeTiO_3$	$LiNbO_3$	$CaTiO_3$
$MnSiO_3$	$FeSiO_3$(h.p.)	$Mg_3Fe^{3+}_2Si_3O_{12}$	$MgTiO_3$	$LiTaO_3$	$SrTiO_3$
	$CoSiO_3$(h.p.)	$Mg_3Cr_2Si_3O_{12}$	$MnTiO_3$	$MnTiO_3$(h.p.)	$BaTiO_3$
	$CaMgSi_2O_6$	$Fe^{3+}_2Al_2Si_3O_{12}$	$CoTiO_3$	$FeTiO_3$(h.p.)	$CaZrO_3$
	$CaFeSi_2O_6$	$Fe^{2+}Fe^{3+}_2Si_3O_{12}$	$MgSnO_3$		$SrZrO_3$
	$CaCoSi_2O_6$	$Fe^{2+}_3Cr_2Si_3O_{12}$	$MgGeO_3$(h.p.)		$BaZrO_3$
	$CaNiSi_2O_6$	$Mn_3Al_2Si_3O_{12}$	$MgSiO_3$(h.p.)		$PbTiO_3$
		$Mn_3Fe^{3+}_2Si_3O_{12}$	$CaGeO_3$(h.p.)		$PbZrO_3$
		$Mn_3Cr_2Si_3O_{12}$	$ZnGeO_3$(h.p.)		$CdTiO_3$(h.p.)
		$Mg_3Al_2Ge_3O_{12}$	$ZnSiO_3$(h.p.)		$MgSiO_3$(h.p.)
		$Mg_3Ga_2Ge_3O_{12}$			$CaGeO_3$(h.p.)
		$Ca_3Al_2Si_3O_{12}$			$MnTiO_3$(h.p.)
		$Ca_3Fe^{2+}_2Si_3O_{12}$			$YAlO_3$
		$Ca_3Cr_2Si_3O_{12}$			$MnTiO_3$(h.p.)
		$MgSiO_3$(h.p.)			$FeTiO_3$(h.p.)
		$CaGeO_3$(h.p.)			
		$CdGeO_3$(h.p.)			
		$Y_3Fe_5O_{12}$			
		$Y_3Al_5O_{12}$			

Note: h.p. = high pressure

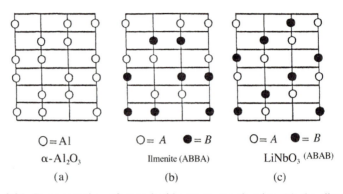

Figure 2.24. Representation of sesquioxide structures showing (a) the disordered (corundum) and the two ordered subfamilies of structures: (b) ilmenite and (c) lithium niobate. The horizontal lines represent the close-packed oxygen planes. The cations occupy the octahedral holes. (After Mehta and Navrotsky 1993.)

aluminum, such as $MgSiO_3$ garnet ($Mg_4Si_4O_{12}$), exist at high pressure. In these latter cases, in which octahedral sites are occupied by cations of different sizes (Y and Fe or Al; Mg and Si), ordering on octahedral sites leads to a reduction of the cubic symmetry to tetragonal.

When both cations in ABO_3 are similar in size and occupy octahedral sites, structures based on ordered derivatives of the corundum structure are formed. The basic structure contains almost hcp oxygen layers with cations defining octahedral layers perpendicular to **c** (see Fig. 2.24a). The occupancy of each layer is identical in the corundum structure for materials such as Al_2O_3, Fe_2O_3, and Cr_2O_3, and each octahedron is slightly distorted as related to the c/a ratio. When two different ions form a ternary ABO_3 structure such as that of ilmenite, $FeTiO_3$, they order in alternate layers (see Fig. 2.24b). The iron and titanium octahedra then have different metal–oxygen bond lengths and different degrees of distortion. The ordered structure is a superstructure of the basic corundum type, and the space group of ilmenite ($R\bar{3}$) is a subgroup of that of corundum ($R\bar{3}$ 2/c). In a solid solution series such as hematite-ilmenite, Fe_2O_3-$FeTiO_3$, order can appear gradually as a higher order phase transition (see Chapter 6).

A different pattern of occupancy of octahedral sites in an hcp oxygen array is encountered in the lithium niobate structure, exemplified by $LiNbO_3$ (see Fig. 2.24c). The octahedral sites occupied are the same ones as in corundum or ilmenite. However, instead of unlike ions segregating into alternate layers as in ilmenite, each layer in the lithium niobate structure contains both species in equal proportions, ordered such that each cation shares three octahedral

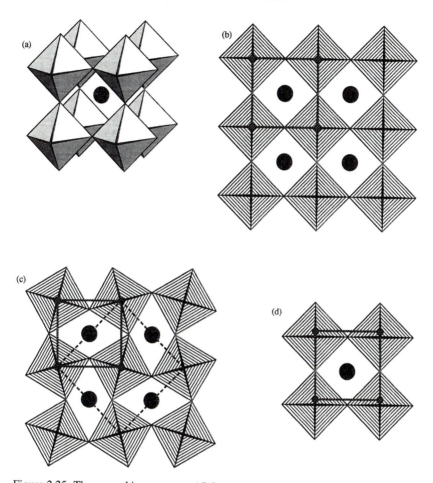

Figure 2.25. The perovskite structure, ABO$_3$, and its distortions. (a) Perspective view of the cubic perovskite structure. It consists of corner-linked BO$_6$ octahedra with the A atom being 12-coordinated at the center of the unit cell. (b) A plan view of the structure looking down on top of the corner-sharing octahedra. The cubic unit cell is outlined with a heavier, shaded line. One possible distortion occurs when the A cation is too small for 12-coordination and the octahedra tilt with respect to one another to accommodate the smaller atom. This is illustrated schematically in (c). The original unit cell is indicated and the new, larger unit cell is shown by the dashed line. (d) A second type of distortion. Here the B cation has moved slightly from the octahedron center resulting in a reduction of symmetry to tetragonal, though the unit cell has the same shape. [(a), (c) and (d) after Putnis 1992.]

edges with one of the opposite type. In this way no cation is bridged through two oxygens to another like cation. The lithium niobate structure can also be described as a very distorted derivative of the perovskite type (see following discussion).

When the A cation requires much longer bond lengths and a higher coordination number than the B cation, the perovskite structure, named after a mineral of idealized composition, $CaTiO_3$, can form. It consists of a framework of virtually regular BO_6 octahedra linked through all their corners (see Fig. 2.25). If the octahedra are linked to form an ideal cubic array, the interstices between them define a site surrounded by twelve equidistant anions that is occupied by the larger A cation. Because the octahedra form an interconnected framework, one can change the bond length in the central site only by perturbing the octahedra. The most common perturbations involve tilting or puckering of the octahedral framework, which leaves the octahedra regular in shape but brings some oxygens closer to the cation in the central site, decreasing its coordination number from twelve to ten or eight. Several different distortions are commonly seen (see Fig. 2.25 and Table 2.19). These distortions allow a much wider range of cation sizes to be accommodated in the perovskite structure than the purely cubic arrangement would permit. For a cubic perovskite, one can define a tolerance factor, t, for a perovskite ABX_3 as

$$t = \frac{r_{AX}}{\sqrt{2}r_{BX}} = \frac{r_A + r_X}{\sqrt{2}(r_B + r_X)} \tag{2.11}$$

where the r terms are bond lengths calculated as sums of ionic radii (see Chapter 5). If $t = 1$, then both cations can be accommodated in the ideal cubic perovskite with their optimum cation–anion bond lengths. If t deviates slightly from unity, typically in the range 0.8 to 1.1, a perovskite may still form, with the degree of distortion from cubic symmetry increasing and the thermodynamic stability decreasing (see Chapter 6) with increasing values of the absolute value of $(1 - t)$. The type and extent of distortion depend sensitively on temperature, pressure, and composition. Perovskites are common in fluorides as well as oxides.

2.6.2 Stoichiometry AB_2X_4

When both cations are small and tetrahedrally coordinated, the phenacite structure, as in Be_2SiO_4 and Zn_2SiO_4, forms (see Fig. 2.26 and Table 2.10). Two

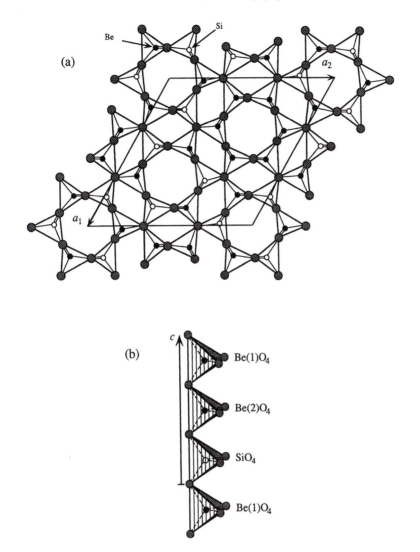

Figure 2.26. The phenacite structure, Be$_2$SiO$_4$. (a) Projection (001). The structure comprises two types of BeO$_4$ tetrahedra and SiO$_4$ tetrahedra. The oxygen atoms (stippled circles) and the metal cations (solid circles = Be, open circles = Si) form continuous chains parallel to the *c*-axis. (b) Schematic representation of the tetrahedral chains running parallel to the *c*-axis (after Zachariasen 1972).

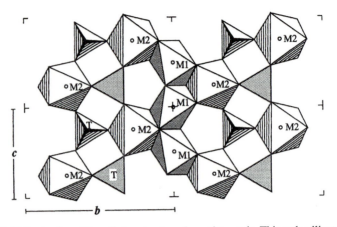

Figure 2.27. Projection of the olivine structure down the *a*-axis. This orthosilicate structure consists of chains of octahedra running parallel to *c*, cross-linked by isolated SiO$_4$
tetrahedra. The tetrahedra share three edges with octahedra in the *b-c* plane, causing
considerable elongation along *a*. There are two distinct octahedral sites, M1 and M2
(after Papike and Cameron 1976). The oxygen array is a distortion of hcp.

structures are common when one cation is tetrahedrally and the other octahedrally coordinated. The less dense of the two is the olivine type, based on a
distorted hexagonal array of oxygens and a pattern of octahedral site occupancy that defines two distinct octahedral sites commonly labeled M1 and M2
(see Fig. 2.27). The denser structure, based on an almost ccp array of oxygens,
is the spinel, described in Section 2.24. The distortion from close packing is
greater in the olivine than the spinel and the olivine contains two distinct octahedra, each more distorted than the one type of octahedron in the spinel. In
general, the difference in tetrahedral and octahedral bond length is greater in
the olivine than in the spinel. Thus cation disorder, common in spinels, does
not occur in olivines. However, varying degrees of ordering between the M1
and M2 octahedra in olivine are commonly seen.

An AB$_2$X$_4$ structure denser than spinel is the K$_2$NiF$_4$ type (see Fig. 2.28).

(a)

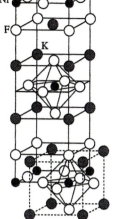

(b)

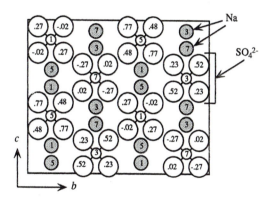

Figure 2.28. (a) A perspective view of the tetragonal K_2NiF_4 structure. This structure is related to the perovskite structure with the Ni in the same octahedral coordination as in perovskite. K, however, is nine-coordinated, instead of twelve. The atomic arrangement in this structure results in a c-axis three times the length of the cubic perovskite a-axis. (b) Projection down the a-axis of the orthorhombic structure of Na_2SO_4. The Na atoms are shown as medium-sized, lightly stippled circles. Each S atom (small circles) is surrounded by four oxygens (large circles). These SO_4^{2-} groups are shown connected by lines. The x-coordinates given for the Na and S atoms are in units of eighths of the a-axis; the x-coordinates of the oxygens are given in decimal form.

2.7 Silicate structures

Silicates are generally classified according to the extent to which their SiO_4 and AlO_4 tetrahedra are cross linked. This in turn depends on their stoichiometry; a phase relatively poor in silica will tend to have its tetrahedra isolated, whereas one that is silica-rich will be more highly polymerized. Thus the term *orthosilicate* refers both to a composition in which the ratio of metal oxide to silica is 2:1 (e.g., Mg_2SiO_4) and, more generally, to a phase containing isolated SiO_4 tetrahedra and no Si–O–Si linkages. The term *metasilicate* means a silicate with a 1:1 metal oxide to silica ratio and a structure containing infinite silicate chains. These can be single chains as in the pyroxenes and pyroxenoids, double chains as in the amphiboles, or even triple and quadruple chains. This compositional and structural use of nomenclature can be confusing at times; for example, the high-pressure polymorph of $MgSiO_3$ having the garnet structure is a metasilicate compositionally but an orthosilicate structurally. The structural classification is summarized in Table 2.11. Several excellent reviews of structures of common rock-forming minerals are available (Griffen 1992; Jaffe 1988; Papike and Cameron 1976; Putnis 1992; Smyth and Bish 1988).

2.7.1 Orthosilicates

Orthosilicates, containing only isolated SiO_4 tetrahedra, constitute a diverse group of structures. The olivine, spinel, phenacite, and garnet structures, already described, are examples. Others of mineralogical importance are zircon, $ZrSiO_4$ (see Fig. 2.29), and the aluminum silicate polymorphs, andalusite, silimanite, and kyanite (see Fig. 2.30). In the latter, although all three structures contain isolated silicate tetrahedra, the coordination number of aluminum increases such that sillimanite is $^{IV}Al\,^{VI}Al\,^{IV}SiO_5$, andalusite is $^{V}Al\,^{VI}Al\,^{IV}SiO_5$, and kyanite is $^{VI}Al_2\,^{IV}SiO_5$. The increasing density of these commonly encountered phases with increasing Al-coordination, and the modest pressure and temperature range in which their transitions occur, make them important indicators of metamorphic grade. The aluminum silicates are reviewed in detail by Kerrick (1990). Orthosilicates are reviewed by Papike and Cameron (1976) and Ribbe (1980).

Table 2.11. *Silicate classification*

Type	Features	Examples
Orthosilicate	Isolated SiO_4 tetrahedra	Olivine, garnet, zircon kyanite, spinel
Pyrosilicate	Si_2O_7 groups	Melilite, wadsleyite
Chain silicate	Chains of linked SiO_4 tetrahedra[a]	Pyroxene (single chains) Amphibole (double chains)
Sheet silicate (phyllosilicate)	Sheets of linked SiO_4 tetrahedra[a]	Mica, clay
Framework silicate (tectosilicate)	3-dimensional framework of linked SiO_4 tettrahedra[a]	Quartz, feldspar, cordierite, zeolites

[a] Al commonly substituting for Si with appropriate charge balance.

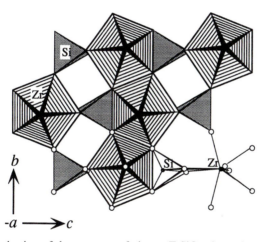

Figure 2.29. Projection of the structure of zircon, $ZrSiO_4$, down the negative *a*-axis. This orthosilicate consists of chains of isolated SiO_4 tetrahedra sharing edges with ZrO_8 triangular dodecahedra in an alternating sequence along *c*. One eight-coordinated Zr site is shown in ball-and-stick representation (after Speer 1982 and after Papike 1987).

(a)

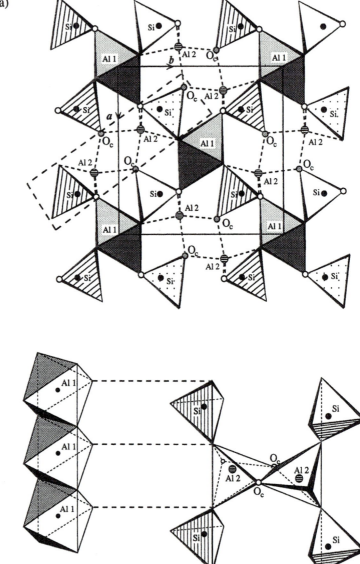

Figure 2.30. The aluminosilicate Al_2SiO_5 structures. (a) Projection of the orthorhombic andalusite structure down the c-axis. Al atoms, Al 1, are in nearly regular octahedral sites that form edge-sharing chains parallel to c. Al 2 atoms occupy unusual five-coordinated sites. The section outlined in a heavy dashed line is shown in projection down a in (b). This direction gives a better view of the Al 2 coordination polyhedron.

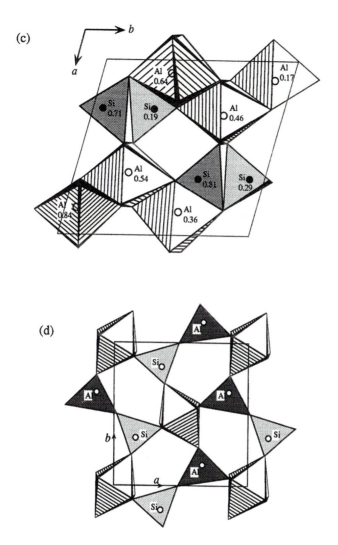

(c) Projection of the kyanite structure down the *c*-axis. This triclinic structure has all the Al in octahedral coordination. Half the Al occurs as isolated AlO$_6$ octahedra and the other half form edge-sharing chains parallel to *c*. (d) Projection of the sillimanite structure down the *c*-axis. The structure is orthorhombic with half the Al in edge-sharing octahedral chains parallel to *c*. The rest of the Al occupy tetrahedra that alternate with SiO$_4$ tetrahedra to form chains parallel to *c* (after Papike and Cameron 1976).

Table 2.12. *Structures of silicates containing* Si_2O_7 *groups or
small rings of tetrahedra*

Structure	Features	Examples
Modified spinel, β, or wadsleyite	Ccp oxygens, Si_2O_7 groups, octahedral M^{2+}	β-Mg_2SiO_4, β-Co_2SiO_4, β-Mn_2GeO_4
Melilite	Si_2O_7 groups plus other tetrahedra and large cations, no octahedra	$CaNaSi_3O_7$ $CaAl_2Si_2O_7$ $CaAlZnSi_2O_7$ $CaAlCoSi_2O_7$
Lawsonite	Hydrous silicate with Si_2O_7 groups	$Ca_2Al_2Si_2O_7(OH)_2H_2O$

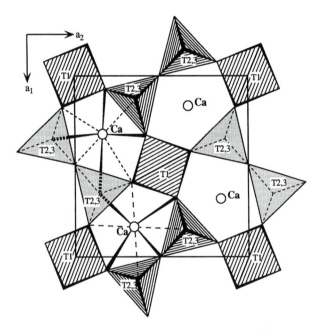

Figure 2.31. Projection of the melilite structure down the *c*-axis. The structure is tetragonal with Ca (or Na) occupying large, eight-coordinated sites in the cavities between
tetrahedra. The tetrahedra may contain Al, Mg, Fe2+, Fe3+, or Si, and various compositions of minerals in this group have different ordering schemes. The T sites form T_2O_7
groups and may be occupied by Si or a 50/50 mixture of Al and Si, depending on
composition (after Papike 1987).

2.7.2 Pyrosilicates

Pyrosilicates, which can be thought of as being formed by the condensation or polymerization of two SiO_4 groups to form an Si_2O_7 group with a bridging oxygen, are relatively uncommon as minerals. The melilite family (see Table 2.12 and Fig. 2.31) are one example. A polymorph of Mg_2SiO_4, called wadsleyite or the β-phase, intermediate between olivine and spinel in density and in its pressure of occurrence, also contains silicate tetrahedra linked through a common oxygen, (also see Sec. 2.8.3 and Chapter 6).

2.7.3 Chain silicates

Silicates containing infinite chains of linked tetrahedra are common at metasilicate stoichiometry. The articulation of the silicate chains determines the space group and the number of distinct cation sites and their symmetry. The tetrahedra lie in chains separated by layers of octahedra (see Fig. 2.32a, c). In true pyroxenes, the tetrahedral chains repeat every two tetrahedra (see Fig. 2.32b), but do not kink. The intertetrahedral angle and rotation of tetrahedra relative to one another can vary to extend or contract the chain, change the unit cell dimensions, and determine the size and distortions of the two inequivalent cation sites, M1 and M2. The space groups and examples of some common pyroxenes are shown in Table 2.13. The orthopyroxenes (Pbnm) common in Ca-poor compositions like enstatite ($MgSiO_3$) and ferrosilite ($FeSiO_3$) and the clinopyroxenes (C2/c) occurring in Ca-rich compositions like diopside ($CaMgSi_2O_6$) and hedenbergite ($CaFeSi_2O_6$) are important examples. In the pyroxenoids, every third, fifth, seventh, or $(2n+1)$st tetrahedron represents a "kink" in the chain (see Fig. 2.32b), providing a greater number of inequivalent cation sites, accommodating larger cations, and resulting in a larger repeat distance along the chain. The pyroxenoids form a *homologous series* (see Sec. 2.8.2) of structures, the chain kink repeat represented by a series of odd integers and the pyroxene itself (no kinks) represented by the end of the series, with n being infinite. Intergrowths and mistakes, having mixtures of different kinds of repeats, are common, especially in phases rich in Mn, Ca, and Fe. The energetics of different pyroxenes and pyroxenoid polymorphs for a given composition are strikingly similar, and because the density generally increases with increasing n, complex polymorphism is often seen as a function of temperature and pressure. The small driving force in terms of overall energetics

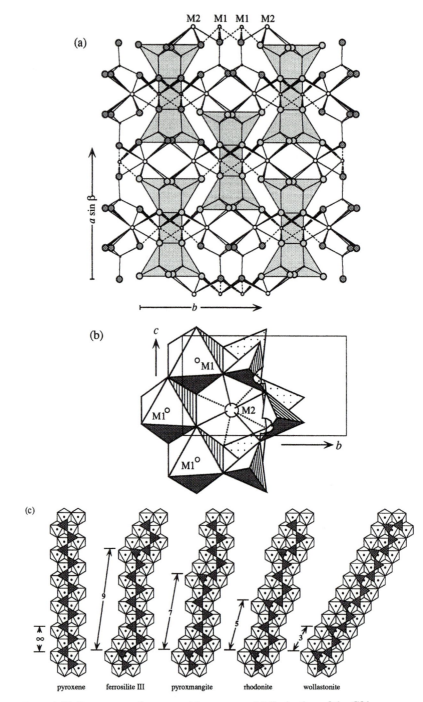

Figure 2.32. Pyroxene and pyroxenoid structures. (a) Projection of the C2/c pyroxene structure down the c-axis. The single silicate chains are normal to the page. There are two distinct octahedral sites, M1 and M2, that form narrow octahedral strips or chains parallel to c. These are sandwiched between opposing silicate chains, forming a struc-

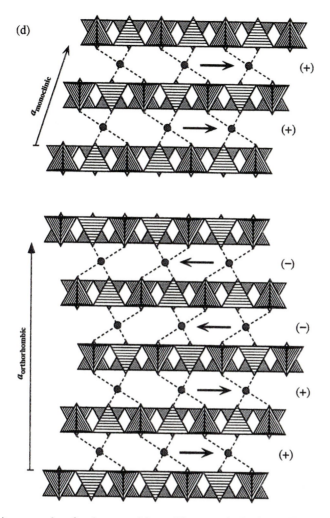

(d)

$a_{\text{monoclinic}}$

(+)

(+)

$a_{\text{orthorhombic}}$

(−)

(−)

(+)

(+)

tural unit commonly refered to as an *I-beam*. These are shaded in (a). The I-beam units are cross-linked by the M2 site. The geometry of the octahedral sites is shown more clearly in (b), which is a view of a portion of the structure projected down the a-axis. The M1 site is nearly regular; M2 can accomodate large cations (e.g., Ca), becoming eight-coordinated, or small cations (e.g., Mg), which would be six-coordinated. (c) Tetrahedral chain and octahedral chain relationships for pyroxenes and pyroxenoids. For all the structures the chains are parallel to c. The numbers indicated for each mineral represent the number of pyroxene-like tetrahedral chain elements present in each structure before the pattern repeats. (d) Schematic diagrams showing the relationship between clinopyroxene and orthopyroxene. The view is parallel to the b-axis. The structures are related by the stacking sequence (+ or −) along a of the octahedral chains. For the clinopyroxenes, the stacking is always in the same direction, symbolized by + + + +. . . , or simply (+). This leads to a monoclinic unit cell. For the orthopyroxenes, the stacking sequence is + + − − + + − − . . . , or (+ −), indicating that the octahedra reverse orientation every other unit cell. This is essentially a twinning operation on (100) at the unit cell scale. The resultant unit cell is orthorhombic with the a-axis being nearly twice as long as in clinopyroxene. [(a) and (c) after Papike and Cameron 1976; (b) after Czank and Liebau 1980.]

Table 2.13. *Structures of chain silicates*

End-member	Solid solution	Composition	Space group	Silicate chain repeat
Pyroxenes				
Mg–Fe^{2+}–Mn pyroxenes				
Orthoenstatite		$Mg_2Si_2O_6$	Pbca	Infinite
Clinoenstatite		$Mg_2Si_2O_6$	P2$_1$/c	
Protoenstatite		$Mg_2Si_2O_6$	Pbcn	
Orthoferrosilite		$Fe_2^{2+}Si_2O_6$	Pbca	
Clinoferrosilite		$Fe_2^{2+}Si_2O_6$	P2$_1$/c	
	Pigeonite	$(Mg,Fe^{2+},Ca)_2Si_2O_6(Ca<0.2)$	P2$_1$/c	
Kanoite		$MnMgSi_2O_6$	P2$_1$/c	
	Donpeacorite	$(Mn,Mg)MgSi_2O_6$	Pbca	
Ca pyroxenes				
Diopside		$CaMgSi_2O_6$	C2/c	
Hedenbergite		$CaFe^{2+}Si_2O_6$	C2/c	
	Augite	$(Ca,Mg,Fe^{2+})_2Si_2O_6(Ca\geq0.2)$	C2/c	
Johannsenite		$CaMnSi_2O_6$	C2/c	
Petedunnite		$CaZnSi_2O_6$	C2/c	
Esseneite		$CaFe^{3+}(AlSi)O_6$	C2/c	

Ca–Na pyroxenes			
Omphacite	$(Ca,Na)(M^{2+},Al)Si_2O_6$	C2/c	
		P2/n	
Aegirine-augite	$(Ca,Na)(M^{2+},Fe^{3+})Si_2O_6$	C2/c	
Na pyroxenes			
Jadeite	$NaAlSi_2O_6$	C2/c	
Aegirine	$NaFe^{3+}Si_2O_6$	C2/c	
Kosmochlor	$NaCr^{3+}Si_2O_6$	C2/c	
Jervisite	$NaSc^{3+}Si_2O_6$	C2/c	
Li pyroxene			
Spodumene	$LiAlSi_2O_6$	C2	
Pyroxenoids			
Wollastonite-Tc	$Ca_2Si_2O_6$	$P\bar{1}$	3
Wollastonite-2M	$Ca_2Si_2O_6$	$P2_1/a$	3
Bustamite	$CaMnSi_2O_6$	$A\bar{1}$	3
Rhodomite	$Mn_2Si_2O_6$	$P\bar{1}$	5
Pyroxmangite	$(Ca,Fe,Mn)_2Si_2O_6$	$P\bar{1}$	7
Ferrosilite III	$Fe_2Si_2O_6$	$P\bar{1}$	9

Source: Modified from Griffen (1992).

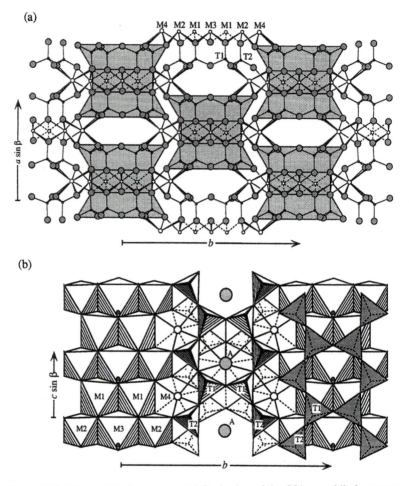

Figure 2.33. The amphibole structure. (a) Projection of the C2/m amphibole structure down the c-axis. The double silicate chains are coming out of the page. There are several octahedral sites, M1, M2, M3, M4, which form an octahedral strip that is sandwiched between two opposing silicate chains. The I-beam structural units for amphibole are shaded. Note they are wider than that of pyroxene (Fig. 2.32a). These I-beams are cross-linked by the M4 site. Amphiboles and related minerals are commonly depicted schematically by simply showing the I-beams (see Fig. 2.38). (b) Projection of the structure down the a-axis. The different octahedral and tetrahedral sites are labeled. The large A site, existing between back-to-back double chains, may be vacant, partially filled, or completely filled. The M4 site is very expandable and may be eight-coordinated for large cations like Ca or Na, or six-coordinated for atoms like Mg (after Papike and Cameron 1976).

Table 2.14. Chemical compositions and names for amphiboles

	A	M	M¹	T	
Tremolite		Ca_2	Mg_5	Si_8	$O_{22}(OH)_2$
Ferro-actinolite		Ca_2	Fe_5^{2+}	Si_8	$O_{22}(OH)_2$
Edenite	Na	Ca_2	Mg_5	Si_7Al	$O_{22}(OH)_2$
Ferro-edenite	Na	Ca_2	Fe_5^{2+}	Si_7Al	$O_{22}(OH)_2$
Pargasite	Na	Ca_2	Mg_4Al	Si_6Al_2	$O_{22}(OH)_2$
Ferro-pargasite	Na	Ca_2	$Fe_4^{2+}Al$	Si_6Al_2	$O_{22}(OH)_2$
Hastingsite	Na	Ca_2	$Fe_4^{2+}Fe^{3+}$	Si_6Al_2	$O_{22}(OH)_2$
Magnesio-hastingsite	Na	Ca_2	Mg_4Fe^{3+}	Si_6Al_2	$O_{22}(OH)_2$
Alumino-tschermakite		Ca_2	Mg_3Al_2	Si_6Al_2	$O_{22}(OH)_2$
Ferro-alumino-tschermakite		Ca_2	$Fe_3^{2+}Al_2$	Si_6Al_2	$O_{22}(OH)_2$
Ferri-tschermakite		Ca_2	$Mg_3Fe_2^{3+}$	Si_6Al_2	$O_{22}(OH)_2$
Alumino-magnesio-hornblende		Ca_2	Mg_4Al	Si_7Al	$O_{22}(OH)_2$
Alumino-ferro-hornblende		Ca_2	$Fe_4^{2+}Al$	Si_7Al	$O_{22}(OH)_2$
Kaersutite	Na	Ca_2	Mg_4Ti	Si_6Al_2	$(O+OH)_{24}$
Ferro-kaersutite	Na	Ca_2	$Fe_4^{2+}Ti$	Si_6Al_2	$(O+OH)_{24}$
Richterite	Na	$CaNa$	Mg_5	Si_8	$O_{22}(OH)_2$
Ferro-richterite	Na	$CaNa$	Fe_5^{2+}	Si_8	$O_{22}(OH)_2$
Ferri-winchite		$CaNa$	Mg_4Fe^{3+}	Si_8	$O_{22}(OH)_2$
Alumino-winchite		$CaNa$	Mg_4Al	Si_8	$O_{22}(OH)_2$
Ferro-alumino-winchite		$CaNa$	$Fe_4^{2+}Al$	Si_8	$O_{22}(OH)_2$
Ferro-ferri-winchite		$CaNa$	$Fe_4^{2+}Fe^{3+}$	Si_8	$O_{22}(OH)_2$
Alumino-barroisite		$CaNa$	Mg_3Al_2	Si_7Al	$O_{22}(OH)_2$

Table 2.14. *Continued*

	A	M	M¹	T	
Ferro-alumino-barroisite		CaNa	$Fe_3^{2+}Al_2$	Si_7Al	$O_{22}(OH)_2$
Ferri-barroisite		CaNa	$Mg_3Fe_2^{3+}$	Si_7Al	$O_{22}(OH)_2$
Ferro-ferri-barroisite		CaNa	$Fe_3^{2+}Fe_2^{3+}$	Si_7Al	$O_{22}(OH)_2$
Magnesio-ferri-katophorite	Na	CaNa	Mg_4Fe^{3+}	Si_7Al	$O_{22}(OH)_2$
Magnesio-alumino-katophorite	Na	CaNa	Mg_4Al	Si_7Al	$O_{22}(OH)_2$
Ferri-katophorite	Na	CaNa	$Fe_4^{2+}Fe^{3+}$	Si_7Al	$O_{22}(OH)_2$
Alumino-katophorite	Na	CaNa	$Fe_4^{2+}Al$	Si_7Al	$O_{22}(OH)_2$
Ferri-taramite	Na	CaNa	$Fe_3^{2+}Fe_2^{3+}$	Si_6Al_2	$O_{22}(OH)_2$
Magnesio-ferri-taramite	Na	CaNa	$Mg_3Fe_2^{3+}$	Si_6Al_2	$O_{22}(OH)_2$
Alumino-taramite	Na	CaNa	$Fe_3^{2+}Al_2$	Si_6Al_2	$O_{22}(OH)_2$
Magnesio-alumino-taramite	Na	CaNa	Mg_3Al_2	Si_6Al_2	$O_{22}(OH)_2$
Glaucophane		Na_2	Mg_3Al_2	Si_8	$O_{22}(OH)_2$
Ferro-glaucophane		Na_2	$Fe_3^{2+}Al_2$	Si_8	$O_{22}(OH)_2$
Magnesio-riebeckite		Na_2	$Mg_3Fe_2^{3+}$	Si_8	$O_{22}(OH)_2$
Riebeckite		Na_2	$Fe_3^{2+}Fe_2^{3+}$	Si_8	$O_{22}(OH)_2$
Eckermannite	Na	Na_2	Mg_4Al	Si_8	$O_{22}(OH)_2$
Ferro-eckermannite	Na	Na_2	$Fe_4^{2+}Al$	Si_8	$O_{22}(OH)_2$
Magnesio-arfvedsonite	Na	Na_2	Mg_4Fe^{3+}	Si_8	$O_{22}(OH)_2$
Arfvedsonite	Na	Na_2	$Fe_4^{2+}Fe^{3+}$	Si_8	$O_{22}(OH)_2$
Kozulite	Na	Na_2	$Mn_4(Fe^{3+},Al)$	Si_8	$O_{22}(OH)_2$

Note: These represent idealized end-member compositions, which seldom exist as such and may not be stable in all cases.

General formula is $A_xM_2^1M_3^1T_8O_{12+y}(OH)_{2-2y}(0 \le x \le 1, 0 \le y \le 1)$

Source: Modified from Veblen (1981) and Griffen (1992).

coupled with high activation energies for rearranging Si–O–Si bonds lead to exceedingly sluggish kinetics for these transformations, and the phase equilibria among chain silicate polymorphs continue to be subjects of debate and confusion. Detailed reviews of pyroxene and pyroxenoids are given by Griffen (1992), Papike and Cameron (1976), and Prewitt (1980).

The amphiboles, general formula $A_aM_2'M_5T_8O_{22}(OH,F)_2$ ($0 < a < 1$) contain double chains of silicate tetrahedra with up to five different coordination sites for cations (see Fig. 2.33). In the preceding formula, three types of cation sites are distinguished: a large A site for alkalis (which may be full, empty, or partly filled), a site (or two distinct sites) M′ for larger divalent ions like Ca, and a set of up to three distinct octahedral M sites for cations like Fe and Mg. The tetrahedral sites contain Si and Al. Like pyroxenes, amphiboles occur in monoclinic and orthorhombic symmetries. Because of this complexity of sites, a large variety of substitutions is possible. Table 2.14 gives a fairly extensive list of amphibole end-member compositions. The number of entries illustrates the richness of substitutions (and nomenclature) typical of chain and sheet silicates. In reality, almost all naturally occurring amphiboles are multicomponent solid solutions seldom near end-member compositions. In addition, the substitution $Fe^{2+} + OH^- = Fe^{3+} + O^{2-}$ results in oxidized species (oxoamphiboles) of general formula $A_aM_2'M_5T_8O_{22+y}(OH)_{2-2y}$. For further discussions, see Veblen (1981).

2.7.4 Sheet silicates

These silicates consist of layers or sheets of tetrahedra linked in a two-dimensional array and possessing a net negative formal charge. The charge balancing di- and trivalent cations in octahedral coordination form an octahedral sheet. Large interlayer cations and water intercalate between the octahedral sheets. The micas and clays are the main members of this family (see Tables 2.15 and 2.16 and Fig. 2.34). Because bonding is strong within each layer but weak between sheets, these minerals easily undergo cation exchange and the intercalation and removal of water molecules (and, in some cases, larger organic molecules as well) from interlayer sites. Polytypism is an additional frequent complication. For details, see Bailey (1984, 1988) and Eslinger and Pevear (1988).

Table 2.15. *Chemical compositions and names for layer silicates*

Mica group	A	M	T $T_4O_{10}(OH)_2$
Common micas			
Dioctahedral micas			
Muscovite	K	Al_2	Si_3Al
Paragonite	Na	Al_2	Si_3Al
Phengite	K	$Al_{1.5}(Mg,Fe^{2+})_{0.5}$	$Si_{3.5}Al_{0.5}$
Celadonite	K	$(Mg,Fe^{2+})(Fe^{3+},Al)$	Si_4
Trioctahedral micas			
Phlogopite	K	Mg_3	Si_3Al
Annite	K	Fe_3^{2+}	Si_3Al
Biotite	K	$(Mg,Fe^{2+})_3$	Si_3Al
Ferriannite	K	Fe_3^{2+}	Si_3Fe^{3+}
Siderophyllite	K	$Fe_2^{2+}Al$	Si_2Al_2
Polylithionite	K	Li_2Al	Si^4
Trilithionite	K	$Li_{1.5}Al_{1.5}$	Si_3Al
Taeniolite	K	Mg_2Li	Si_4
Zinnwuldite	K	$Fe^{2+}Li(Al,Fe^{3+})$	Si_3Al
Ephesite	Na	$LiAl_2$	Si_2Al_2
Brittle micas			
Dioctahedral			
Margarite	Ca	Al_2	Si_2Al_2
Trioctahedral			
Clintonite	Ca	Mg_2Al	Si_2Al_2
Bityite	Ca	Al_2Li	Si_2AlBe
Anandite	Ba	Fe_3^{3+}	Si_3Fe^{3+}
Clays			
Dioctahedral			
Pyrophyllite		Al_2	Si_4
Trioctahedral			
Talc		Mg_3	Si_4
Other layer silicates (clay minerals)			
Dioctahedral			
Kaolinite		$Al_2Si_2O_5(OH)_4$	
Trioctahedral			
Serpentine		$Mg_3Si_2O_5(OH)_4$	

Note: Many clays derived by substituting cations $A_A^+ + Al_T$ for Si_T in these structures (see Table 2.16). Idealized end-member compositions not common or always stable.
Source: Modified from Bailey (1988).

Table 2.16. *Classification of layer silicates*

Layer type	Interlayer material	Group	Subgroup	Species (examples)
1:1	None or H_2O only	Serpentine-kaoline ($x \sim 0$)[a]	Serpentine	Chrysotile, lizardite, berthierine
			Kaolin	Kaolinite, dickite, nacrite, halloysite
2:1	None	Talc-pyrophyllite ($x \sim 0$)	Talc	Talc, willemseite
			Pyrophyllite	Pyrophyllite
	Hydrated exchangeable cations	Smectite ($x \sim 0.2–0.6$)	Saponite	Saponite, hectorite, sauconite, stevensite, etc.
			Montmorillonite	Montmorilloite, beidellite, nontronite
	Hydrated exchangeable cations	Vermiculite ($x \sim 0.6–0.9$)	Trioctahedral vermiculite	Trioctahedral vermiculite
			Dioctahedral vermiculite	Dioctahedral vermiculite
	Nonhydrated cations	True mica ($x \sim 0.5–1.0$)	Trioctahedral true mica	Phlogopite, biotite, lepidolite, annite
			Dioctahedral true mica	Muscovite, illite, glauconite, paragonite, celadonite
	Nonhydrated cations	Brittle mica ($x \sim 2.0$)	Trioctahedral brittle mica	Clintonite
			Dioctahedral brittle mica	Margarite
	Hydroxide	Chlorite (x variable)	Trioctahedral chlorite	Clinochlore, chamosite, nimite, pennantite
			Dioctahedral chlorite	Donbassite
2:1 mixed-layer (regular)	Variable	None	None	Hydrobiotite, rectorite, corrensite, aliettite, tosudite, kulkeite
Modulated 1:1 layer	None	No group name ($x \sim 0$)	No subgroup name	Antigorite, greenalite
Modulated 2:1 layer	Hydrated exchangeable cations	Sepiolite-palygorskite ($x \sim$ variable)	Sepiolite	Sepiolite, loughlinite
			Palygorskite	Palygorskite
	Variable	No group name ($x \sim$ variable)	No subgroup name	Minnesotaite, stilpnomelane, zussmanite

[a] x = layer change, whose absolute value = $\sum M^+ + 2M^{2+}$ concentrations where M are interlayer cations.
Source: Eslinger and Pevear (1988)

(a)

(b)

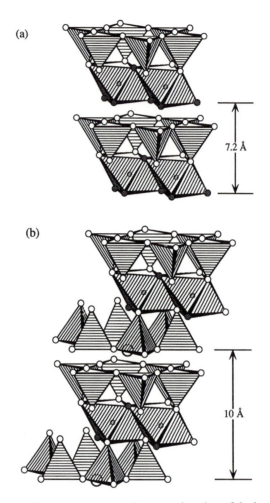

7.2 Å

10 Å

Figure 2.34. Layer silicate structures. (a) A perspective view of the kaolinite structure, a 1:1 layer silicate. The structure consists of layers of corner-sharing SiO_4 tetrahedra bonded to layers of MO_6 octahedra. These T–O packages, which usually have a basal repeat of about 7.2 Å, are held together by weak van der Waals forces. Hydroxyl groups are indicated by solid circles in the figure. (b) Perspective view of the mica structure. This group of layer silicates has a T-layer to O-layer ratio of 2:1, with the octahedral sheet sandwiched between two opposing tetrahedral sheets. In the true micas, the inter-layer region between T–O–T sandwiches houses large alkali cations such as K or Na. This interlayer site may also be empty, as in pyrophyllite, or contain H_2O, as in vermicu-lite. The basal repeat for most 2:1 layer silicates is about 10 Å, although there are many different polytypes. Hydroxyl groups are shown as solid circles (after Klein and Hurlbut 1985).

Table 2.17. *Feldspar structures*

Composition	Structure type	Space group	Characteristics
$NaAlSi_3O_8$	Low albite	C2/m	Complete Al-Si order
	Intermediate albite	C2/m	Al-Si partially disordered
	Monalbite (high albite)	C2/m	Al-Si disordered
	Analbite	C$\bar{1}$	Al-Si disordered, displacive transformation to this structure at 1223 K
$KAlSi_3O_8$	Maximum microcline	C$\bar{1}$	Complete Al-Si order
	Microcline	C$\bar{1}$	Intermediate temperature form, partial Al-Si disorder
	Orthoclase	C2/m	Intermediate temperature form, partial Al-Si disorder
	Sanidine	C2/m	High-temperature form, maximum Al-Si disorder
$CaAl_2Si_2O_8$	Anorthite	P$\bar{1}$	Al-Si ordered
		I$\bar{1}$	Al-Si ordered, above 518 K (may have some disorder above 1600 K)
$BaAl_2Si_2O_8$	Celsian	I$\bar{1}$/c	Al-Si ordered

2.7.5 Framework silicates

Tetrahedra linked in three dimensions form framework silicates (see Tables 2.9 and 2.17). The quartz structure (see Fig. 2.21a, b) is the most compact of these, containing rings of six tetrahedra. It undergoes an unquenchable displacive transition (α- to β-quartz) near 846 K. Alpha-quartz allows very little ionic substitution, but β-quartz, with more open cavities within the rings, permits a "stuffing" substitution (see Sec. 2.8.1), with aluminum replacing silicon and a small cation (Li or Mg) filling a vacant interstitial A-site to form a solid solution series $Li_xAl_xSi_{1-x}O_2$ or $Mg_{x/2}Al_xSi_{1-x}O_2$. Cristobalite and tridymite (see Fig. 2.21) are high-temperature polymorphs of lower density than quartz. Each can also undergo displacive transitions (α to β) analogous to those in quartz. The tridymite type can undergo extensive substitution of Al in the framework, with larger cations, especially Na and K, substituting for charge balance in the interstitial sites. When this substitution extends to $Na_{0.5}Al_{0.5}Si_{0.5}O_2$ or

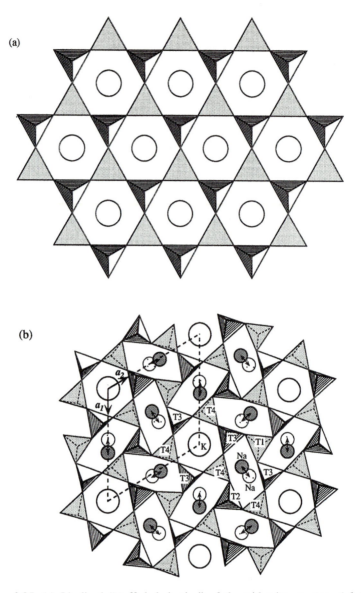

Figure 2.35. (a) Idealized "stuffed derivative" of the tridymite structure (cf. Fig. 2.21c, d) in which Al replaces about half of the Si in the tetrahedral sites, and large cations such as Na, K, and Ca fill the interstitial voids to maintain charge balance. (b) Projection of the nepheline structure down the c-axis. There are four distinct tetrahedral sites, T1, T2, T3, T4, with Al usually ordering into T1 and T4 and Si ordering into T2 and T3. Note that although some of the six-membered rings of tetrahedra retain their hexagonal shape, many are severely distorted into oval-shaped rings. Within the oval rings are shown two Na atoms, which is meant to represent the uncertainty in the actual position of Na (after Papike and Cameron 1976).

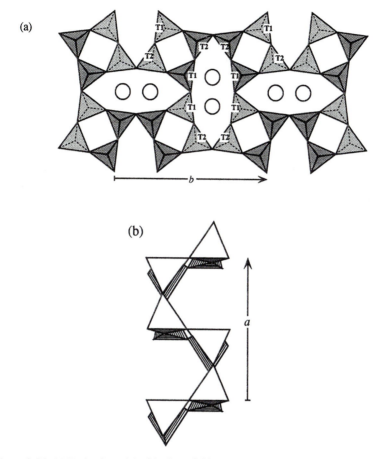

Figure 2.36. (a) Projection of the idealized feldspar structure normal to the *b*-axis. The fundamental structural units are four-membered rings of tetrahedra, with two pointing up and two pointing down. Cations such as K, Na, or Ca fill the large oval-shaped voids between the rings. Note there is mirror symmetry normal to *b* and two-fold rotational symmetry parallel to *b*. There are two distinct tetrahedral sites, T1 and T2, in which Si and Al can order. (b) When viewed normal to *a*, the linkage between rings above and below form a chain resembling a crankshaft (after Papike and Cameron 1976).

NaAlSiO$_4$, the feldspathoid mineral nepheline results. This is often called a stuffed-silica derivative structure. It is shown as an idealized packing of rings in Figure 2.35a and as a projection of the real structure in Figure 2.35b. Coesite, a high-pressure form denser than quartz, contains both eight- and four-membered rings (see Fig. 2.21). This ring arrangement is formally similar to that in the feldspar structure. That structure contains "double crankshafts" or spirals of linked tetrahedra (see Fig. 2.36) that define a pattern of four- and eight-membered rings, with large cations filling the interstitial voids. Commonly occurring feldspars and some synthetic analogues are shown in Table 2.17. A crucial feature of the feldspar structure especially, and of aluminosilicate framework, sheet, and chain structures in general, is aluminum-silicon order–disorder. In low albite and maximum microcline minerals, characteristic of metamorphic environments involving annealing at 600–1000 K for millions of years, complete Al-Si ordering takes place. Alkali feldspars characteristic of higher temperature environments, either natural igneous associations or laboratory heating experiments, show Al-Si disorder and associated symmetry changes. The complex phase relations in ternary Na, K, and Ca feldspars largely reflect order–disorder phenomena. The physical and thermodynamic descriptions of these phenomena are discussed in Section 6.8. Feldspars are discussed in detail in books by Griffen (1992), Ribbe (1975), and Smith (1974).

2.8 Relating different structures

2.8.1 Sites and substitutions

One can identify different crystallographic sites, each with fairly constant geometry, that occur in a variety of structures. These include tetrahedral sites (T), usually occupied by Si or Al, sites of approximately octahedral symmetry (M) occupied by cations of intermediate size, large sites with seven or higher coordination (A) filled by large cations, and others. In addition, hydroxyl or halogen (Y) may be present. Though the crystallographic details vary from structure to structure, the general similarity of coordination environments and average bond lengths for each type of site makes this categorization extremely useful. One can then write a mineral structural formula in terms of these sites as $A_a M_m T_t O_o Y_y$. The overall stoichiometry of the phase is then fixed by the structural formula and the likely compositions that minerals can have are determined by the possible substitutions onto the available sites, bearing in mind that each site will accept only a certain range of bond lengths and, therefore, of ions. Most mineral phases can accept a variety of ionic substitutions, leading to the general observation that minerals seldom have ideal end-member compositions but are complex solid solutions. Substitutions can occur involving (1) ions of the same charge on one set of sites, for example, $Si_T = Ge_T$ (or

Table 2.18. *Charge coupled substitutions in minerals*

Substitution	Examples
Same charge	
$Mg_M^{2+} = Fe_M^{2+}$	Pyroxene, olivine
$Al_T^{3+} = Fe_T^{3+}$	Mica, clay, epidote, feldspar, tourmaline
$Al_M^{3+} = Fe_M^{3+}$	Mica, clay, amphibole, garnet, spinel, pyroxene
$Al_T^{3+} = B_T^{3+}$	Feldspar, tourmaline
$Si_T^{4+} = Ge_T^{4+}$	Olivine, garnet, spinel, pyroxene, feldspar
$Na^+ = K^+$	Feldspar, mica, clay
Different charges	
$Si_T^{4+} + Na_A^+ = Al_T^{3+} + Ca_A^{2+}$	Plagiclase feldspar, pyroxene, amphibole, mica, clay
$Si_T^{4+} = Na_A^+ + Al_T^{3+}$	Stuffed silica derivative, amphibole, mica, clay
$Si_T^{4+} = K_A^+ + Al_T^{3+}$	Stuffed silica derivative, mica, clay
$Ca_M^{2+} + Si_T^{4+} = Al_M^{3+} + Al_T^{3+}$	Pyroxene, amphibole, mica, clay
$Mg_M^{2+} + Si_T^{4+} = Al_M^{3+} + Al_T^{3+}$	Pyroxene, amphibole, mica, clay
$3Fe_M^{2+} = vacancy_M + 2Fe_M^{3+}$	Defects in olivine, spinel
$Si_T^{4+} = 4H_T^+$	Hydrogarnet

$GeSi_{-1}$); (2) the coupled substitution of two ions of different charges on the same sublattice, for example, $2Mg_M = Li_M + Al_M$ (or $(Li, Al) (2Mg)_{-1}$); (3) the coupled substitution of altervalent cations on different sublattices, for example, $Na_A + Si_T = Ca_A + Al_T$ (or $(Ca, Al) (Na, Si)_{-1}$); (4) or substitutions involving vacant sites, for example, $Si_T = Na_A + Al_T$ (or $(Na, Al) Si_{-1}$). Various common substitutions are shown in Table 2.18.

One phase may be derived from another by a set of coupled substitutions. Thus anorthite ($CaAl_2Si_2O_8$) and albite ($NaAlSi_3O_8$) are related by the substitution $Na_A + Si_T = Ca_A + Al_T$ ((Ca, Al)(Na, Si)$_{-1}$). Diopside ($CaMgSi_2O_6$) and calcium Tschermak's pyroxene ($CaAl_2SiO_6$) are related by $Mg_M + Si_T = Al_M + Al_T$ or $(Al,Al)(Mg,Si)_{-1}$. That same substitution in micas relates phlogopite, $KMg_3Si_3AlO_{10}(OH)_2$, and eastonite, $KMg_2Si_2Al_3O_{10}(OH)_2$. A question currently under study is whether the same type of substitution has comparable energetics in different structures.

Structures may also be derived by considering ordered coupled substitutions. Ordered substitution of two different atoms into diamond produces the

Table 2.19. *Structures related to perovskite*

Type	Space group	Relationship	Examples
Perovskite			
ABO_3	Pm3m	Parent cubic perovskite	$BaTiO_3$ above 393 K SrTiO_3
	P4mm	Displacement of cations (Ba)	$BaTiO_3$ below 393 K
	Pbnm	Tilting of octahedra	$CaTiO_3$, $NaMgF_3$, $MgSiO_3$
	Pnmm	Cation displacement (A-site) and tiled octahedra	$NaNbO_3$, $GdFeO_3$
	R3c	Cation displacement (both sites) and tilted octahedra	$PbZr_{0.9}Ti_{0.9}O_3$ (PZT)
Lithium niobate ABO_3	R3c	Large displacements and tilts, also considered an ordered derivative of corundum	$LiNbO_3$
Ordered perovskites			
$AB_{0.5}B'_{0.5}O_3$	Fm3m	Ordering of octahedral cations	$BaFe_{0.5}Mo_{0.5}O_3$
$AB_{0.33}B'_{0.67}O_3$	P3m1	Ordering of octahedral cations	$BaSr_{0.33}Ta_{0.65}O_3$
Different stoichiometries			
ReO_3	Pm3m	Central cation absent	ReO_3
K_2NiF_4	I4/mmm	Perovskite layers with extra A-layer, CN of A = 9	K_2NiF_4, La_2CuO_4
YBCO type	Pmmm	Ordering of cations and vacancies	$YBa_2Cu_3O_7$ (oxide super-conductors)
Brownmillerite	Pnma	Ordered oxygen vacancies	$Sr_2Fe_2O_5$

wurtzite and sphalerite structures. Ordered substitution of two unlike atoms into bcc produces the cesium chloride structure. Substitution of Na and Al into cristobalite produces nepheline, and both the ilmenite and lithium niobate structures can be viewed as ordered substitutions into the parent corundum type. The perovskite structure may itself be related to a parent ReO_3 structure type that contains the octahedral framework with no cations in the large central site. It may also be related to the cesium chloride type by intercalculating another set of cations between the chlorine sites.

Further substitutions involving more ions are also possible. The ordered $LiFeO_2$ type is produced by substituting two different cations into sodium chlo-

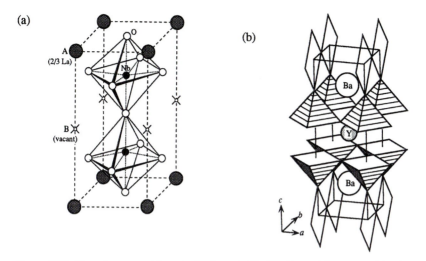

Figure 2.37. Complex perovskite substitutions. (a) $La_2Nb_3O_9$, a defect perovskite. The 12-coordinated sites are two-thirds filled and B sites are empty. (b) The superconducting 1:2:3 phase $YBa_2Cu_3O_7$ has a tripled *c*-axis as a consequence of barium (large circles) and yttrium (small circles) ordering in the sequence Ba–Y–Ba, Ba–Y–Ba. Two of every nine perovskite oxygens are absent, giving the superconductor a uniquely ordered arrangement of oxygen atoms.

ride such that each cation has the opposite type as its next-nearest-neighbor. The perovskite structure can be further modified by ordered substitutions to form many new ordered ternary and quaternary perovskites (see Table 2.19 and Fig. 2.37). In general, ordered substitutions of ions with unequal bond lengths in an ordered arrangement will lead to a lowering of symmetry to optimize the structure. Ionic substitutions, order–disorder transitions, and changes in symmetry are therefore frequently coupled.

The perovskite structure's propensity to deviations from cubic symmetry (Glazer 1972), to cation substitution and ordering, and to anion vacancies make this simple structure type show a rich complexity of behavior in terms of phase transitions and physical properties. Noncentrosymmetric structures can be materials of high dielectric constant, and ferroelectrics and materials like $BaTiO_3$ and lead zirconate titanate are commercially very important. Perovskites with a high valence state and both ionic and electronic conductivity, such as $LaCrO_3$, $LaMnO_3$, and $LaNiO_3$, are potential fuel cell electrodes and interconnectors. The excitement of high-temperature superconductors is based on defect perovskites and complex layer structures derived from perovskite (see Fig. 2.37).

The structures related by octahedral tilts and off-centering of cations can often be transformed one to another by mechanisms involving short-range correlated motions of atoms and no long-range diffusion. Thus, such transitions are often rapid and nonquenchable and sometimes involve the "softening" of vibrational modes related to those displacements (see Chapter 6).

2.8.2 *Structural elements, intergrowths, and homologous series*

Frequently, structural elements can be identified whose repetition makes up the structure. If their sequence or frequency can be varied, a series of structures, characterized as members of a homologous series, can be formed. An example from the preceding discussion is the relation of pyroxenes and pyroxenoids, where the frequency of chain kinking is the structural element. In this case, all the structures could conceivably exist at the same composition. Another example arises in relating the chain and sheet silicates. The double chains of amphiboles are separated by cation-rich regions. Occasionally, a triple or quadruple chain is found (see Fig. 2.38), and if these multiple chains occur in an ordered sequence, a new ordered structure is found, such as chesterite (see Fig. 2.38b). Because the composition of the chains is different (silica rich) from that of the interchain regions (cation and water rich), a change in the thickness of the chains implies a change in overall composition. In this series, the sheet silicate structure can be thought of as an end-member of infinite chain thickness. This relationship also suggests a possible mechanism for the transformation, and regions that look partly transformed from amphibole to clay contain such intermediate sequences.

The spinelloids (see Table 2.20) are a third illustration of a homologous series of structures. The cation positions in the spinel structure (see Section 2.4) , projected onto the (110) plane of the face-centered cubic unit cell (or the (100) plane of a body-centered tetragonal unit cell of half the volume) are shown in Figure 2.39a. The lines outline identical building blocks from which the entire structure can be constructed. One can create an antiphase boundary by moving one portion of the crystal relative to the other by half a building block distance (see Fig. 2.39b). If this twinning occurs periodically, a new structure is formed (Hyde et al. 1982). A number of spinel-related phases, called spinelloids, found in Mg_2SiO_4, Co_2SiO_4, and in $NiAl_2O_4$–Ni_2SiO_4, $MgGa_2O_4$–Mg_2GeO_4, and other systems have structures that can be represented in this way (see Fig. 2.39c). The twinning produces linked tetrahedra. Table 2.20 gives some examples of spinelloid structures. A particularly interesting case is the hypothetical extreme in which twinning in this omega structure

Table 2.20. *Spinelloids*

Phase	Tetrahedral groups				Examples
	TO_4	T_2O_7	T_3O_{10}	T_4O_{12}	
Spinel (γ)	x				γ-Mg_2SiO_4,$NiAl_2O_4$,γ-Ni_2SiO_4
V	x	x			$0.5Ni_2SiO_4$–$0.5NiAl_2O_4$
IV	x	x			$0.5Ni_2SiO_4$–$0.5NiAl_2O_4$
					$0.5Ni_2SiO_4$–$0.5NiAl_3O_4$
III(β)		x			β-Mg_2SiO_4,β-Co_2SiO_4
I	x		x		$0.25Ni_2SiO_4$–$0.75NiAl_2O_4$
II			x		$0.40Ni_2SiO_4$–$0.60NiAl_2O_4$
Li₂WO₄-II				x	Li_2WO_4

Source: Classification after Horiuchi et al. (1982).

occurs at every building block. Although the oxygens in this omega phase remain in cubic close packing, the cation positions are analogous to those in olivine. By shearing along the oxygen planes to convert an ABAB hcp oxygen arrangement to an ABCABC ccp oxygen array, one can envision a transformation from olivine to the omega phase as an intermediate, possibly low energy, pathway for the conversion of olivine to spinel. This initial step, involving no cation migration, would be followed by movement of cations to their appropriate positions to form the spinelloid or spinel. In addition, if a structure accommodated periodic "mistakes" in oxygen layer sequence as well as the twinning just described, a complex series of intermediates involving elements of both olivine and spinel structures might be generated. They would all conform to the AB_2O_4 stoichiometry.

A fourth way of building related structures is through the intergrowth of structural elements of different compositions. The K_2NiF_4 structure (see Fig. 2.28) may be viewed as an intergrowth of perovskite and rocksalt units, and other more complex sequences, called Ruddlesden-Popper phases, exist. A series of complex germanates involving an intergrowth of spinel-like and rocksalt layers has been identified (Barbier and Hyde 1987). Other examples abound.

A fifth theme, common in oxidation-reduction series, is the creation of oxygen vacancies and the collapse of the structure to eliminate them by a shear mechanism. This is seen most clearly in the ReO_3 structure, Figure 2.40a, where reduction leads to the formation of shear planes as shown in Figure

(a)

(b)

(c)

Figure 2.38. Schematic diagrams showing several members of the polysomatic series that exists between the double-chain silicates (amphiboles) and the layer silicates. All these diagrams are I-beam representations (compare Fig. 2.33a). (a) I-beam representation of the amphibole structure. This consists only of a sequence of double chains along *b*. This chain repeat sequence is, therefore, 222 . . . , or simply (2). (b) I-beam representation of the chesterite structure that has a regular alternation of double chain and triple chain material along *b*. The chain repeat sequence is 232323 . . . , or (23). (c) I-beam

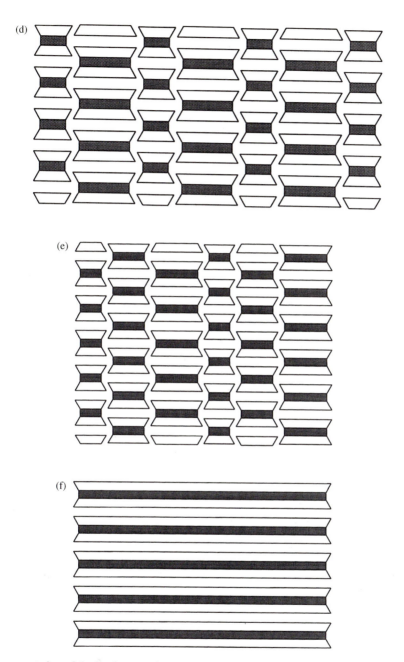

representation of the jimthompsonite structure that is comprised only of triple chains, so the chain repeat is given by (3). (d) and (e) These two diagrams are I-beam representations of two unusually ordered polysomes recently reported by Grobety (1992); see Veblen (1992). In (d) the chain repeat sequence is 242424 . . . , or just (24), and in (e) the sequence is (234). These last two polysomes have only been observed by transmission electron microscopy. (f) The end of this polysomatic series, namely, a layer silicate, is shown.

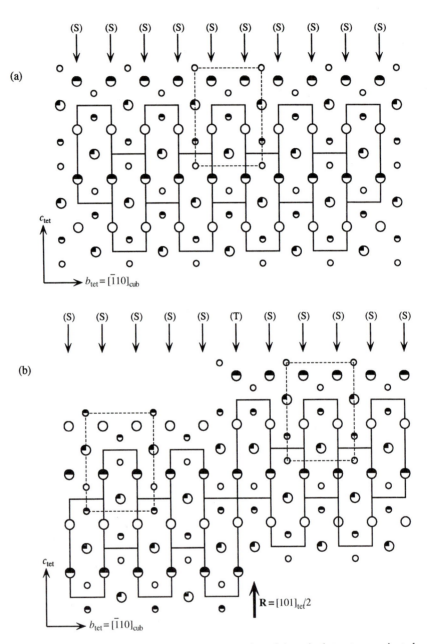

Figure 2.39. Spinelloids. (a) Schematic representation of the spinel structure projected onto (100) of the tetragonal unit cell, which is equivalent to (110) of the cubic unit cell. In these diagrams only the cation positions are shown. The tetrahedral (A) cations are shown as small circles and the octahedral (B) cations are shown as larger circles. Open symbols have x-coordinates of 0, half-filled symbols are cations at $x = 1/2$, and symbols that are a quarter full have x-coordinates of $\pm 1/4$. The structure may be divided up into fundamental rectangular building blocks, separated by boundaries labeled (S). (b) The

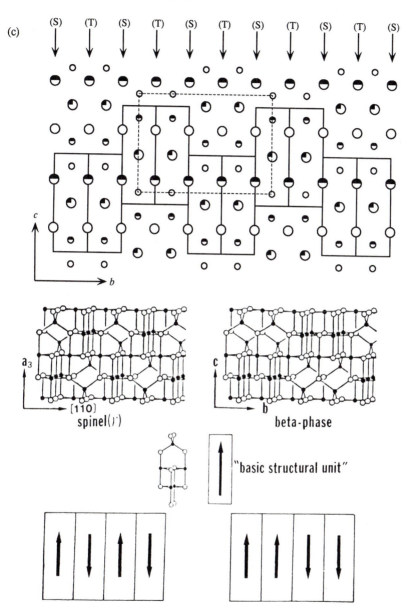

formation of a single antiphase boundary (apb) by applying a slip vector of [101]/2. The apb is a twin plane and is labeled (*T*). Systematic twinning at the unit-cell scale gives rise to a number of related structures. Shown in (c) is the structure of the beta-phase, known to occur at high pressure, which consists of a series of fundamental building blocks separated by a regular alternation of (S) and (T) type boundaries (after Hyde et al. 1982). (d) another view of the relation of spinel and β-phase (from Horiuchi, Akaogi, and Sawamoto 1982).

(a)

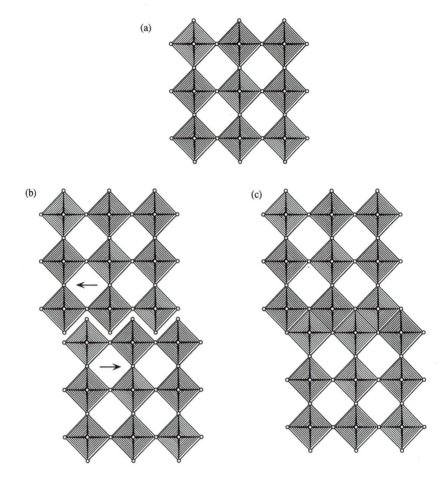

(b)   (c)

(d)

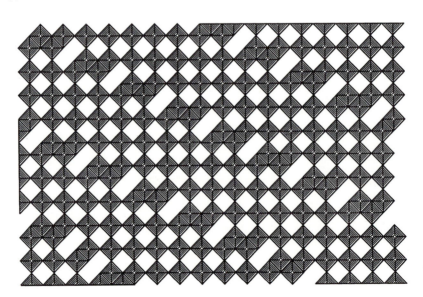

2.40b. If these form at ordered intervals, a series of phases related by the frequency and orientation of these shear planes can form, for example, in WO_{3-x} phases (see Fig. 2.40c). Similar behavior is seen in reduced rutile (TiO_2), where a homologous series of phases, Ti_nO_{2n-1}, forms. Such series are often called Magneli phases. Introduction of other trivalent ions (Cr, Fe) into TiO_2 may produce similar structural effects as reduction of Ti^{4+} to Ti^{3+}. These various homologous series are summarized in Table 2.21 (p. 88).

Structural relations such as those described in this section may be conceptual or they may be actual in the sense of invoking mechanisms of formation or transformation. The more closely one looks at phases, especially by high-resolution transmission electron microscopy (see Section 3.4), the more one realizes the wealth of complexity even an ostensibly simple crystal structure can hold. This complexity is all-important for chemical reactivity, rates of solid-state reactions, and transport properties. The detailed structures of minerals at the level of order–disorder, intergrowths, microstructures, and defects record their geologic history.

2.9 References

2.9.1 General references and bibliography

Adams, D. M. (1974). *Inorganic solids, an introduction to concepts in solid-state structural chemistry.* New York: Wiley.

Bailey, S. W., ed. (1984). *Micas.* Reviews in Mineralogy, vol. 13. Washington, DC: Mineralogical Society of America.

Bailey, S. W., ed. (1988). *Hydrous phyllosilicates (exclusive of micas).* Reviews in Mineralogy, vol. 19. Washington, DC: Mineralogical Society of America.

Boisen, M. B., Jr., and G. V. Gibbs. (1985). *Mathematical crystallography, an introduction to the mathematical foundations of crystallography.* Reviews in Mineralogy, vol. 15. Washington, DC: Mineralogical Society of America.

Burns, G., and A. M. Glazer. (1978). *Space groups for solid state scientists.* New York: Academic Press.

Eslinger, E., and D. Pevear. (1988). *Clay minerals for petroleum geologists and engineers.* SEPM Short Course Notes No. 22. Tulsa, OK: Society of Economic Paleontologists and Mineralogists.

Figure 2.40. (*Facing page*) (a) Plan view of the ReO_3 structure, composed of a three-dimensional network of corner-sharing octahedra. This structure may be sheared, as shown in (b) and (c). Here the shear plane is (100) and eliminates oxygen vacancies that would occur during reduction. These shear planes may occur in an ordered fashion, leading to a wide range of related structures known as shear structures. The example shown in (d) is a plan view of the Mo_9O_{26} structure that results from regular shearing of the ReO_3 structure along (102) planes.

Table 2.21. *Homologous series of structures*

Group	Stoichiometry	Structural elements and end-members	Examples	References
Pyroxenoids	ABO_3	Kinking of tetrahedral chains every $2n + 1$ tetrahedra $n = \infty$, pyroxene $n = 1$, wollastonite	$MgSiO_3$ $CaSiO_3$ $MnSiO_3$ $FeSiO_3$ III	Papike and Cameron (1976)
Spinelloids	AB_2O_4	Twin planes on (010) every n unit cells $n = \infty$, spinel $n = 1$, hypothetical ω structure	β-Mg_2SiO_4 $NiAl_2O_4$–Ni_2SiO_4 phases $MgGa_2O_4$–Mg_2GeO_4 phases	Hyde et al. (1982) Horiuchi et al. (1982) Barbier and Hyde (1986)
Biophyriboles	amphibole-mica intergrowths	Triple and multiple (n) chains in double-chain amphibole matrix, can be ordered or disordered $n = 1$, pyroxene $n = 2$, amphibole $n = \infty$, mica	Chesterite Jimthompsonite	Veblen (1981)
Ruddlesden-Popper phases	$A_{n+1}B_nO_{3n+1}$	Perovskite layers and A-containing cubes $n = 1$, K_2NiF_4-type $n = \infty$, perovskite-type	La_2NiO_4, Sr_2TiO_4, $Sr_3Ti_2O_7$, $Sr_4Ti_3O_{10}$	Ruddlesden and Popper (1958) Hyde and Andersson (1989)
Structures with crystallographic shear planes (CS)	W_nO_{3n-1} Ti_nO_{2n-1}	Removal of oxygen and "shear" of structure to eliminate defects $n = \infty$, TiO_2 rutile $n = 3$, Ti_3O_5	Reduced rutiles, reduced tungsten and vanadium oxides, niobium oxides, and oxyfluorides	Hyde and Andersson (1989)

Evans, R. C. (1966). *An introduction to crystal chemistry.* 2nd ed. Cambridge: Cambridge University Press.
Griffen, D. T. (1992). *Silicate crystal chemistry.* New York: Oxford University Press.
Hahn, T. (1983). *International tables for crystallography.* Dordrecht, Holland: Reidel.
Hume-Rothery, W., and G. V. Raynor. (1944). *The structure of metals and alloy.* 2nd ed. London: Institute of Metals.
Hyde, B. G., and S. Andersson. (1989). *Inorganic crystal structures.* New York: Wiley.
Jaffe, H. W. (1988). *Introduction to crystal chemistry.* Student ed. New York: Cambridge University Press.
Kerrick, D. M., ed. (1990). *The Al₂SiO₅ polymorphs.* Reviews in Mineralogy, vol. 22. Washington, DC: Mineralogical Society of America.
Lindsley, D. H., ed. (1991). *Oxide minerals: petrologic and magnetic significance.* Reviews in Mineralogy, vol. 25. Washington, DC: Mineralogical Society of America.
Papike, J. J., and M. Cameron. (1976). Crystal chemistry of silicate minerals of geophysical interest. *Rev. Geophys. and Space Phys. 14*, no. 1, 37–80.
Pauling, L. (1960). *The nature of the chemical bond.* 3rd ed. Ithaca, NY: Cornell University Press.
Pearson, W. B. (1972). *The crystal chemistry and physics of metals and alloys.* New York: Wiley.
Prewitt, C. T., ed. (1980). *Pyroxenes.* Reviews in Mineralogy, vol. 7. Washington, DC: Mineralogical Society of America.
Putnis, A. (1992). *Introduction to mineral sciences.* Cambridge: Cambridge University Press.
Reeder, R. J., ed. (1983).*Carbonates: mineralogy and chemistry.* Reviews in Mineralogy, vol. 11. Washington, DC: Mineralogical Society of America.
Ribbe, P. H., ed. (1975). *Feldspar mineralogy.* Reviews in Mineralogy, vol. 2. Washington, DC: Mineralogical Society of America.
Ribbe, P. H., ed. (1980). *Ortho-silicates.* Reviews in Mineralogy, vol. 5. Washington, DC: Mineralogical Society of America.
Smith, J. V. (1974). *Feldspar mineralogy.* Berlin: Springer-Verlag.
Smyth, J. R., and D. L. Bish. (1988). *Crystal structures and cation sites of the rock-forming minerals.* New York: Allen and Unwin.
Veblen, D. R., ed. (1981). *Amphiboles and other hydrous pyriboles-mineralogy.* Reviews in Mineralogy, vol. 9A. Washington, DC: Mineralogical Society of America.
Wells, A. F. (1984). *Structural inorganic chemistry.* Oxford: Oxford University Press.
Wyckoff, R. W. G. (1986). *Crystal structures.* 2nd ed. Vol. 2, *Inorganic Compounds RXₙ, RₙMX₂, RₙMX₃.* Malabar, FL: Krieger.

2.9.2 Specific references

Barbier, J., and B. G. Hyde. (1986). Spinelloid phases in the system MgGa₂O₄–Mg₂GeO₄. *Phys. Chem. Minerals 13*, 383–92.
Barbier, J., and B. G. Hyde. (1987). Mg₇Ga₂GeO₁₂, a new spinelloid-related compound, and the structural relations between spinelloids (including spinel) and the β-Ga₂O₃ and NaCl types. *Acta Cryst. B43*, 34–40.
Czank, M., and F. Liebau. (1980). Periodicity faults in chain silicates: A new type of planar lattice fault observed with high resolution electron microscopy. *Phys. Chem. Minerals 6*, 85–93.

Dollase, W. A. (1967). The crystal structure at 220° C of orthorhombic high tridymite from the and Steinbach meteorite. *Acta Cryst. 23*, 617–23.

Giammattei, V. D., and N. G. Reichert. (1975). *Art of a vanished race, the Mimbres classic black-on-white.* Woodland, CA: Dillon-Tyler Publishers.

Glazer, A. M. (1972). The classification of tilted octahedra in perovskites. *Acta Cryst. B28*, 3384–92.

Grobety, B. (1992). *Electron microscopy of mineral intergrowths in metamorphic rocks.* Ph.D. diss., Eidgennosische Technische Universität, Zürich, Switzerland.

Horiuchi, H., M. Akaogi, and H. Sawamoto. (1982). Crystal structure studies on spinel-related phases, spinelloids: Implications to olivine-spinel phase transformation and systematics. *Adv. Earth and Planet. Sci. 12*, 391–403.

Horiuchi, H., and H. Sawamoto. (1981). β-Mg_2SiO_4: Single-crystal X-ray diffraction study. *Amer. Mineral. 66*, 568–75.

Hyde, B. G., T. J. White, M. O'Keeffe, and A. W. S. Johnson. (1982). Structures related to those of spinel and the β-phase, and a possible mechanism for the transformation olivine $\leftrightarrow$ spinel. *Zeitschrift für Kristallographie 160*, 53–62.

James, H. L. (1976). *Posts and rugs, the story of Navajo rugs and their homes.* Popular Series No. 15. Globe AZ: Southwest Parks and Monuments Association.

Klein, C., and C. S. Hurlbut, Jr. (1985). *Manual of mineralogy.* 20th ed. New York: Wiley.

Mehta, A., and A. Navrotsky. (1993). Structural transitions in $LiNbO_3$ and $NaNbO_3$. *J. Solid State Chem. 102*, 213–25.

Novak, G. A., and G. V. Gibbs. (1971). The crystal chemistry of the silicate garnets. *Amer. Mineral. 56*, 791–825.

Papike, J. J. (1987). Chemistry of the rock-forming silicates: Ortho, ring, and single-chain structures. *Rev. Geophys. 25*, 1483–526.

Peacor, D. R. (1973). High-temperature single crystal study of the cristobalite inversion, *Z. Krist. 138*, 274–98.

Ruddlesden, S. N., and P. Popper. (1958). The compound $Sr_3Ti_2O_7$ and its structure. *Acta Cryst. 11*, 54–5.

Smith, J. V. (1982). *Geometrical and structural crystallography.* New York: Wiley.

Speer, J. A. (1982). *Zircon.* Reviews in Mineralogy, vol. 5, 67–112. Washington, DC: Mineralogical Society of America.

Veblen, D. R. (1992). *Electron microscopy applied to nonstoichiometry, polysomatism, and replacement reactions in minerals.* Reviews in Mineralogy, vol. 22, 181–229. Washington, DC: Mineralogical Society of America.

Zachariasen, W. H. (1972). Refined crystal structure of phenacite Be_2SiO_4. *Sov. Phys. Cryst. 16*, 1021–5.

Zoltai, T., and M. J. Buerger. (1959). The crystal structure of coesite, the dense, high-pressure form of silica. *Z. Krist. 3*, 129–41.

3

Experimental methods for studying structure

3.1 Interaction of radiation with matter

Radiation is the major probe by which one obtains information on the structure of matter. Table 3.1 shows the electromagnetic spectrum and indicates the energy regions relevant to various processes in solids. The variation of intensity with wavelength of solar radiation striking the surface of the earth is shown in Figure 3.1. It is similar to that emitted by a black body at about 1000 K, but with "bites" out of the spectrum due to absorption by the sun's upper atmosphere and the Earth's atmosphere. The radiation we detect as color, ranging from red to violet, is near the peak in solar intensity.

Quantum mechanics defines the energy states of a bound system, such as a crystal, as being, not a continuum including all energies, but a set of discrete allowable energies that correspond to solutions of the Schrödinger equation. A transition between two such states is accomplished by the absorption or emission of that quantum of energy. Measuring these energies by direct observation forms the basis for most spectroscopic techniques. The energy, wavelength, and frequency are related by

$$E = h\nu = hc/\lambda \qquad (3.1)$$

where E is energy (per quantum absorbed, i.e., per molecule), ν is frequency, λ is wavelength (wavenumber in cm^{-1} is $1/\lambda$), and h is Planck's constant.

Absorption or emission spectra in isolated atoms generally consist of a series of sharp lines from which both the nature of the element and the spacing of energy levels can be deduced. Such transitions are governed by quantum mechanical selection rules that determine which transitions are "allowed," that is, what changes in quantum numbers can lead to absorption or emission of radiation. In molecules, and even more so in condensed matter, the sharp lines are broadened into absorption bands because of the interaction of each atom with its neighbors. In general, the more variable the local environment of an atom, the broader the peak, so peak width gives information on disorder.

91

Table 3.1. *The electromagnetic spectrum and characteristic frequencies, energies, and wavelengths for processes in solids*

Energy $\log_{10} E$ (kJ)	Wavelength $\log_{10} \lambda$ (m)	Frequency $\log_{10} \nu$ (Hz)	Regions	Phenomena causing absorption
		20		
7	−11		γ radiation	Nuclear transitions
		19		
6	−10		X radiation	Core electron transitions
		18		
5	−9			
		17		
4	−8		Vacuum ultraviolet	Loss of valence electrons
		16		
3	−7			
		15	Ultraviolet visible	Valence electron transitions
2	−6			
		14		
1	−5		Infrared	Molecular vibrations
		13		
0	−4		Far infrared	
		12		
−1	−3			
		11		Molecular rotations
−2	−2			
		10	Microwave	
−3	−1			Electron spin resonance
		9		
−4	0			
		8		
−5	+1			
		7	Radio frequency	Nuclear spin resonance
−6	+2			
		6		
−7	+3			Nuclear quadrupole resonance
		5		
−8	+4			
		4		

Source: After Calas and Hawthorne (1988).

Increasing temperature also generally broadens peaks because of increased vibrations and disorder. Thus to enhance resolution, it is often desirable to perform spectroscopy and diffraction at low temperature (liquid nitrogen, 77 K, or liquid helium, 4 K).

Table 3.1 also shows some characteristic frequencies, energies, and wave-

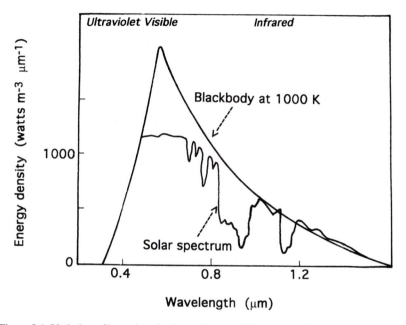

Figure 3.1. Variation of intensity of solar radiation striking the Earth's surface compared to black body radiation at 1000 K, semischematic.

lengths for processes occurring in solids. Virtually all methods used to probe the structure of matter rely on the interaction of radiation or of elementary particles (electrons, neutrons, protons), each having appropriate wavelengths and energies. The radiation can be absorbed, emitted, reflected, or diffracted. Strong interaction will occur when the incoming energy matches the spacing of appropriate energy levels, or, in the case of diffraction, the interplanar spacing. The intensity will be governed by the number of quanta of incident radiation, the amount of matter present, and by a property intrinsic to the atoms involved, called variously a cross section, absorption coefficient, or scattering factor. The change in intensity and/or wavelength of the incident radiation is then detected optically, photographically, or by appropriate scintillation counters, phosphors, or position-sensitive detectors. The fundamental data are in the form of a spectrum of intensity versus energy (or corresponding wavelength, wave number, or frequency).

The answers to three questions can be used to characterize all spectroscopic (and diffraction) techniques: (1) What radiation is used to probe the sample? (2) What sample response (diffraction, intensity change, emitted radiation) is detected? (3) What is the technique good for? As an overview, Table 3.2 presents a summary of major techniques characterized by this scheme.

Table 3.2. *Classification of spectroscopic and diffraction techniques*

Method	Exciting radiation	Analyzed radiation	Applications
X-ray diffraction	X rays	Diffracted X rays	Crystal structure determination
Neutron diffraction	Neutrons	Diffracted neutrons	Crystal structure determination, especially for light atoms, magnetic structure
Inelastic neutron scattering (INS)	Neutrons	Energy loss of neutrons due to scattering by phonons and magnons	Vibrational and magnetic structure
Electron microscopy	Focused beam of electrons		
Scanning (SEM)	Focused beam of electrons	Electrons reflected from sample surface	Sample morphology, phase separation
Transmission (TEM) and high resolution (HRTEM)	Focused beam of electrons	Electrons passing through sample	Structure on an atomic scale, "lattice images"
Electron diffraction	Focused beam of electrons	Diffracted electrons	Structure on an atomic scale, especially useful for ordering and superstructures
Electron energy loss spectroscopy (EELS)	Focused beam of electrons	Loss of energy of reflected electrons	Chemical analysis
X-ray microanalysis	Focused beam of electrons	X rays generated by electron bombardment	Chemical analysis, also basis of electron microprobe (EPMA)
Infrared absorption spectroscopy	IR, vary wavelength	Decrease in intensity due to absorption	Lattice vibrations
Raman spectroscopy	Single wavelength, visible light from laser	Visible light with shift of frequency due to vibrational transition corresponding to a vibrational frequency	Lattice vibrations

Technique	Excitation	Signal	Information
Photo acoustic infrared spectroscopy	IR, vary wavelength	Change in transducer signal related to change in sample temperature due to absorption at vibrational frequencies	Lattice vibrations
Optical spectroscopy	UV or visible radiation	Decrease in intensity due to absorption	Electronic transitions, oxidation states
XANES	Intense X rays	Fine structure near absorption edge	Local coordination, oxidation state
EXAFS	Intense X rays	Fine structure far beyond absorption edge	Local coordination, mid- and long-range order, chemical analysis
XRF	X rays	Fluorescent X-rays of different energy	Chemical analysis
XPS	X rays	Emitted photoelectrons	Surface structure
UPS	Ultraviolet	Emitted photoelectrons	Surface structure
Auger spectroscopy	X rays or electrons	Electronic transitions	Surface structure
NMR	rf in a magnetic field excites nuclear spin transitions	rf pulse as function of time	Local environment, dynamics, chemical analysis, order–disorder; especially Si, Al, P, O, C, H, F, Na
ESR	Microwave in a magnetic field excites electron spin transitions	Microwave absorption	Unpaired electrons, local environment, magnetic interactions, chemical analysis, order–disorder, mainly for transition metals
Mössbauer	Gamma ray emitted by nuclear energy level transition	Gamma ray absorbed by sample, energy shifted by Doppler shift due to moving sample	Local environment, oxidation state, order–disorder, magnetic ordering, chemical analysis; especially for Fe
STM, AFM	Ultrathin needle probe above surface	Current induced in probe by "tunneling" or repulsion of probe from surface	Surface morphology and structure at atomic level

It is instructive to consider timescales. Normal electronic transitions occur "instantaneously," that is, at 10^{-15} to 10^{-20} second timescales, though fluorescence and phosphorescence can have timescales of 10^{-12} to 10^2 seconds. Vibrational transitions occur typically on the 10^{-12} to 10^{-13} timescale, while the nuclear magnetic resonance timescale is 10^{-3} to 10^{-8} seconds. Thus each method, in addition to probing different transitions, probes excitations that occur and relax at very different rates. This is important to bear in mind when comparing structural information obtained by different methods, for it gives clues as to how long a given structural unit stays intact in a dynamic system such as a surface or a liquid.

Similarly, the distance scales of various techniques differ. Diffraction probes long-range order and crystallinity over hundreds of angstroms. Nuclear magnetic resonance (NMR), electron spin resonance (ESR), and Mössbauer spectroscopy are sensitive primarily to short-range interactions, the nearest and next-nearest neighbor environments. Vibrational spectra are sensitive to local environments and somewhat (the exact range of interaction is still debated) to longer range order. No spectroscopic method is ideally suited to looking at order in the 20–100 Å scale, which may be exactly the critical range for amorphous solids, glasses, and liquids.

This chapter is intended to be an overview of experimental techniques. The reference texts and monographs listed as "General References and Bibliography" in Section 3.12.1 provide excellent reviews, "Spectroscopic Methods in Mineralogy and Geology" (Hawthorne 1988) is especially recommended. These general references should be consulted for further information; this chapter is not heavily peppered with specific references because of the availability of these high-quality reviews.

3.2 X-ray diffraction

X rays are produced when electrons strike a metal anode, exciting electrons in the atoms to orbitals of higher principal quantum number. When these electrons return to their ground states, radiation in the X-ray range is emitted. A typical X-ray emission spectrum is shown in Figure 3.2, with the peaks marked as to their origin. In addition to these sharp lines, a broad background is produced. Synchrotron radiation, producing X rays several orders of magnitude more intense than in a normal X-ray tube, is becoming very useful for diffraction and spectroscopy of solids and has opened new vistas for structural work, including studies at high pressure and on very small samples (also see Sec. 3.7).

Because X-ray wavelengths are of the magnitude of interplanar spacings in

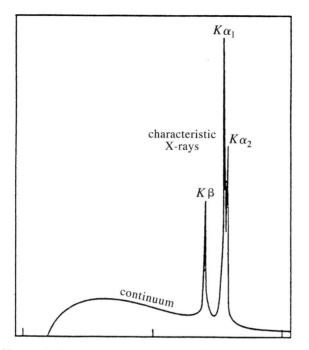

Figure 3.2. X-ray emission spectrum of copper, semischematic.

crystals, X rays interact strongly with the periodic lattice. Although essentially a scattering phenomenon between the radiation and the electrons in the solid (see Fig. 3.3), X-ray diffraction can be viewed as analogous to diffraction by an optical grating, with the periodic spacing of atoms in the crystal structure providing the "grating" on an atomic scale (see Fig. 3.3). This analogy provides an intuitive justification of the fundamental relation known as Bragg's law,

$$n\lambda = 2d \sin \theta \tag{3.2}$$

where λ is the X-ray wavelength, n the order of the diffraction (usually one), and θ the angle of incidence of X rays on the diffracting planes. When the path length of diffracting X rays is increased by an integral number of wavelengths as they diffract from parallel planes, they remain in phase and a peak in intensity (diffraction peak) occurs.

A number of geometrical arrangements can be used to study crystal structure by X-ray diffraction. Polychromatic (white) radiation can impinge on a stationary single crystal and the angles of diffracted beams of different wavelengths used to obtain the interplanar spacings. This method, originated in the early 1900s by Laue, has come back into vogue for use with synchrotron radiation and with neutrons, where obtaining monochromatic radiation is not always easy. If a single energy can be selected (and monochromators are frequently themselves diffracting crystals), then the single crystal being studied can be rotated to bring different interplanar spacing into the diffraction condition. Commercially available, computer controlled, fully automated four-circle single crystal diffractometers greatly ease the drudgery of data collection and analysis. X-ray crystallography has been revolutionized by modern computerization.

The relation of interplanar spacings for diffraction to the crystallographic parameters for the structure is governed by Eqs. 2.2, 2.3, and 3.2. Whether a given *hkl* value can give a peak is governed by the symmetry of the crystal, and the detection of systematic absences of reflections is essential in determining the space group. The reader is referred to standard crystallography texts and to the listings in the *International Tables for Crystallography* for detailed discussion of this important point (see References to Chapters 2 and 3). The intensity of a symmetry-allowed diffraction peak depends on the electron density in the diffracting plane; heavy atoms with many electrons are far more effective diffractors than light atoms. This monotonic increase of X-ray scattering factor with atomic number Z (see Fig. 3.4) places two important limitations on X-ray structural studies. First, it is hard to get good information about the location of light atoms, especially hydrogen. Information about OH group location, orientation, and hydrogen bonding in hydrous minerals is almost impossible to obtain by conventional X-ray techniques. Second, atoms adjacent in the periodic table (e.g., Fe and Mn) will be virtually indistinguishable in normal diffraction experiments, so that, for example, Fe-Mn order–disorder in carbonates or pyroxenes or Al-Si order–disorder in framework silicates can not be inferred from X-ray diffraction intensities. *Anomalous scattering* using radiation near the absorption edge, has been used to attempt to distinguish Mn and Fe and determine site occupancies, but the results have sometimes been equivocal.

The diffraction information (positions and intensities of individual reflections) is used to determine the crystal structure (space group and positional parameters for all atoms). This is often done in a series of iterative cycles (refinement). Although one speaks of solving a crystal structure, what one usually does is to propose a model for that structure and calculate a quality of fit (reliability factor, *R*). A small *R* factor ($<$ about 3%) argues for the proposed structure being correct. Nevertheless, a different model might fit the data even better, or seemingly minor misfits in the proposed model may be indicative of

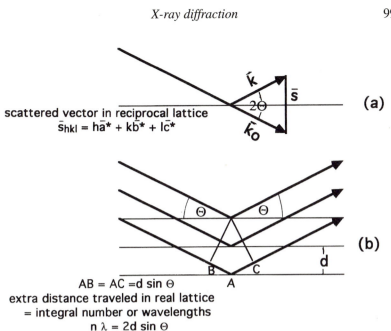

scattered vector in reciprocal lattice
$\bar{s}_{hkl} = h\bar{a}* + k\bar{b}* + l\bar{c}*$

(a)

(b)

$AB = AC = d \sin \Theta$

extra distance traveled in real lattice
= integral number or wavelengths

$n \lambda = 2d \sin \Theta$

Figure 3.3. X-ray diffraction schematic representation of Bragg's law.

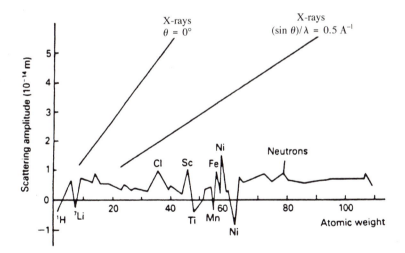

Figure 3.4. Variation of X-ray scattering factor and neutron scattering cross section, with atomic number Z. Note the monotonic increase with Z for electrons and the irregular behavior for neutrons. A negative neutron cross section means that neutrons are scattered with a 180° change of phase (from Bacon 1975).

some phenomenon not properly considered (e.g., order–disorder, twinning, etc.). As our concepts of structural chemistry and especially of defect chemistry evolve, many complex structures will need to be reinvestigated.

Once the structure is known, the calculated interatomic distances provide additional information. Bond length variations can pinpoint ordering. For example, although Si and Al can not be distinguished by their diffraction intensities, the difference in tetrahedral Si–O distance (1.62 Å) and Al–O (1.73 Å) makes it possible to infer the ordering state from the bond lengths (see Fig. 3.5).

Maps of electron density within the unit cell from single crystal X-ray diffraction studies provide direct experimental information to compare with models of chemical bonding (see Chap. 5). Anomalies in distributions can indicate that the proposed structure is "not quite right." An example of such a case are the amphiboles, with A-sites filled by alkali ions. Studies have repeatedly suggested, but have been unable to prove definitively, the hypothesis of a *split A-site*, in which the alkali takes up positions somewhat away from the center of the large cavity (see Fig. 3.6).

Although powder diffraction patterns provide less readily analyzed structural information than single crystal diffraction studies, in many cases, powder is all that is available, and useful structural studies are possible. Powder diffraction is a very useful fingerprinting technique; it can identify phases, detect impurity phases down to 3–5% abundances, and, through variation of lattice parameters with composition, estimate the composition of solid solutions. If the space group is known with some confidence, powder diffraction data can be refined to give accurate lattice parameters. The variation of lattice parameters with composition, with temperature (T), and with pressure (P) (see Fig. 3.7) gives valuable insight into chemical bonding, order–disorder, and the effect of P, T, and chemical environment on individual bond lengths (also see Chap. 6). Powder diffraction, especially in situ at high P and/or T, allows direct observation of phase transitions.

A number of geometries is possible for powder diffraction studies, but the automated powder diffractometer has gained popularity because of convenience and versatility. Figure 3.8 shows powder patterns of a cubic and triclinic phase. The forest of overlapping peaks seen in a sample of low symmetry can present problems in distinguishing and assigning individual peaks. An alternative approach, now increasingly popular, is the Rietveld method, in which the entire pattern is modeled as a continuous profile of intensity versus diffraction angle. This does not require the direct deconvolution of the experimental data into individual peaks, but one has to take considerable care in making assump-

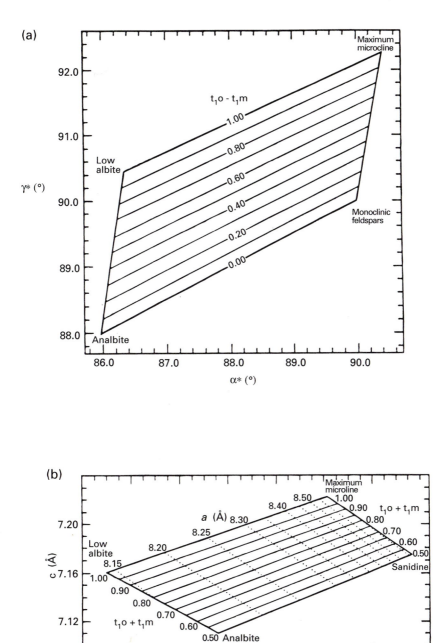

Figure 3.5. (a) The α^*-γ^* plot for alkali feldspars. It relates the reciprocal space angles to contours of the difference in aluminum content between T_1O and T_1m sites. (b) The b-c plot for alkali feldspars. It relates the b- and c-axis lengths to contours of the total Al content of T_1 sites, $T_1O + T_1m$. (After Stewart and Wright 1974.)

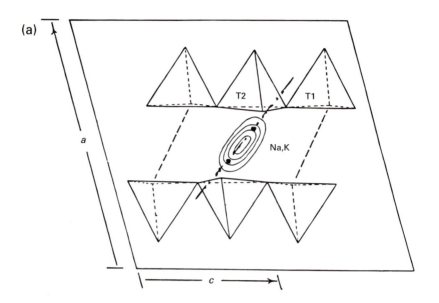

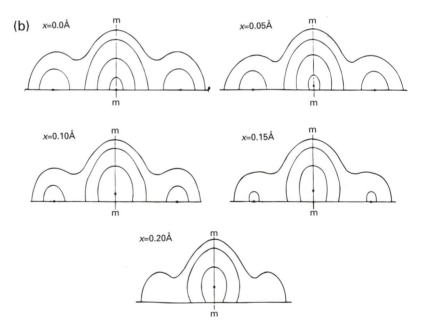

Figure 3.6. Split A-site in amphibole. (a) The hornblende A-site viewed along *b*. The contour lines (contour interval 4 electrons/Å³) represent the (Na, K) contribution to the electron density. Black dots mark the "split-atom" positions for (Na, K) (from Papike, Ross, and Clark 1969). Dot-dashed line shows axis corresponding to contours in (b). (b) A series of Fourier sections taken in the vicinity of the A-site perpendicular to the normal to (100). X distance along the *a* direction. Section X = 0·0 Å is through the A-site; m is the mirror plane; small circle shows locations of maxima in electron density (from Hawthorne and Grundy 1973).

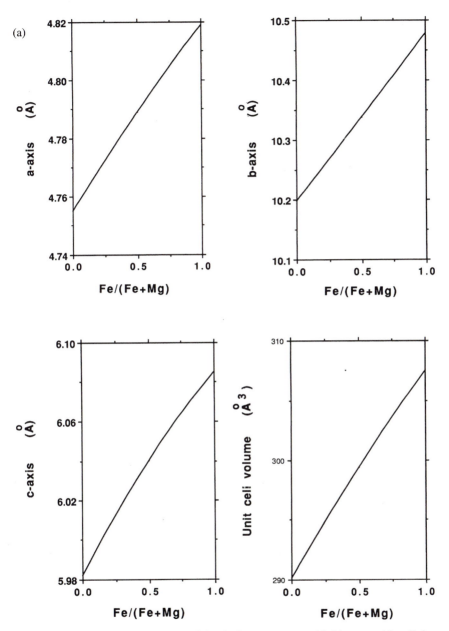

Figure 3.7. (*pp. 103–5*) Variation of olivine lattice parameters with (a) composition, Fe/(Fe + Mg) at ambient conditions (data from Matsui and Syono 1968), (b) temperature for Mg_2SiO_4, and (c) pressure for Mg_2SiO_4 (data from Hazen 1976).

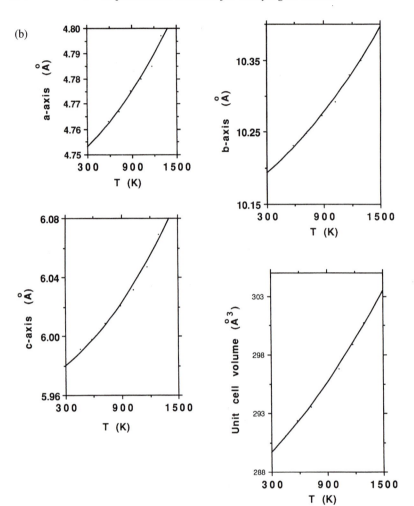

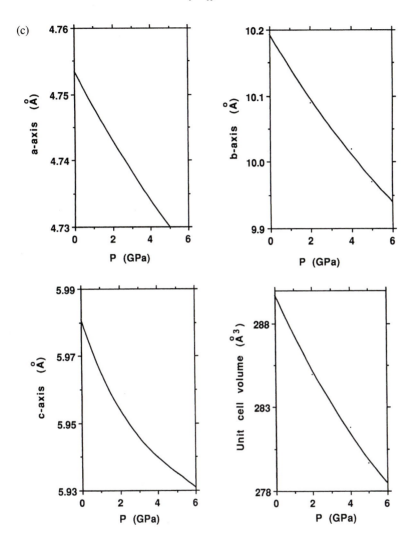

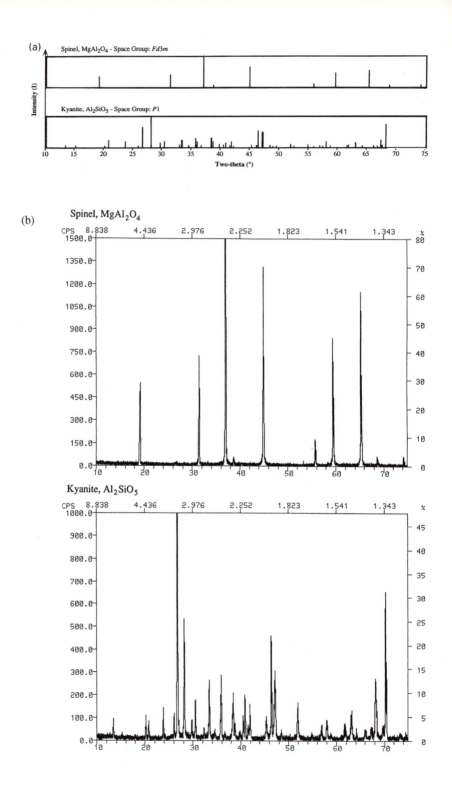

tions about peak shapes and thermal parameters. The Rietveld method has met considerable success in dealing with phases of low symmetry, with lattice parameter refinements for multiphase mixtures, and with site occupancy studies. It is also well suited to studies with a polychromatic beam of synchrotron X rays or of neutrons.

The broadening and disappearance of X-ray diffraction peaks is a sign of decreasing crystallinity of a sample. Figure 3.9 shows the powder pattern of talc of several grain sizes. When individual particles are smaller than several hundred angstroms, the pattern broadens significantly. This is the same size range for which surface effects and poor crystallinity begin to seriously affect thermodynamic properties. Such fine-grained materials are important in near surface geologic environments and may play an important role in the trapping and transport of trace elements, radioactive waste, and other environmental pollutants. They are also crucial in ceramic processing and in the emerging field of "nanoscale" composites.

Amorphous materials will give a broad "hump" as a powder diffraction pattern (see Fig. 3.9) rather than sharp peaks. This pattern can be analyzed to give the most likely average interatomic distances and their dispersion through the construction of a radial distribution function. This is an important means of studying mid-range structure in glasses and will be discussed further in Chapter 8. Radiation can induce amorphization (metamict minerals), and, recently, pressure-induced amorphization is proving to be a fairly widespread phenomenon. X-ray diffraction is essential in the study of such behavior.

In powder diffraction studies, preferred orientation of crystals is often a problem. This is especially true for clays, which show almost complete orientation when deposited in a typical glass diffraction slide. The diffraction pattern then reflects a nonrandom averaging of crystal orientations, with certain peaks enhanced and others essentially missing. Patterns of the same sample but with different degrees of preferred orientation can look deceptively different.

Because atoms vibrate about their equilibrium positions with corresponding dynamic displacement of their electron clouds, a diffraction experiment provides information on lattice vibrations. Often such information is disregarded or simply used to parameterize "thermal vibration ellipsoid" parameters, but closer links between diffraction studies and lattice dynamics (see Chap. 6) are desirable.

Figure 3.8. (*Facing page*) Powder X-ray diffraction profiles. Intensity (I) versus diffraction angle (2θ) for spinel, $MgAl_2O_4$ (cubic), and kyanite, Al_2SiO_5 (triclinic). (a) Schematic patterns based solely on data from the X-ray powder diffraction file (JCPDS) cards. Only peak positions and relative peaks' heights are shown. (b) Actual diffraction patterns.

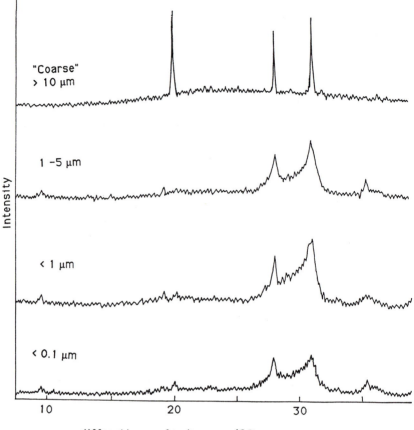

diffraction angle, degrees (2Θ)

Figure 3.9. Powder diffraction patterns of talc of several grain sizes. Note extreme broadening for < 1μm samples. (From K. Bose, pers. comm.)

A frontier of X-ray diffraction lies in the study of very small samples and materials at very high pressure and temperature (Fei et al. 1992; Finger 1992; Navrotsky et al. 1992).

3.3 Neutron diffraction

The advantages of neutrons for diffraction studies lie in two realms. First, the neutron-scattering cross section, in contrast to that for X rays, varies in a complex way with atomic number (see Fig. 3.4). Thus, adjacent elements in the periodic table can be distinguished, and hydrogen is a good scatterer. The positions of OH groups and waters of hydration can be well established. Second, neutrons interact with the magnetic moments of the atoms and can be used to determine magnetic ordering. Inelastic neutron-scattering experiments are the most rigorous experimental approach for getting the complete vibrational density of states of a solid (see Chap. 6).

Fairly large samples of powder (2–10 g) or large single crystals (centimeter sized) are generally required for neutron diffraction. This limitation, coupled with the need to collaborate with specialized facilities, makes this technique less readily available than X-ray diffraction. Nevertheless, it has probably been underutilized in the earth sciences relative to materials science.

3.4 Electron microscopy and electron diffraction

The electron microscope uses a finely focused beam of electrons to interact with a sample in a high-vacuum environment. In scanning electron microscopy (SEM) the electrons form an image analogous to the reflected light image in an optical microscope, but magnifications from several thousand to several million times and resolution down to several hundred angstroms are possible. Thus particle morphology, phase separation, intergrowths, and heterogeneity on a very fine scale can be studied (see Fig. 3.10). The electron image can be analyzed for variations in brightness related to variations in the mean atomic number of the materials under the beam. The X rays generated by the impinging electrons can provide for both qualitative and quantitative chemical analysis of the elements present in the sample and their spatial distribution. When this is done with micron spatial resolution (limited by the larger volume of sample that emits X rays even when excited by a smaller electron beam) and the X rays are analyzed by several wavelength-selective spectrometers, the resulting instrument is the modern electron microprobe (or electron probe microanalyzer, EPMA). Modern image analysis and computerization melds these techniques to provide maps, on a micron to millimeter scale, of sample composition (see Fig. 3.10).

When electrons transmitted through a thin sample are analyzed, the technique is called transmission electron microscopy (TEM) or, when resolution in the 2–10 Å range is achieved, high-resolution transmission electron microscopy (HRTEM). The method then resolves structural features on the atomic scale and provides images related to electron density within the unit cell (see Fig. 3.11). Electron diffraction patterns of the region under the beam are also obtained (see Fig. 3.11). Although HRTEM images and diffraction patterns generate data that are averages over many unit cells, they detect local variations in structure that can not be seen by X-ray diffraction. Thus modulated structures in feldspars, chain sequence faults in amphiboles, and the progress of chemical reactions on the unit cell scale can be studied (see Fig. 3.11). Indeed,

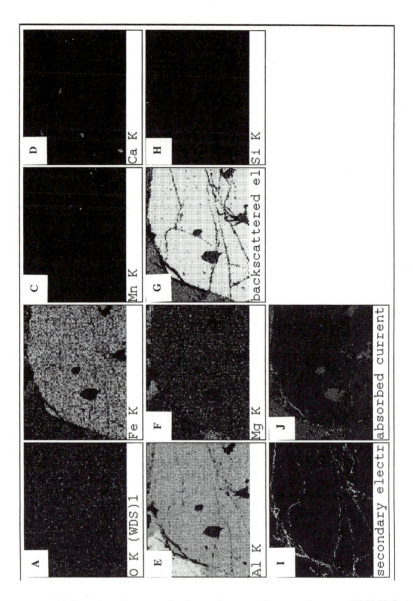

Figure 3.10. Scanning electron microbeam images of a natural garnet ($X_3Y_2Si_3O_{12}$) crystal with inclusions of quarts (SiO_2). Figures A–F, and H are X-ray intensity maps for K_α lines in which the white regions represent high concentrations of a given element and darker areas represent progressively lower concentrations. (A) Oxygen K_α, X rays collected with a wavelength spectrometer using a multilayer pseudo crystal (effective $2d = 60$ Å). (B–F, H) K_α lines for Fe, Mn, Ca, Al, Mg, and Si, respectively, collected with an energy dispersive spectrometer. (G) Backscattered electron image in which greyscale is proportional to the average atomic number. (I) Secondary electron image reflects topographic relief and hence surface features. (J) Absorbed current image represents the proportion of electron beam current "taken in" by the sample, and for a polished sample will reflect the inverse of the reflected electron (or backscattered) image. The field of view for each image is 300 microns. (These images were taken by E. Vicenzi at the Princeton Materials Institute.)

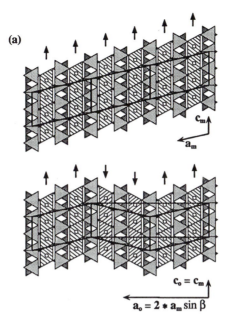

(a)

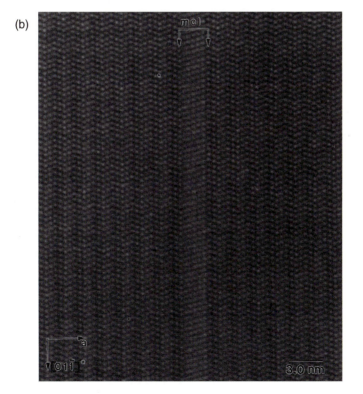

(b)

almost no mineral or ceramic looks "simple" and "single phase" after a careful HRTEM study.

Analytical capabilities of modern HRTEMs allow chemical analysis using emitted X rays and electrons. Analysis of the energy of transmitted electrons forms the basis of the analytical technique known as EELS (electron energy loss spectroscopy). Combinations of scanning and transmission capabilities are available in the scanning transmission electron microscope (STEM). The field is evolving rapidly and new specialized techniques are being developed. Modern electron microscopy is indispensable to materials characterization and to the study of the intricacies of crystal chemistry.

3.5 Vibrational spectroscopy

Both in isolated molecules and in solids, vibrations have energies corresponding to the infrared. These can be studied by several techniques. Infrared absorption spectroscopy directly measures the fraction of energy absorbed when radiation of a given frequency impinges on a sample (Gadsden 1975, McMillan and Hofmeister 1988). It is customary to plot percent transmission versus wavenumber (see Fig. 3.12). Conventional infrared absorption spectroscopy uses a single crystal, a polished piece of glass, or a powder, the latter often pressed in a pellet of a soft material like KBr to minimize scattering from surfaces. Data from single crystals, in which anisotropy and polarization effects can be studied, are more informative than data from powders, but the latter are nevertheless very useful, both as fingerprints and for fundamental understanding of structural and vibrational features.

There are several recent advances in infrared spectroscopy. Beamlines on the synchrotron, operating in the infrared, give advantages of enhanced inten-

Figure 3.11. (*p. 111*) (a) Schematic diagrams showing the relation between the monoclinic and orthorhombic amphibole structures. In both parts of (a) the tetrahedral chains are running from top to bottom. In the monoclinic C2/m structure (top half of (a)), the octahedra are all stacked in the same direction parallel to a, as shown by arrows and symbolized by $(++++)$. Three monoclinic unit cells are outlined. The Pnma orthorhombic structure results from periodic twinning along (100) such that each monoclinic cell is related to its neighbor by a mirror reflection across (100) (bottom half of (a)). This makes the sequence $(++--++--)$. This stacking gives an orthorhombic unit cell with an a-axis nearly twice as long as that of the monoclinic structure. (b) High-resolution transmission electron microscope (HRTEM) image of the orthoamphibole anthophyllite down [011], showing a direct picture of this relationship. This homogeneous area contains a single stacking fault parallel to (100). The break in the regular $(++--++--)$ stacking in the orthorhombic structure gives rise to a narrow strip of monoclinic material with the $(++++)$ stacking sequence. This strip of monoclinic amphibole is only two unit cells thick (from Smelik and Veblen 1993).

sity analogous to those for X-ray studies. IR spectroscopy through an optical microscope allows the study of small single crystals and of individual grains in a thin section of rock or composite. High surface area powders and optically opaque materials can be studied by photoacoustic infrared spectroscopy (PAS-IR), without the need to embed the materials in a pellet of KBr (Benziger, Royce, and McGovern 1985). Spectroscopy in reflection rather than transmission mode has advantages for fine-grained and strongly absorbing samples (McMillan and Hofmeister 1988). Fast Fourier transform techniques

(a)

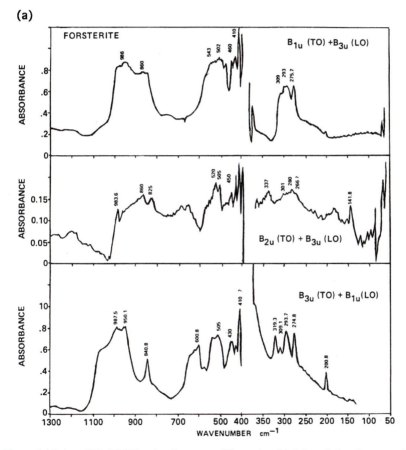

Figure 3.12. (*pp. 113–14*) Vibrational spectra of forsterite. (a) Infrared absorbance and (b) infrared reflectance (both from Hofmeister 1987). (c) Raman (from Iishi 1978). The *B* symbols refer to light polarization directions relative to crystal symmetry; the variation of intensity with polarization direction gives additional information about the nature and symmetry of vibrational modes.

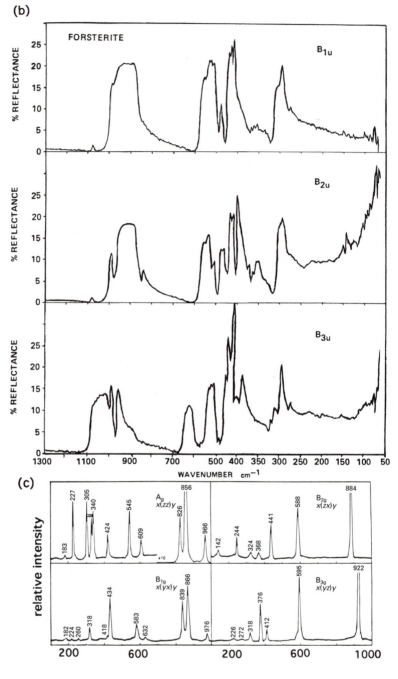

have speeded up data acquisition, improved resolution and sensitivity, and enabled the study of samples that decompose on prolonged irradiation.

In Raman spectroscopy, the sample is irradiated by intense monochromatic visible radiation, generally from a laser (McMillan and Hofmeister 1988). Most of this light is transmitted or scattered with no change in energy, but a small proportion (generally less than a percent) interacts with the solid, is absorbed, causes vibrational excitation, and is re-emitted at a lower energy (lower frequency, longer wavelength). This Raman-shifted radiation is then observed as a function of its energy. Micro-Raman spectroscopy, with a resolution of less than 5 microns, is commercially available.

Figure 3.12 shows Raman spectra, and both reflectance and transmission IR spectra of Mg_2SiO_4 forsterite single crystals. Several points are evident. Raman bands are generally sharper than IR bonds. The IR spectra in reflection and transmission mode are somewhat different. The intensity of a given peak depends strongly on the orientation of the single crystal with respect to the polarization of the light. These points are important to proper interpretation of spectra and assignment of peaks (see McMillan and Hofmeister 1988). For forsterite, the IR and Raman peaks above 900 cm^{-1} are generally assigned to Si–O stretching vibrations and those in the 200–800 cm^{-1} region to bending modes and vibrations involving the lattice as a whole.

For a vibrational mode to be active in the infrared, the motion must change the dipole moment of the molecule or crystal. For Raman activity, the polarizability of the excited state must be different from that of the ground state. Symmetry considerations, through group theory and normal mode analysis, enable one to predict the number of IR and Raman active modes for a given crystal structure. Some vibrational modes are active in the IR, others in the Raman, a few in both, and some in neither. Thus IR and Raman spectroscopy are complementary. The lower the symmetry of the space group, the greater the number of different observable vibrational frequencies. Disorder, both in crystalline and in amorphous phases, leads to severe peak broadening. Thus vibrational spectroscopy is a useful qualitative, and sometimes quantitative, tool for the study of symmetry and positional disorder.

Certain frequencies correspond to specific localized vibrations, for example, an O–H stretch in a hydroxyl group, an Si–O stretch in an SiO_4 tetrahedron, or a C–O stretch in a carbonate (see Table 3.3). However, much of the richness of a vibrational spectrum of a solid arises from lattice modes not assignable to single bond vibrations but involving cooperative motion of several neighboring atoms. The proper assignment of vibrational bands is an involved and often controversial topic, especially in glass science.

There is a large literature using vibrational spectroscopy to elucidate the

Table 3.3. *Vibrational modes (cm^{-1}) in solids*

Mode	Wavenumber (cm^{-1})	Activity
Lattice modes		
Si–O bend	400–800	IR and Raman
octahedral cation M–O	500–800	IR and Raman
(including stretching Si)		
modes involving weakly	200–500	IR and Raman
bonded cations		
Internal vibrations of complex ions		
Carbonate, CO$_3^{2-}$		
asymmetric stretch	1400	Strong IR, weak Raman
symmetric stretch	1000–1100	Strong IR, weak Raman
Silicate, SiO$_4^{4-}$		
stretching	900–1200	IR and Raman
bending	400–600	IR and Raman
(but modes not really isolated,		
strongly coupled)		
Water, H$_2$O		
HOH bend	1630	Strong IR
OH stretch	3400	Strong IR
Hydroxyl, OH$^-$		
isolated ion stretch	3735	Strong IR
hydrogen bonded stretch	3400–3500	Strong IR

structure of aluminosilicate glasses in terms of species of differing degrees of polymerization (see Chap. 8). There are inherent difficulties in assigning peaks to vibrations of specific species or clusters, in deconvoluting overlapping peaks, and in relating peak intensities (or areas) to species concentrations. Because the distance scale on which atomic motions in a glass are correlated to produce vibrations of a given frequency is still somewhat controversial, the use of vibrational spectroscopy as an analytical tool for silicate group speciation is theoretically less straightforward than the use of methods such as NMR that are clearly sensitive almost exclusively to nearest and next-nearest neighbor configurations.

Vibrational spectra are sensitive to defects and to atomic motions at surfaces and interfaces. Thus, for the study of surface adsorbed atoms, of mineral dissolution and precipitation, and of surface and interface structure, Raman and IR spectroscopy are important techniques. Under certain circumstances, the Raman signal from surfaces can be enhanced in intensity, and this forms the basis for a growing field of research. The description and analysis of the vibrations of a crystal (lattice dynamics) are discussed in Chapter 6.

3.6 Optical spectroscopy

Because visible and ultraviolet (UV) radiation is higher in frequency than infrared, it probes transitions of higher energy, typically those among electronic energy levels. In minerals and ceramics, these are mainly of four types: transitions between *d*- or *f*-electronic levels in transition metals and rare earths, transitions among levels associated with a defect or "color center," charge transfer transitions, and strong UV absorption related to *s* and *p* electron levels. These phenomena, plus diffraction effects from closely spaced exsolution lamellae (as in labradorite feldspar) are the main sources of color in minerals, a complex subject about which much continues to be written (Nassau 1983). Because of the dominant role of impurities and defects, a mineral's color – though of considerable esthetic, gemological, and sometimes economic importance – is not an intrinsic and immutable property of its structure and major element composition, but is influenced by trace impurities and thermal history. This sensitivity accounts for the rainbow of colors attainable in natural and synthetic gems and for the use of heat treatment, irradiation, and other methods to change or enhance color.

Spectroscopy in the UV and visible can be done in either transmission or reflection mode (Rossman 1988). In the former, the light transmitted through the sample is monitored, and the absorption as a function of wavelength recorded. In reflection spectroscopy, the light reflected from a polished surface is collected. Absorption results in a diminution of intensity of the reflected light of the given energy. The color seen by the eye is the color of the remaining transmitted or reflected light; thus a mineral that appears blue does so because it absorbs strongly in the red. As in IR spectroscopy, computerization, fast Fourier transform methods, and more intense and stable light sources have brought about advances in the field. Spectroscopy through a microscope and at high pressure in a diamond anvil cell also extend the usefulness of this technique.

Data from UV-visible spectroscopy can be used to determine crystal field splitting (see Sec. 5.4). Spectra of Ni^{2+} in olivine are shown in Figure 3.13. Each peak corresponds to a specific *d*-electron transition, and Ni^{2+} in M1 and M2 sites can be resolved.

A spectrum involving charge transfer is shown in Figure 3.14. The colors attained are often more intense than those due to *d-d* transitions, because the transitions in the former are quantum mechanically "allowed", whereas those among *d* or *f* levels are "forbidden" and occur only weakly. Indeed, in minerals containing ions in several oxidation states (typically, Fe^{2+} and Fe^{3+} or Fe^{2+} and Ti^{4+}), charge transfer bands are so strong that the material is often black.

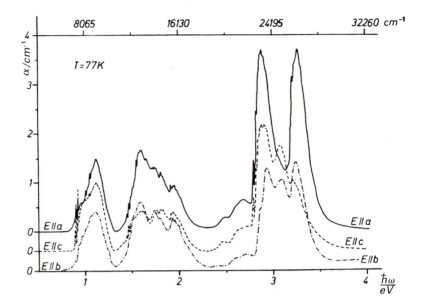

Figure 3.13. Polarized optical absorption spectra of the synthetic Ni doped forsterite at 77 K. The resolution is roughly 4 cm^{-1}. For clarity the curves have been displaced along the ordinate (from Rager, Hosoya, and Weiser 1988).

The intensity of absorption is related to other physical parameters by the Beer Lambert law, with I the observed intensity, I_o the initial intensity, and ε an absorption coefficient.

$$\text{Absorbance} = \log (I/I_o) = \varepsilon \cdot \text{concentration} \cdot \text{pathlength} \qquad (3.3)$$

By controlling the pathlength, by a cuvette of known thickness holding a solution or by a crystal or polished section of known thickness, optical spectroscopy becomes a very useful analytical tool for determining concentrations of transition metals or other species, providing the absorption coefficient can be determined by a set of standards whose concentrations bracket those in the unknown. Because absorption spectra are characteristic of a given species, oxidation state, coordination number, and bond distance, they can be used to identify transition metals and their environment. Since such spectra are generally fairly broad, peak overlap, line shape, and proper subtraction of baseline can be serious challenges.

The fate of the absorbed radiation yields further information on structure

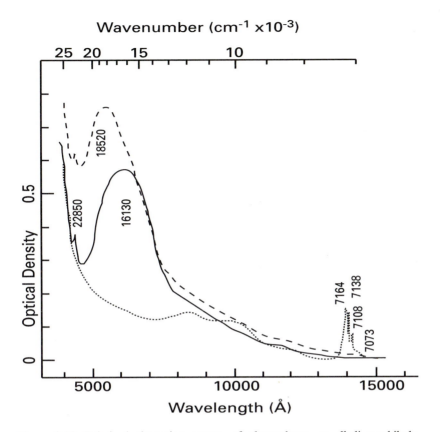

Figure 3.14. Polarized absorption spectra of glaucophane, an alkali amphibole, $Na_2(Fe^{2+}, Mg)_3$ $(Fe^{3+}, Al)_2Si_8O_{22}(OH)_2$. When light is polarized in the b-c plane of linked octahedra, strong Fe^{2+}-Fe^{3+} charge-transfer bands at 16130 and 18520 cm^{-1} are seen. When light is polarized perpendicular to this plane, little absorption occurs. The mineral is blue-violet in color and strongly pleochroic. (From Burns et al. 1980.)

and bonding. Because an electronic transition is rapid relative to nuclear motion, it may be considered to occur vertically between two states as depicted on a potential energy curve (see Fig. 3.15). If the excited electronic state has a longer equilibrium bond length than the ground state (as is usually the case), this electronic excitation places the system in a vibrationally excited state. The vibrational excitation may dissipate through equipartition among vibrational modes, that is, as heat. The electronically excited ground state may then emit UV or visible radiation of a lower energy, putting the system in a vibrationally excited (too long bond length) ground electronic state, which then decays

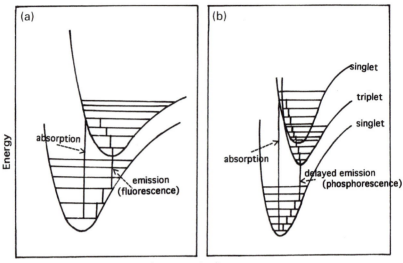

interatomic distance

Figure 3.15. Sources of absorption, emission, fluorescence, and phosphorescence depicted on potential energy curves. (a) A symmetry allowed transition from electronic ground to excited state. The equilibrium-internuclear separation is the ground electronic state, r_0, which corresponds to a contracted bond in the excited electronic state. Horizontal lines correspond to vibrational energy levels. Energy is re-emitted at a lower frequency (fluorescence) from the ground vibrational state of the excited electronic state to an excited vibrational state (lengthened bond) of the ground electronic state. Vibrational energy is dissipated as heat. (b) Fluorescence – if the allowed excited electronic state potential energy curve crosses that of a lower lying excited electronic state that can not be accessed from the ground state (is symmetry forbidden), a radiationless transition between the two states can occur. Decay from the forbidden excited state to the ground electronic state is then slow, resulting in fluorescence or, if very slow, phosphorescence, a delayed glow visible when the exciting radiation is turned off.

by heat dissipation to the original state. This emission of red-shifted light is fluorescence. If the emission is delayed such that one detects a glow after the exciting radiation is turned off, this emission is called phosphorescence. The time lag may occur when the excited state can undergo a radiationless transition (because the energy level manifolds cross) to a set of states whose symmetry makes decay to the ground state forbidden. Both fluorescence and phosphorescence are observed in minerals; famous examples are the beautifully green-fluorescing willemites (Zn_2SiO_4) and red-fluorescing calcites ($CaCO_3$) from Franklin, New Jersey. The electronic transitions are probably those of small amounts of Mn^{2+} substituted for Zn in tetrahedral coordination in

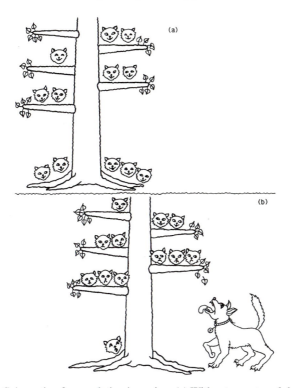

Figure 3.16. Schematic of a population inversion. (a) Without an external driving force, the particles (cats) are distributed in the lower-lying energy levels (branches) according to the Boltzmann distributions. (b) With external pumping of energy (dog), the particles populate higher energy states (higher branches) more than lower ones. Lasing action may be likened to an upper branch breaking, returning its occupants to their ground state in a coherent pulse.

willemite and for Ca in octahedral coordination in calcite. The contrasting colors of fluorescence provide a vivid example of the importance of coordination number on optical properties (see also Sec. 5.6).

Fluorescence and phosphorescence are technologically important. Fluorescent screens, TV tubes, and other optical devices use electron-excited optical transitions to generate light of specific colors. The best materials (phosphors) for such applications are frequently complex oxides analogous to minerals and containing transition metals, especially Mn^{2+}. Low symmetry sites generally enhance fluorescent intensity by relaxing selection rules.

While discussing optical phenomena, a few words need to be said about lasers. Excitation with a bright source can cause a large number of atoms to achieve electronically excited states. If these states decay relatively slowly, then a "population inversion" occurs in which a far greater than thermodynamically stable population of excited species exists (see Fig. 3.16). Under proper conditions, this population can decay to the ground state together, resulting in an intense pulse of coherent light, a so-called laser (light amplification by stimulated emission of radiation) pulse. Such pulses can be reflected by mirrors, their frequencies doubled, and other optical games played. Of particular interest is their short duration (now it is possible to get pulses in the 10^{-12}–10^{-15} s timescale), which enables time-resolved spectroscopy of vibrational and electronic excitations.

3.7 X-ray and electron spectroscopy

In addition to being diffracted, X rays interacting with a solid may be absorbed and their energy later re-emitted as X rays or lower energy (UV or visible) photons. X rays may cause electronic excitation followed by emission of the excited electron. These interactions form the basis for a variety of spectroscopic techniques that probe geometry and bonding, in some cases of the surface, in others of the bulk material. These processes and their corresponding spectroscopies are summarized in Figure 3.17.

Because such interactions generally involve only a small fraction of the incident X rays, effective spectroscopy has been advanced by the development of stable and intense X-ray sources and of associated detector and computerized data analysis technologies. The use of synchrotron radiation is essential to X-ray absorption spectroscopy. The synchrotron was developed initially by high-energy physicists to study the interactions of electrons accelerated along a circular path to relativistic speeds. It produces, almost as a by-product, radiation whose frequency is related to the design parameters. Thus synchrotron rings that produce primarily X rays, primarily vacuum UV radiation, and primarily IR radiation have been constructed. Research at the synchrotron is essentially a collaborative team effort, and users bring samples to the facility, work around the clock, and return home with data to analyze.

Synchrotron radiation has a number of unique properties. First, it is of high and relatively uniform intensity or flux (number of photons emitted per second per certain bandwidth of energy) over a wide range of energies (see Fig. 3.18). The gain in intensity over laboratory X-ray sources is four to eight orders of magnitude. Second, the beam has small geometrical divergence, producing high brilliance (flux per unit area or solid angle). The resulting high collimation enables high spatial control and resolution for the beam interacting with

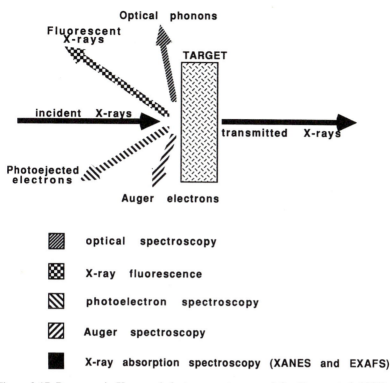

Figure 3.17. Processes in X-ray and electron spectroscopy (after Brown et al. 1988).

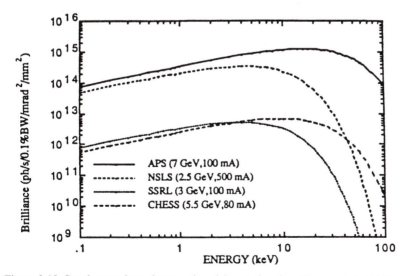

Figure 3.18. Synchrotron intensity at various laboratories (from Brown et al. 1988).

the sample. Third, because of the way in which electrons are accelerated and injected into the ring, the radiation actually comes in time-resolved bunches or pulses of several tenths of a nanosecond duration, enabling time-resolved spectroscopy of rapid chemical processes. Fourth, the radiation is horizontally polarized in the orbital plane and elliptically polarized perpendicular to the plane. This allows polarized absorption spectroscopy on single crystals, which can yield detailed information about structure and bonding.

X-ray absorption spectra are shown in Figure 3.19. At low energies the absorption is low, and this is the region favorable for diffraction experiments. The absorption edge, characteristic for each element, marks the onset of strong absorption. This region is characterized by fine structure both before and after the absorption edge (Waychunas 1988). The characterization of these oscillations is the realm of *X-ray absorption near edge structure* spectroscopy, XANES. The relatively sharp features seen are associated with strong multiple scattering of emitted photoelectrons by nearest and next-nearest neighbor atoms. Thus XANES is a good probe of local coordination number, coordination geometry, and bond angles. It is beginning to yield information about bond lengths as well. However, the physics of multiple scattering is complex and interpretation of XANES spectra is still controversial. The region well above the absorption edge (from 50 to perhaps 1000 eV higher) contains weaker and more spread-out oscillations resulting from single scattering events. These are known as the *extended X-ray absorption fine structure*, EXAFS. The spectra contain information about local coordination and mid- and long-range order.

XANES and EXAFS are applicable to solutions, amorphous materials, and crystalline materials. Theory is complex and evolving rapidly. Interpretation is currently a mixture of first-principles considerations and empirical comparisons to the spectra of materials with known structures. Because of high sensitivity and because the environment of only one kind of atom is probed, these techniques can give information on the local coordination of minor or trace elements, as well as of one element in the presence of others. For example, though X-ray diffraction will give average bond length information for an $(Mg,Fe)_2SiO_4$ olivine solid solution, including the site occupancies and average bond geometries of the M1 and M2 sites, X-ray absorption techniques have the capability to tell whether the coordination about the iron atom (bond lengths and angles) is the same as or different from the average. EXAFS and XANES

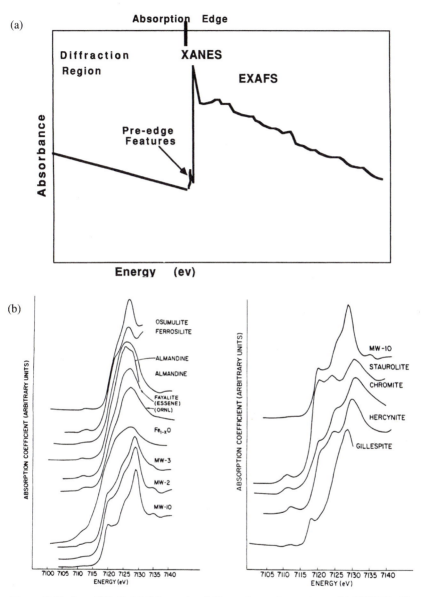

Figure 3.19. (*pp. 125–6*) (a) Schematic of X-ray absorption regions. (b) XANES, X-ray absorption near edge structure, of minerals containing Fe^{2+} in octahedral (left side) and tetrahedral (right side) coordination. Note sensitivity to local environment. (From Waychunas, Apted, and Brown 1983.) (c) EXAFS data analysis for Co(II) K-edge in $Na_2O–2SiO_2$ glass (4 wt% CoO). (Top) Experimental absorption spectrum; (middle) EXAFS obtained after background substraction; (bottom) Fourier transform gives peaks that correspond to interatomic distances (Calas et al. 1987).

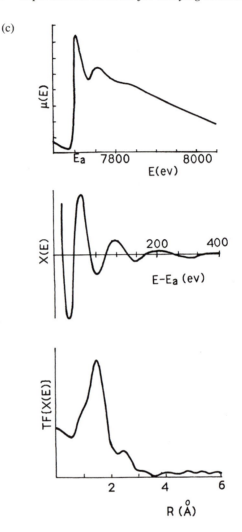

can probe the coordination environments of specific atoms in glasses. The techniques can also distinguish different oxidation states. At present X-ray absorption spectroscopy is generally limited to heavier elements (roughly, calcium and higher).

When X rays interact with a solid, they can produce a spectrum of other X rays. This occurs if the exciting X ray is absorbed, causing the excitation of an electron to a higher energy level, followed by emission of the remaining energy as an X ray of longer wavelength. This X-ray fluorescence (XRF) is analogous to fluorescence in optical spectroscopy. Because specific fluorescence can be attributed to particular elements (although problems of overlapping peaks exist), and because the intensity of fluorescence is proportional to the concentration of that element (in a constant matrix), XRF is a commonly used analytical technique that lends itself easily to automation and to processing multiple samples in an industrial environment as well as in academic research. An XRF microprobe with 5 micron spatial resolution and trace element detection limits of ppm to ppb is operating using synchrotron radiation at Brookhaven National Laboratory.

If an X ray is absorbed, it can cause the ejection of an electron from a bound orbital. Ejected photoelectrons can also be produced by electron bombardment and by UV irradiation. The detection and analysis of these ejected photoelectrons form the basis for a family of photoelectron spectroscopies: X-ray photoelectron spectroscopy (XPS), ultraviolet photoelectron spectroscopy (UPS), and Auger electron spectroscopy (AES) (Hochella 1988). In general, because electrons interact strongly with a solid, only those electrons generated near the surface (i.e., in the first few hundred Å) will emerge and be detected. Thus the electron spectroscopies probe primarily the surface and near surface environment, whereas the X-ray spectroscopies (and electron microprobe) investigate the bulk.

3.8 Nuclear magnetic resonance

Atomic nuclei have quantized energy levels characterized by quantum numbers representing major shells, angular momentum, magnetic properties, and spin, analogous in many ways to energy levels occupied by electrons around an atom. Protons and neutrons each occupy a series of levels; thus the properties of a nucleus depend both on its atomic number and its mass. Nuclear spin is the crucial parameter in nuclear magnetic resonance (NMR); nuclides with even mass number and even charge (e.g., ^{12}C) have zero spin and do not show NMR spectra. Those with even mass number and odd charge (e.g., ^{14}N) have integral spin and are relatively difficult to study. Nuclides with odd mass number and odd charge (e.g., ^{1}H, ^{13}C, ^{17}O, ^{27}Al, ^{29}Si, ^{31}P) have half integral spin ($I = 1/2, 3/2, 5/2, \ldots$) and are most useful for NMR (see Table 3.4). In a magnetic field, these spin levels are split into $2I + 1$ states (see Fig. 3.20a). This splitting is in the radio frequency (rf) region, with megahertz (MHz) frequencies.

The magnetic moment of an isolated nucleus is a fundamental physical property, thus its frequency of absorption in a magnetic field of given strength is also a fixed parameter. However, a nucleus surrounded by other atoms in a

Table 3.4. *Characteristics of NMR nuclei*

Nucleus	Readily observed	Spin	Quadrupole moment (10^{-24} cm^2)	Natural abundance (%)	Frequency MHz (at 11.7 T)
H-1	yes	1/2	—	99.985	500
H-2	yes	1	0.0028	0.015	76.8
Li-7	yes	3/2	−0.03	92.58	194.3
Be-9	yes	3/2	0.0512	100	70.3
B-10	yes	3	0.074	19.58	53.7
B-11	yes	3/2	0.0355	80.42	160.4
C-13	yes	1/2	—	1.1	125.7
N-14	yes	1	0.016	99.6	36.1
N-15	yes	1/2	—	0.37	50.7
O-17	yes	5/2	−0.026	0.037	67.8
F-19	yes	1/2	—	100	470.4
Na-23	yes	3/2	0.14	100	132.3
Mg-25	yes	5/2	N.D.	10.1	30.6
Al-27	yes	5/2	0.149	100	130.3
Si-29	yes	1/2	—	4.7	99.3
P-31	yes	1/2	—	100	202.4
S-33	no	3/2	−0.064	0.76	38.4
Cl-35	yes	3/2	−0.0789	75.5	49.0
K-39	yes	3/2	0.11	93.1	23.3
Sc-45	yes	7/2	−0.22	100	121.5
Ti-49	no	7/2	N.D.	5.5	28.2
V-51	yes	7/2	−0.04	99.76	131.4
Cu-63	yes	3/2	0.16	69.1	132.5
Zn-67	yes	5/2	0.15	4.1	31.3
Ga-71	yes	3/2	0.112	39.6	152.5
Ge-73	yes	9/2	−0.2	7.8	17.4
Se-77	no	1/2	—	7.6	95.3
Br-79	yes	3/2	0.33	50.5	125.3
Rb-85	yes	5/2	0.27	71.25	48.3
Sr-87	no	9/2	0.2	7.0	21.7
Y-89	yes	1/2	—	100	24.5
Zr-91	no	5/2	N.D.	11.2	46.7
Nb-93	yes	9/2	−0.2	100	122.2
Mo-95	yes	5/2	0.12	15.7	32.6
Ag-109	yes	1/2	—	48.18	23.3
Cd-113	yes	1/2	—	12.26	110.9
In-115	yes	9/2	1.14	95.72	109.6
Sn-119	yes	1/2	—	8.58	186.4
Te-125	no	1/2	—	6.99	158.0
Cs-133	yes	7/2	−0.003	100	65.6
Ba-137	no	3/2	0.2	11.3	55.6
La-139	yes	7/2	0.21	99.9	70.6
Yb-171	no	1/2	—	14.3	88.1
W-183	yes	1/2	—	14.4	20.8
Pt-195	no	1/2	—	33.8	107.5
Hg-199	yes	1/2	—	16.8	89.1
Ti-205	yes	1/2	—	70.5	288.5
Pb-207	yes	1/2	—	22.6	104.6

Source: Kirkpatrick (1988)

NUCLEAR ENERGY LEVELS IN A MAGNETIC FIELD

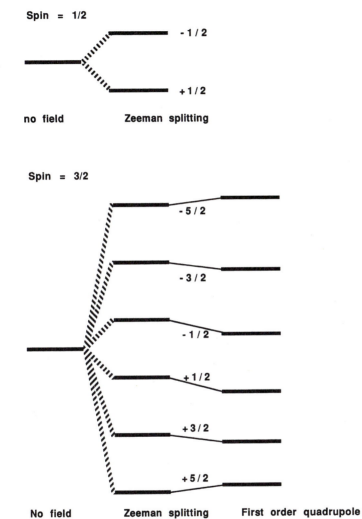

Figure 3.20. Splitting of nuclear spin levels in a magnetic field (after Kirkpatrick 1988).

solid, glass, or liquid feels not the externally applied magnetic field, but an effective field due to the diminution in the condensed phase of the applied field by shielding by the electrons. That is, the actual position of the resonance is sensitive to chemical effects, making NMR a useful probe of local environment (Kirkpatrick 1988; Stebbins 1988).

These concepts are embodied in the following equations. The angular momentum, $\mathbf{J}$, of a spinning nucleus is related to its quantized spin, I, by

$$\mathbf{J} = h \left[(I \, (I + 1)) \right]^{1/2} \qquad (3.4)$$

The magnetic moment, μ, is given by

$$\mu = \gamma I \qquad (3.5)$$

where γ is a constant, the gyromagnetic ratio.

The resonance frequency, ν, with $\mathbf{H}$ the magnetic field, is given by

$$\nu = \frac{\gamma}{2\pi} \mathbf{H} \qquad (3.6)$$

However, because the nucleus sees the effective and not the applied field,

$$\mathbf{H} = \mathbf{H}_o \, (1 - \sigma) \qquad (3.7)$$

where σ is a shielding constant.

NMR is particularly useful for studying structure and dynamics in mineralogical systems because several NMR active nuclides (notably Si, Al, O, B, P, Na) are elements for which few other spectroscopic probes exist. Si and Al NMR is very useful in elucidating questions of Al–Si order–disorder.

When an rf pulse of the appropriate energy is applied, the population of excited states is increased; when the pulse is turned off, this population decays back into thermal equilibrium, emitting rf radiation, which is detected by an antenna. Thus a simple "free induction decay" experiment consists of applying a known magnetic field, pulsing a sample with rf, and detecting the decay of these pulses. Fourier transformation of the decay into the frequency domain gives the normal NMR spectrum, which consists of a set of peaks. These are usually plotted, not against absolute frequency, but against a relative change known as the chemical shift, δ, and given by

$$\delta = \left(\frac{\nu_{\text{sample}} - \nu_{\text{standard}}}{\nu_{\text{standard}}} \right) \times 10^6 \qquad (3.8)$$

A typical spectrum of a solid is shown in Figure 3.21a. Rather broad peaks are seen; this is sometimes called wide-line NMR. Several processes result in different effective magnetic fields being felt by a given nucleus in otherwise crystallographically identical positions. These include interaction of nearby nuclear dipoles (dipole–dipole interaction), anisotropy of the electronic shielding at various sites (chemical shift anisotropy), and interaction of the quadrupole moment of a nucleus with the electric field gradient at the nucleus.

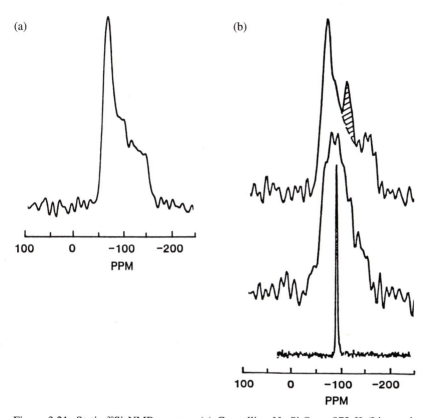

Figure 3.21. Static ^{29}Si NMR spectra. (a) Crystalline $Na_2Si_2O_5$ at 973 K (Liu et al. 1988). (b) $Na_2Si_2O_5$ liquid above the glass transition. Cross-hatched areas are due to the presence of Q^4 sites. (Top) 822 K, motion is too slow to cause significant line narrowing. (Middle) 921 K, chemical exchange and motional averaging begin to affect line shape. (Bottom) 1077 K, structural rearrangement takes place much more rapidly than the inverse of the static line width (from Brandriss and Stebbins 1988).

Additional broadening occurs in disordered solids and amorphous materials because these have a distribution of different chemical environments caused by differences in nearest neighbor and next-nearest neighbor environments.

In liquids and solutions of low viscosity, very sharp NMR peaks are generally seen because atomic motion is so rapid that all nuclei in a given chemical environment sense the same effective magnetic field. An exciting use of NMR, long familiar to organic chemistry and recently applied to silicate melts, is the study of motional narrowing of NMR peaks as a function of temperature (see

Fig. 3.21b). A broad solid-like spectrum for the glass narrows to a single narrow line as motion in the liquid increases as viscosity decreases above the glass transition. This gives direct information on rates of structural rearrangements and atomic motion in the liquid (also see Chap. 8).

The peak broadening not related to structural disorder can be overcome to a great extent by the application of magic angle spinning (MAS). For complex reasons emerging from analysis of the time-dependent Hamiltonian describing the spin system, the magnitude of the dipole–dipole, chemical shift anisotropy, and quadrupolar interactions depend on the angle between the applied magnetic field and the principal axis of the interaction. When this angle, θ, is such that $3\cos^2\theta - 1 = 0$, namely, $\theta = 54.7°$, these interactions are minimized. Thus physically spinning a solid sample at an angle of $54.7°$ to the field results in dramatic line narrowing (see Fig. 3.22) and improved signal-to-noise ratio. Spinning side bands also appear but can be distinguished from true peaks.

MAS NMR has revolutionized the study of aluminosilicates and aluminophosphates. The ^{29}Si spectra are easier to interpret than the ^{27}Al spectra because ^{29}Si has spin 1/2, whereas ^{27}Al has spin 5/2 and its spectra are complicated by quadrupolar effects. On the other hand, ^{27}Al occurs in a natural abundance of 100%, but ^{29}Si is present at only 4.7%. Thus for certain experiments, synthesis of samples enriched in ^{29}Si is essential; this is both tedious and expensive. ^{31}P has spin 1/2 and is present in 100% abundance, so it is an ideal NMR nucleus. ^{31}P NMR has greatly aided the understanding of the structural chemistry of phosphate glasses, aluminophosphate zeolites, and phosphate minerals. Because living organisms use phosphate linkages (largely in adenosine triphosphate, ATP) to store energy, ^{31}P NMR (known in the medical community as magnetic resonance imaging, MRI, to avoid the frightening term *nuclear*) can detect rapidly metabolizing (i.e., malignant) regions in the body. NMR imaging techniques are beginning to be used to scan rocks, for example, drill cores, to detect porosity and specific minerals. As imaging technology becomes more readily available and its resolution improves to the micron scale, MRI will become a valuable technique in mineral physics and materials science.

Figure 3.22 shows ^{29}Si MAS NMR spectra, a series of $Mg_4Si_4O_{12}$–$Mg_3Al_2Si_3O_{12}$ garnets. The chemical shift is sensitive to coordination number. Phases with octahedral silicon have chemical shifts in the -180 to -220 ppm range and tetrahedral silicon in the -60 to -126 ppm range (all chemical shifts are relative to tetramethylsilane liquid as an arbitrary but accepted standard). Within the tetrahedral range, the chemical shift depends on the degree of polymerization (whether an SiO_4 tetrahedron is joined to four (Q^4), three (Q^3), two (Q^2), one (Q^1), or zero (Q^0) other SiO_4 tetrahedra). Thus tectosili-

Garnets on Mg$_4$Si$_4$O$_{12}$ (En) - Mg$_3$Al$_2$Si$_3$O$_{12}$ (Py) Join

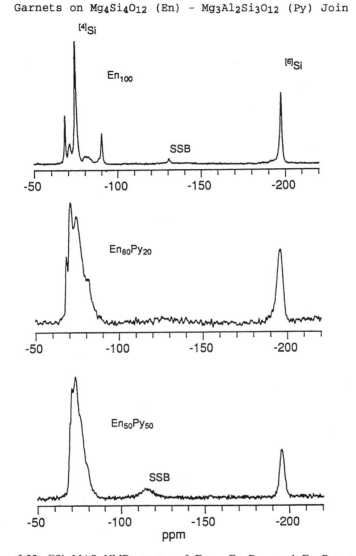

Figure 3.22. ^{29}Si MAS NMR spectra of En$_{100}$, En$_{80}$Py$_{20}$, and En$_{50}$Py$_{50}$ garnets (Mg$_4$Si$_4$O$_{12}$–Mg$_3$Al$_2$Si$_3$O$_{12}$) showing both ^{4}Si and ^{6}Si peaks. The chemical shifts are reported relative to tetramethylsilane (TMS). The increasing peak width on pyrope substitution reflects the variety of next-nearest neighbor environments as Al and Si mix on tetrahedral sites and Mg and Si mix on octahedral sites. Note that octahedral and tetrahedral Si are clearly distinguished. SSB = spinning side bond. (Phillips et al. 1992.)

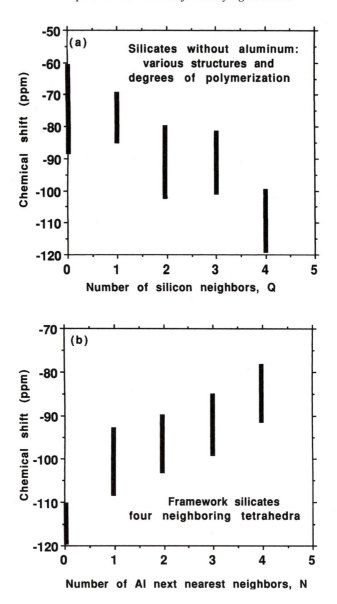

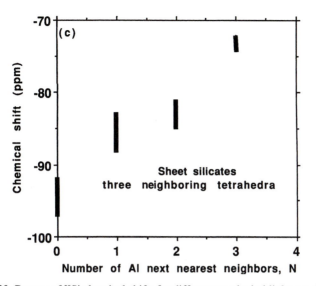

Figure 3.23. Ranges of ^{29}Si chemical shifts for different tetrahedral linkages. (a) Variation with degree of polymerization for Al-free silicates. (b) Variation with nuclear of next-nearest neighbor Al for tectosilicates. (c) Variation with number of next-nearest neighbor Al for phyllosilicates. All chemical shifts relative to tetramethylsilane. (Data from Kirkpatrick 1988.)

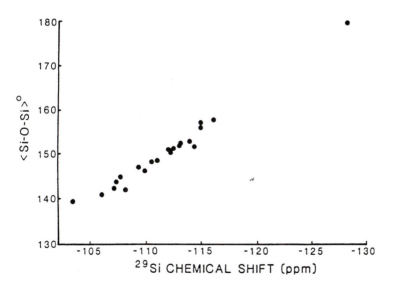

Figure 3.24. Correlation of ^{29}Si chemical shift with Si–O–Si angle in tectosilicates (Kirkpatrick 1988).

cates, phyllosilicates, chain silicates, pyrosilicates, and orthosilicates each show distinct silicon NMR signatures (see Fig. 3.23). Extending these correlations to glasses can deduce their degree of polymerization (see Chap. 8). Within a given degree of polymerization, the chemical shift is sensitive to the number of aluminum next-nearest neighbors (see Fig. 3.23) and thus has been used to quantify short range Al–Si order in feldspars, clays, zeolites, and other minerals (see Sec. 6.6). A correlation also exists between ^{29}Si chemical shift and average intertetrahedral angle (Si–O–Si angle) (see Fig. 3.24). Analogous correlations can be found for aluminum and phosphorus chemical shifts.

More complex NMR experiments involve controlling the rf pulse sequence to obtain further information on structure and dynamics. As both computer software and NMR users become more sophisticated, these methods will be more widely used in mineral physics.

3.9 Electron spin resonance

An unpaired electron has spin $\pm 1/2$. At low concentrations in the absence of an external magnetic field, such spins, generally arising from unpaired d or f electrons, are aligned at random in a solid, and the solid is paramagnetic. In a magnetic field, the two levels are split by

$$\Delta E = -g\beta \mathbf{H} \qquad (3.9)$$

where $\mathbf{H}$ is the applied magnetic field, β the magnetic moment of the electron, and g a proportionality constant, the gyromagnetic ratio. Transitions between these two levels can occur when the appropriate radiation is applied (see Fig. 3.20b). The electron spin resonance (ESR; also called electron paramagnetic resonance, EPR) experiment is then analogous to an NMR experiment (Calas 1988). However, for similar applied fields of several kGauss, ESR occurs at GHz frequencies (10^9/sec) while NMR is seen at MHz (10^6/sec). If more than one unpaired electron is present on a given atom (e.g., the five d electrons in Fe^{3+}), their spins interact to produce further splitting, known as fine structure. If nuclear spins are also present, they interact with the electron spins, causing further, so-called hyperfine, splitting. The magnitude of the g-factor reflects how strongly the electrons interact with each other. The fine structure is a probe of local environment, and the hyperfine structure is an indirect measure of covalent interactions.

Figure 3.25 shows an ESR spectrum of a synthetic forsterite, Mg_2SiO_4, containing trace amounts of Mn^{2+}, Fe^{3+}, and Ni^{2+}. Details of the splittings of the peaks for each paramagnetic ion are well resolved because these resonances occur at different fields.

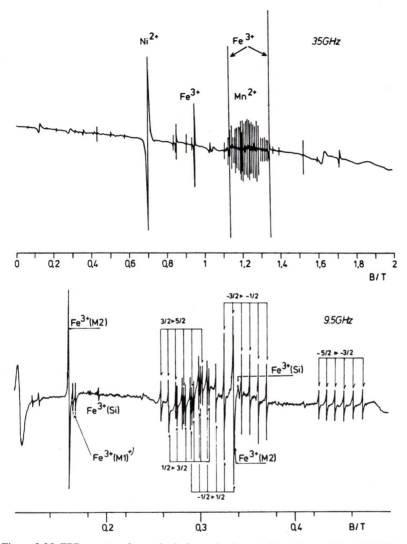

Figure 3.25. EPR spectra of a synthetic forsterite doped with nickel at 35 and 9.5 GHz. The external magnetic field **B** was parallel to **c** (space group D_{2h}^{16}/Pbnm). The main resonance signals are due to Mn^{2+} at M2 and Fe^{3+} at M1, M2, and Si positions. Only the 35 GHz EPR spectrum shows the Ni^{2+} signal with a line width of about $50 \cdot 10^{-4}T(50G)$. The five EPR transitions of Mn^{2+} between the spin sublevels, $M_s = 5/2, 3/2, 1/2, -1/2, -3/2, -5/2$, are denoted according to the selection rule $\Delta M_s = \pm 1$. Each transition is further split into six components by the hyperfine interaction between the electron spin $S(5/2)$ and nuclear spin $I(5/2)$ of the isotope ^{55}Mn (100% natural abundance). The splitting of the Fe^{3+} (M1) EPR signal is due to a misorientation of **c** with respect to **B** of less than 0.5°. (From Rager et al. 1988.)

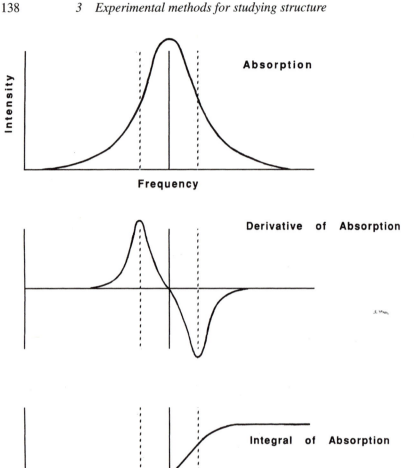

Figure 3.26. Relation of peak, its derivative, and its area or integral as commonly encountered in ESR and NMR spectra.

Uses of ESR can be grouped into analytical and structural applications. The number of spins in a sample is proportional to the area under the ESR peak. As commonly collected, the ESR signal is the derivative of the peak (see Fig. 3.26), therefore finding the area involves a double integration. In general, the proportionality constant must be determined through known standards. Because ESR easily detects transition metal ions present at the ppb to ppm level, ESR is potentially an excellent trace element analytical technique. If a trace element is present in several oxidation states, these can be distinguished and quantified.

A variety of structural studies can be performed by ESR. Distributions of trace elements among inequivalent cation sites can be studied. These tracers can also be used to probe their next nearest environment and give information on the degree of disorder of major elements (for an application to $MgAl_2O_4$ spinel see Sec. 7.3.2). The evolution of local site geometry in a glass with composition, or during phase separation or crystallization, can be investigated using low concentrations of a paramagnetic ion as a probe of local structure (see Chap. 8). Site symmetry associated with phase transitions and order–disorder can be detected. ESR is also a major tool for studying paramagnetic defects produced by radiation damage. Relaxation times associated with ESR transitions give information on dynamics.

3.10 Mössbauer spectroscopy

Mössbauer spectroscopy involves the emission and subsequent absorption of gamma rays by atomic nuclei (Hawthorne 1988). When nuclear energy levels are appropriate, radioactive decay involves the emission of a gamma ray. An example is

$$^{57}Co_{27} + {}^0e_{-1} = {}^{57}Fe^*_{26} = {}^{57}Fe_{26} + \gamma \qquad (3.10)$$

The ^{57}Fe nucleus is initially in an excited state; its return to the ground state produces the gamma ray. These gamma rays from the source can be absorbed by a different sample, the absorber, containing ^{57}Fe. If the source and absorber are chemically the same, the energy of the emitted gamma will be right for resonant absorption. If the chemical environments are different, for example, iron in a steel emitter versus iron in a silicate absorber, the energies will not match by a small amount because of interactions with the electric and magnetic fields produced by the surrounding atoms (in some ways analogous to the chemical shift in NMR). Physically moving the sample and source relative to each other at a velocity of a few millimeters per second changes the energy by a Doppler shift and brings the absorption into resonance.

Not all nuclei have energy level schemes appropriate for Mössbauer spectroscopy. For mineralogical purposes, it is fortunate that iron is suitable. Because ^{57}Fe is naturally present at only 2% abundance, synthesizing a sample enriched in this isotope improves sensitivity, especially for systems in which iron is a minor or trace element. Table 3.2 lists some properties of Mössbauer nuclei.

A Mössbauer spectrum of andradite, a mineral that has only one type of site

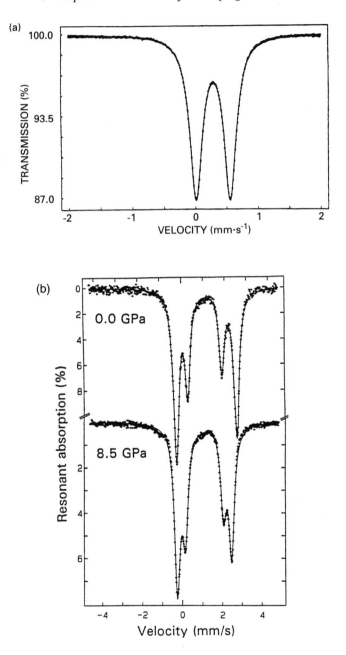

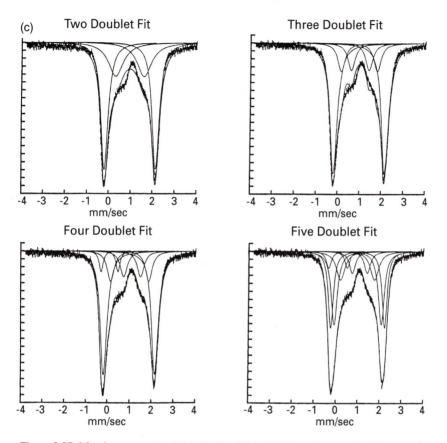

Figure 3.27. Mössbauer spectra. (a) Andradite (Murad 1984), (b) grunerite (Zhang and Hafner 1992), (c) staurolite (Dyar et al. 1991).

containing iron, is shown in Figure 3.27a. Two peaks of equal intensity are seen. The splitting is caused by the interaction of the nuclear spin and the electric field gradient at the nucleus. The latter is directly related to crystal symmetry, nearest and next-nearest neighbor environment, and electron distribution, and thus this quadrupole splitting is a very sensitive chemical probe. The actual location of the doublet (i.e., the average position of the two peaks related by quadrupole splitting) is called the isomer shift, and is usually taken relative to an arbitrary standard material, commonly iron foil or sodium nitroprusside. In addition, the peaks may be characterized by their shape and their intensity. The latter, proportional to absorber concentration, provides the basis for use of Mössbauer as an analytical tool.

Mössbauer peaks may be split further by several additional interactions. One source is magnetic interaction, which occurs differently for paramagnetic, ferromagnetic, and antiferromagnetic minerals. Thus Mössbauer spectroscopy is a probe of magnetic ordering. A number of other factors, grouped together as hyperfine effects, involve the coupling of various electric and magnetic interactions and cause further splittings. Thus, the Mössbauer spectrum can be quite complex (see Fig. 3.27).

When several distinct sites occur for the Mössbauer nucleus in a mineral, the observed spectrum is a superposition of spectra for each environment. The largest differences in isomer shift occur for iron in different valence states (2+ and 3+) and in very different coordinations (octahedral and tetrahedral). Mössbauer spectroscopy has been applied extensively to determining iron oxidation states in minerals, glasses, and rocks, and indeed it has been shown to compare favorably with standard wet chemical analysis. It has also been used to study cation distributions in spinels, pyroxenes, amphiboles, and other minerals. Figure 3.27b shows ^{57}Fe Mössbauer spectra for an amphibole (grunerite) at atmospheric and high pressure (Zhang and Hafner 1992). The spectrum is relatively simple because there are only two well-resolved doublets, the outer one corresponding to Fe^{2+} in the M1, M2, and M3 sites (which have similar environments) and the inner one to Fe^{2+} in M4. The changes of isomer shift and quadruple coupling constant with pressure probes changes in site geometry and bonding.

A more complex case, as is frequently encountered, is shown in Fig. 3.27c, namely, several fits to the spectrum of a staurolite, a complex silicate containing layers of kyanite composition alternating with layers of iron aluminum hydroxide (Dyar et al. 1991). The Mössbauer spectrum has a complex shape, indicating the presence of a number of overlapping peaks. As more doublets are used to fit the spectrum, the quality of fit improves, and the authors argue that a five-doublet fit is justified in terms of both statistics and crystal chemistry. Figure 3.27c illustrates a limitation of Mössbauer spectroscopy; when a mineral contains a large number of similar sites, the spectrum can become very difficult to interpret, and unique deconvolution to give a well-defined number of assignable peaks is no easy task. There continue to be controversies in the literature about the extent to which spectral deconvolution can be pushed and the uniqueness and robustness of the results. Some users of Mössbauer spectroscopy probably tend to overinterpret the results, and appropriate caution should be exercised when reading the literature.

3.11 Scanning tunneling and scanning force microscopies

Scanning tunneling and scanning force microscopies (STM and SFM, respectively) are revolutionary new tools for the study of surfaces (Hochella 1990). With both techniques, sharp tips are used in conjunction with piezoelectric scanners. For STM, the tip is rastered by the scanner just above the surface (within a few Å) so that when a small bias voltage is applied, electrons can "tunnel" across the gap. The tunneling current is sensitive to the chemical and electronic nature of the surface and varies exponentially with the distance between tip and surface. For SFM, the piezoelectric scanner rasters the sample under a tip that is mounted on the end of a microcantilever. The tip is actually in contact with the sample surface, but exceptionally small tracking forces are used. As the tip rides over the sample, deflection of the microcantilever due to surface roughness is detected with a laser optical system.

The theory behind both techniques, at least for initial understanding and use, is moderately straightforward. Both STM and SFM are relatively inexpensive, especially in comparison with a high-resolution transmission electron microscope, which provides similar spatial resolution, and STM/SFM instrumentation is relatively uncomplicated and easy to maintain. Both methods have excellent adaptability and versatility. They both can be operated with a sample in vacuum, air, or solution. All three environments are important depending on the experimental and/or spectroscopic information needed. In particular, characterization of the sample-air or sample-solution interface offers new views of mineral reaction mechanisms under realistic conditions. Finally, and perhaps most important of all, both methods have unparalleled spatial resolution. STM can be used to "image" atoms on conducting and semiconducting surfaces (actually, the technique is probing electronic tunneling characteristics on an atom by atom basis), and one can easily observe topographic features as small as monatomic steps. In fact, STM has been used to obtain electronic spectra (i.e., electron state distributions in both the valence and conductions bands) of individual atoms on surfaces before, during, and after interaction with gas molecules in vacuum. SFM has better than angstrom resolution in the z direction (i.e., perpendicular to the surface), but its lateral resolution is generally considered to be molecular rather than atomic. Because SFM does not require electron conduction, only mechanical contact, it can be used on any solid surface, including insulating oxides and silicates.

Modern image-processing techniques and high-speed laboratory computers provide intriguing pictures obtained from the STM and SFM techniques. Yet quantitative interpretation, especially of STM images, is still in its infancy, much as HRTEM was in 1970. Both techniques encompass growing and prom-

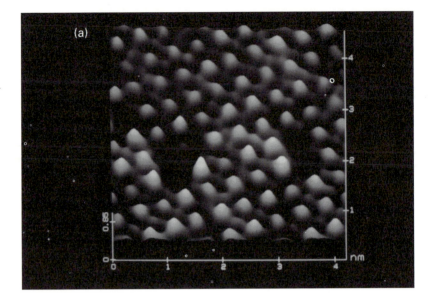

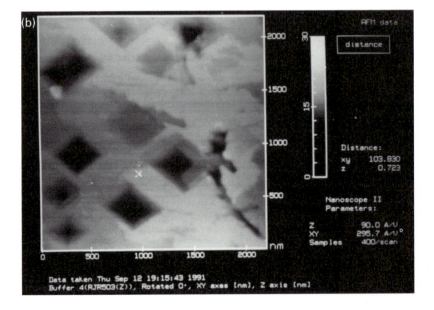

ising fields, with major applications in mineral physics, surface science, and environmental chemistry.

Figure 3.28a shows an STM image of a freshly cleaved galena (PbS) surface that had been exposed to water at room temperature for only one minute. Figure 3.28b shows a SFM image of a freshly cleaved calcite surface after only twenty seconds of water exposure. In both cases, substantial reaction with water, not detectable by other means, is clearly documented.

3.12 References

3.12.1 General references and bibliography

Bacon, G. E. (1975). *Neutron Diffraction*. Oxford: Clarendon Press.

Bish, D. L., and J. E. Post, eds. (1989). *Modern powder diffraction*. Reviews in Mineralogy, vol. 20. Washington, DC: Mineralogical Society of America.

Brown, G. E., Jr., G. Calas, G. A. Waychunas, and J. Petiau. (1988). *X-ray absorption spectroscopy: Applications in mineralogy and geochemistry*. Reviews in Mineralogy, vol. 18, 431–512. Washington, DC: Mineralogical Society of America.

Buseck, P. R., ed. (1992). *Minerals and reactions at the atomic scale: Transmission electron microscopy*. Reviews in Mineralogy, vol. 27. Washington, DC: Mineralogical Society of America.

Calas, G. (1988). *Electron paramagnetic resonance*. Reviews in Mineralogy, vol. 18, 513–71. Washington, DC: Mineralogical Society of America.

Calas, G., and F. C. Hawthorne. (1988). *Introduction to spectroscopic methods*. Reviews in Mineralogy, vol. 18, 1–9. Washington, DC: Mineralogical Society of America.

Cheetham, A. K., and P. Day, eds. (1987). *Solid state chemistry techniques*. Oxford: Oxford Science Publications, Oxford University Press.

Gadsden, J. A. (1975). *Infrared spectra of minerals and related inorganic compounds*. Boston: Butterworth.

Ghose, S. (1988). *Inelastic neutron scattering*. Reviews in Mineralogy, vol. 18, 161–92. Washington, DC: Mineralogical Society of America.

Griffen, D. T. (1992). *Silicate crystal chemistry*. New York: Oxford University Press.

Hawthorne, F. C. (1988). *Mössbauer spectroscopy*. Reviews in Mineralogy, vol. 18, 255–340. Washington, DC: Mineralogical Society of America.

Hawthorne, F. C., ed. (1988). *Spectroscopic methods in mineralogy and geology.*

Figure 3.28. (*Facing page*) (a) STM image of a galena (001) surface after exposure to water for one minute. Image collected at $+200$ mV sample bias and 1.8 nA tunneling current in the constant height mode. The high bumps represent surface sulfur atoms; the low bumps in between represent lead atoms. The sulfur–lead distance is approximtely 3 Å. The apparent vacancies are interpreted as oxidized sulfur sites. All scales are in nanometers. (Courtesy of C. M. Eggleston and M. F. Hochella, Jr., Stanford University). (b) SFM image of a rhombohedral cleavage surface of calcite after 20 seconds of water exposure. Image collected in air in the constant deflection mode. Etch pits are up to 20 nm deep. The step marked by the whie x is 7 Å in height, equivalent to two carbonate layers. All scales are in nanometers. (Courtesy of M. F. Hochella, Jr., Stanford University).

Reviews in Mineralogy, vol. 18. Washington, DC: Mineralogical Society of America.

Hawthorne, F. C., and G. A. Waychunas. (1988). *Spectrum-fitting methods.* Reviews in Mineralogy, vol. 18, 63–98. Washington, DC: Mineralogical Society of America.

Hochella, M. F., Jr. (1988). *Auger electron and X-ray photoelectron spectroscopies.* Reviews in Mineralogy, vol. 18, 573–637. Washington, DC: Mineralogical Society of America.

Hochella, M. F., Jr., and A. F. White, eds. (1990). *Mineral-water interface geochemistry.* Reviews in Mineralogy, vol. 23. Washington, DC: Mineralogical Society of America.

Hofmeister, A. M. (1987). Single crystal absorption and reflection infrared spectroscopy of forsterite and fayalite. *Phys. Chem. Minerals 14*, 499–513.

Kirkpatrick, R. J. (1988). *MAS NMR spectroscopy of minerals and glasses.* Reviews in Mineralogy, vol. 18, 341–403. Washington, DC: Mineralogical Society of America.

McMillan, P. F., and A. C. Hess. (1988). *Symmetry, group theory, and quantum mechanics.* Reviews in Mineralogy, vol. 18, 11–61. Washington, DC: Mineralogical Society of America.

McMillan, P. F., and A. M. Hofmeister. (1988). *Infrared and Raman spectroscopy.* Reviews in Mineralogy, vol. 18, 99–159. Washington, DC: Mineralogical Society of America.

Nassau, K. (1983). *The physics and chemistry of color: The fifteen causes of color.* New York: Wiley.

Putnis, A. (1992). *Introduction to mineral sciences.* Cambridge: Cambridge University Press.

Rossman, G. R. (1988). *Vibrational spectroscopy of hydrous components.* Reviews in Mineralogy, vol. 18, 193–206. Washington, DC: Mineralogical Society of America.

Rossman, R. G. (1988). *Optical spectroscopy.* Reviews in Mineralogy, vol. 18, 207–54. Washington, DC: Mineralogical Society of America.

Stebbins, J. F. (1988). *NMR spectroscopy and dynamic processes in mineralogy and geochemistry.* Reviews in Mineralogy, vol. 18, 405–29. Washington, DC: Mineralogical Society of America.

Waychunas, G. A. (1988). *Luminescence, X-ray emission and new spectroscopies.* Reviews in Mineralogy, vol. 18, 639–98. Washington, DC: Mineralogical Society of America.

White, J. C., ed. (1985). *Applications of electron microscopy in the earth sciences.* Short Course Handbook, vol. 11. Toronto: Mineralogical Association of Canada.

3.12.2 Specific references

Benziger, J. B., B S. H. Royce, and S. J. McGovern. (1985). IR photoacoustic spectroscopy of silica and aluminum oxide. In *Catalyst characterization science,* ACS Symposium Series No. 288, ed. M. L. Diveney and J. L. Gland. Washington, DC: American Chemical Society.

Brandriss, M. E., and J. F. Stebbins. (1988). Effects of temperature on the structure of silicate liquids: ^{29}Si NMR P results. *Geochim. Cosmochim. Acta 52*, 2659–70.

Burns, R. G., D. A. Nolet, K. M. Parkin, C. A. McCammon, and K. B. Schwartz. (1980). Mixed-valence minerals of iron and titanium: Correlations of structural, Mössbauer and electronic spectral data. In *Mixed-valence compounds: Theory and application in chemistry, physics, biology,* ed. D. B. Brown, 295–336. Boston: Reidel.

Calas, G., G. E. Brown, Jr., G. A. Waychunas, and J. Petiau. (1987). X-ray absorption spectroscopic studies of silicate glasses and minerals. *Phys. Chem. Minerals 15*, 19–29.

Dyar, M. D., C. L. Perry, C. R. Rebbert, B. L. Dutrow, M. J. Holdaway, and H. M. Lang. (1991). Mössbauer spectroscopy of synthetic and naturally occurring staurolite. *Amer. Mineral. 76*, 27–41.

Fei, Y., H. K. Mao, J. Shu, G. Parthasarathy, W. A. Bassett, and J. Ko. (1992). Simultaneous high-*P*, high-*T* X-ray diffraction study of β-$(Mg,Fe)_2SiO_4$ to 26 GPA and 900 K. *J. Geophys. Res. 97*, B4, 4489–95.

Finger, L. W. (1992). Instrumentation for studies at the powder/single-crystal boundary. Reprinted from *Accuracy in Powder Diffraction II, NIST Special Publication 846*, 183–8.

Hawthorne, F. C., and R. Grundy. (1973). The crystal chemistry of the amphiboles I: Refinement of the crystal structure of ferrotschermakite. *Min. Mag. 39*, 6–48.

Hazen, R. M. (1976). Effects of temperature and pressure on the crystal structure of forsterite. *Amer. Mineral. 61*, 1280–93.

Iishi, K. (1978). Lattice dynamics of forsterite. *Amer. Mineral. 63*, 1198–208.

Liu, S.-B., J. F. Stebbins, R. Schneider, and A. Pines. (1988). Diffusive motion in alkali silicate melts: An NMR study at high temperature. *Geochim. Cosmochim. Acta 52*, 527–38.

Matsui, Y., and Y. Syono. (1968). Unit cell parameters of some synthetic olivine group solid solutions. *Geochem. Jour. 2*, 51–9.

Murad, E. (1984). Magnetic ordering in andradite. *Amer. Mineral. 69*, 722–4.

Navrotsky, A., D. J. Weidner, R. C. Liebermann, and C. T. Prewitt. (1992). Materials science of the earth's deep interior. *MRS Bull. 17*, 30–7.

Papike, J. J., M. Ross, and J. M. Clark. (1969). Crystal-chemical characterization of clinoamphiboles based on five new structure refinements. Special Paper 2, pp. 117–36. Washington, DC: Mineralogical Society of America.

Phillips, B. L., D. A. Howell, R. J. Kirkpatrick, and T. Gasparik. (1992). Investigation of cation order in $MgSiO_3$-rich garnet using ^{29}Si and ^{27}Al MAS NMR spectroscopy. *Amer. Mineral. 77*, 704–12.

Rager, H., S. Hosoya, and G. Weiser. (1988). Electron paramagnetic resonance and polarized optical absorption spectra of Ni^{2+} in synthetic forsterite. *Phys. Chem. Minerals 15*, 383–9.

Smelik, E. A., and D. R. Veblen. (1993). A transmission and analytical electron microscope study of exsolution microstructures and mechanisms in the orthoamphiboles anthophyllite and gedrite. *Amer. Mineral. 78*, 511–32.

Stewart, D. R., and T. L. Wright. (1974). Al/Si order and symmetry of natural alkali feldspars and relationship of strained parameters to bulk composition cell. *Bull. Soc. France Mineral. Cristallog. 97*, 356–77.

Waychunas, G. A., M. J. Apted, G. E. Brown, Jr. (1983). X-ray K-edge absorption spectra of Fe minerals and model compounds: Near-edge structure. *Phys. Chem. Minerals 10*, 1–9.

Zhang, L., and S. S. Hafner. (1992). Gamma resonance of ^{57}Fe in grunerite at high pressures. *Amer. Mineral. 77*, 474–9.

4

Methods for studying thermodynamic properties

4.1 Thermochemistry

The thermodynamics of mineral reactions can be studied by a variety of approaches. Because minerals are structurally complex and their stability and phase relations are governed by a delicate balance of energetic, entropic, and volumetric factors that depend on temperature, pressure, and composition, such work must pay careful attention to sample characterization. One must be concerned with sample purity (single phase, stoichiometric, unzoned material), structural state (order–disorder, polymorphism), careful experimental technique, and equilibrium (or lack of it) in the experiments. In analyzing the data, one must make the punishment fit the crime by using thermodynamic models that are appropriate to the processes involved, especially when the possibility of order–disorder exists. These models must be complex enough to incorporate the essential physics and chemistry and yet simple enough to apply to data frequently of limited extent and accuracy. It is essential to avoid too many fitting parameters – with enough parameters you can fit an elephant, but not grasp its essential nature. Thus although mineral thermodynamics might be regarded as a sub-branch of solid-state physics, the complexity of the structures and inherent messiness of natural systems, as well as the difficulty with which definitive experiments can be made gives this field its own flavor. In this chapter, experimental approaches are summarized. Chapter 6 and 7 deal with the systematics of mineral thermodynamics, especially linking macroscopic thermodynamics and physical properties to microscopic descriptions of structure and bonding.

4.1.1 Experimental measurement of heat capacity

To obtain the standard entropy (S°_{298}) of a phase, its heat capacity must be measured down to temperatures below 10 K, or lower for systems with magnetic

148

transitions. Adiabatic calorimeters (Robie and Hemingway 1972, Westrum 1984) exist in a number of laboratories for work between liquid helium temperature (4 K) and 350–400 K (see Fig. 4.1). These calorimeters are usually custom built. Especially for the lower temperature range, a relatively large amount of sample, 1–10 g, is generally required. Because of this and because measurements are time-consuming and require painstaking attention, calorimetric entropies are not routinely available for many substances, especially for exotic materials that can be synthesized in only small amounts (e.g., high-pressure phases). Indeed there is concern that this somewhat unglamorous, old-fashioned but essential experimentation may be dying out.

In the range 200–1000 K, differential scanning calorimetry (DSC) (Callanan and Sullivan 1986; Stebbins, Weill, and Carmichael, 1982; Stebbins, Carmichael, and Moret 1984) has made great headway in obtaining heat capacities with an accuracy of about ±1% on samples as small as a few milligrams. Above 1000 K, heat contents are generally measured by drop calorimetry, with a sample in a furnace at high temperature dropped into a calorimeter at room temperature (Stebbins et al. 1982). More recently, samples have been dropped into calorimeters operating at high temperatures (up to 1773 K) (Ziegler and Navrotsky 1986) and high-temperature scanning calorimetry (Lange, DeYoreo, and Navrotsky 1991) has also been used (see Fig. 4.2). The advantages of using a calorimeter operating at high temperature is that the final structural state of the sample can be an equilibrium rather than a metastable quench state. The disadvantages stem from greater experimental difficulty. These measurement methods are summarized in Table 4.1.

4.1.2 Enthalpies of mineral reactions

Heat capacity measurements permit the characterization of the thermal properties of a single substance. The measurement of the enthalpy of a chemical reaction can be used to relate the thermodynamic properties of a given substance to those of other substances, and thus to obtain heats of formation, of mixing, and of solid-state reactions. These calorimetric methods can be classified into two groups: *direct* methods in which the reaction studied by calorimetry is the one of interest and *indirect* methods in which a thermochemical cycle is used to obtain the enthalpy of interest. In the former category is combustion calorimetry, in which the enthalpy of formation of an oxide is obtained by burning the metal or an oxide in oxygen (Charlu and Kleppa 1973), or the heat of formation of a fluoride is measured by burning the metal in a fluorine atmosphere. The absorption or release of oxygen by nonstoichiometric oxides – for example, "FeO" (Marucco, Gerdanian, and Dode 1970) or oxide super-

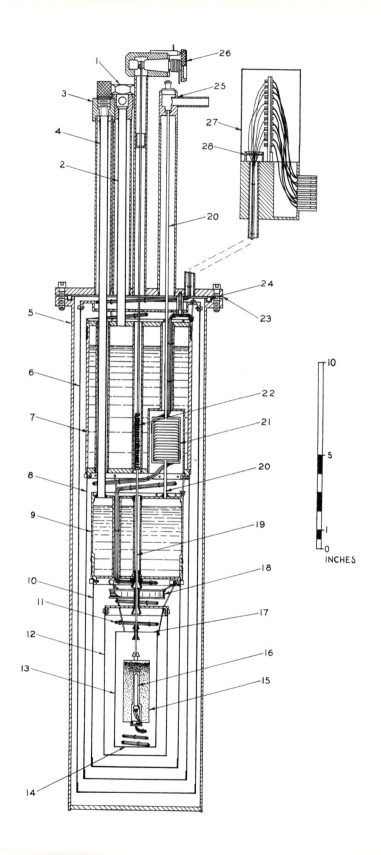

INCHES

Table 4.1. *Calorimetric methods for heat capacity measurements*

Method	Basic measurement	Sample required	Approximate temp. range (K)
Adiabatic low-temperature calorimetry	Temperature change for known heat input under adiabatic conditions	1–10 g	5–~500
Differential scanning calorimetry	Heat flow temperature difference or power difference between sample and standard during controlled heating conditions	0.5–100 mg	200–1000
Scanning calorimetry	Enthalpy required to raise temperature of sample by a 5–10 K increment	1–3 g	900–1800
Drop calorimetry	Heat released when hot sample is dropped into calorimeter at room temperature	1–10 g	300–2500
Transposed temperature drop calorimetry	Heat absorbed when sample is dropped from room temperature into calorimeter at high temperature	10–500 mg	300–1800

conductors (Parks et al. 1989) – can also be studied by high-temperature reaction calorimetry. By combining such calorimetric measurements with free energy measurements and a knowledge of the composition of the phase as a function of the partial pressure of oxygen, the partial molar entropy of solution of the gas in the nonstoichiometric oxide can be obtained with fairly high accuracy. This lends insight into solubility mechanisms and defect chemistry.

Figure 4.1 (*Facing page*). Adiabatic calorimeter for heat capacity measurements at 4–350 K. (1) Liquid nitrogen inlet and outlet connector, (2) liquid nitrogen filling tube, (3) compression fitting to seal inlet to liquid helium tank, (4) liquid helium filling tube, (5) brass vacuum jacket, (6) outer "floating" radiation shield, (7) liquid nitrogen tank, (8) nitrogen tank radiation shield, (9) liquid helium tank, (10) helium tank radiation shield, (11) bundle of lead wires, (12) guard adiabatic shield, (13) main adiabatic shield, (14) bottom adiabatic shield, (15) calorimeter vessel, (16) calorimeter heater/thermometer assembly, (17) top adiabatic shield, (18) temperature-controlled ring, (19) supporting braided silk line, (20) helium exit tube, (21) economizer (effluent helium vapor heat exchanger), (22) coil spring, (23) cover plate, (24) O-ring gasket, (25) helium exit connector, (26) windlass, (27) copper shield for terminal block, (28) vacuum seal. (Provided by J. Boerio-Goates, after Woodfield 1988.)

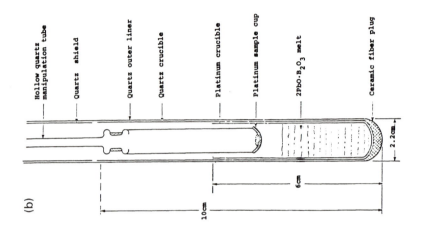

(b)

Hollow quartz manipulation tube

Quartz shield

Quartz outer liner

Quartz crucible

Platinum crucible

Platinum sample cup

2PbO·B₂O₃ melt

Ceramic fiber plug

10cm

6cm

2.2cm

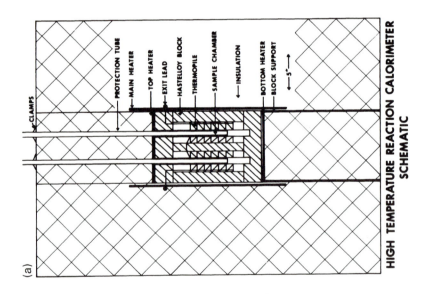

(a)

CLAMPS

PROTECTION TUBE

MAIN HEATER

TOP HEATER

EXIT LEAD

HASTELLOY BLOCK

THERMOPILE

SAMPLE CHAMBER

INSULATION

BOTTOM HEATER

BLOCK SUPPORT

5"

HIGH TEMPERATURE REACTION CALORIMETER
SCHEMATIC

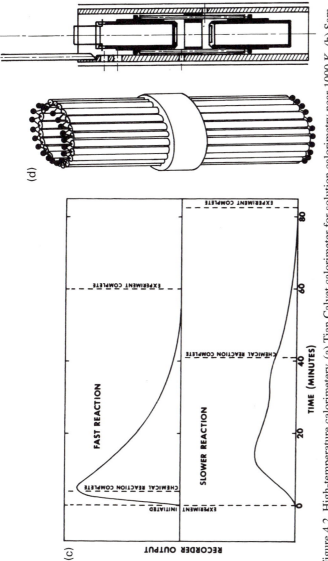

Figure 4.2. High-temperature calorimetry. (a) Tian Calvet calorimeter for solution calorimetry near 1000 K. (b) Sample assembly for solution calorimetry of oxides in molten $2PbO \cdot B_2O_3$. (c) Typical calorimetric curves for fast and slower reactions. (d) Detector for Setaram-type calorimeter used for transposed-temperature drop and scanning calorimetry to 1773 K. (e) Schematic of calorimetric arrangement within detector in (d). The upper crucible contains the sample, the lower contains the reference.

The heats of mixing in molten salts, some oxide melts, and metal alloys can be measured directly by calorimetry at high temperature (Ostvold and Kleppa 1969). Viscous melts, glasses, and many crystalline solid solutions or compounds cannot be studied by direct reaction calorimetry because they react far too slowly. Instead, indirect methods, generally solution calorimetry, are used. The solvent in such measurements can be an aqueous solution, for example, hydrofluoric acid or mixtures of HF, HCl, and/or HNO_3 (Robie and Hemingway 1972) for oxides and silicates, an acidified brine for carbonates (Capobianco and Navrotsky 1987), or a molten salt or oxide melt (lead borate, alkali borate, or alkali molybdate or tungstate for oxides and silicates) (Navrotsky 1977). Recent advances in high-temperature oxide melt solution calorimetry are the use of low oxygen fugacity for the study of Fe^{2+}-bearing phases (Akaogi and Navrotsky 1989; Chatillon-Colinet et al. 1983) and the development of solution calorimetric techniques for sulfides (Cemic and Kleppa 1986). In addition, progress in high-temperature solution calorimetry of hydrous minerals (Pawley, Graham, and Navrotsky 1993) and carbonates (Chai and Navrotsky 1993) has been made. These methods are summarized in Table 4.2.

Thermochemical cycles can be devised to deal with phases that decompose if left to equilibrate in a hot (~1000 K) calorimeter prior to dissolution. One method, transposed temperature-drop calorimetry, takes advantage of the rapid decomposition itself, directly measuring its enthalpy by dropping a sample from room temperature into a hot calorimeter with no solvent (Parks et al. 1989). A second method, drop solution calorimetry, drops the metastable sample into the solvent, where it dissolves (Chai and Navrotsky 1993).

As examples, the heats of solution of a number of oxides in molten $2PbO \cdot B_2O_3$ (near 1000 K) are given in Table 4.3. Using these values we can construct the following thermodynamic cycle:

$$SiO_2 \text{ (quartz)} \rightarrow \text{dilute solution, } \Delta H = -3.5 \pm 0.2 \text{ kJ mol}^{-1} \quad (4.1)$$

$$Mg_3SiO_4 \text{ (olivine)} \rightarrow \text{dilute solution, } \Delta H = +67.1 \pm 0.1 \text{ kJ mol}^{-1} \quad (4.2)$$

$$\text{Dilute solution} \rightarrow 2MgSiO_3 \text{ (orthoenstatite)}, \quad (4.3)$$
$$\Delta H = -72.6 \pm 0.6 \text{ kJ mol}^{-1}$$

Therefore, for Mg_2SiO_4 (olivine) + SiO_2 (quartz) = $2MgSiO_3$ (orthoenstatite),

$$\Delta H = -3.5 + 67.1 - 72.6 = -9.0 \pm 0.6 \text{ kJ mol}^{-1} \quad (4.4)$$

Similarly, for the phase transformation Mg_2SiO_4 (olivine) = Mg_2SiO_4 (modified spinel),

$$\Delta H = 67.1 - 37.1 = 30.0 \pm 0.7 \text{ kJ mol}^{-1} \quad (4.5)$$

Table 4.2. *Methods of high-temperature reaction calorimetry*

Method	Process	Application
Solution calorimetry	Solute (T) + solvent (T) = dilute solution (T)	Sample must be able to equilibrate at calorimetric T
lead borate (973–1073 K)		Oxides containing MgO, CaO, alkali, alkaline earth, Fe_2O_3, CoO, NiO, ZnO, GeO_2, Ga_2O_3
alkali molydbate or tungstate (973 K)		Titanates, vanadates
alkali borate (1100–1200 K)		Titanates, zirconates, stannates, Fe^{2+}-compounds under controlled fO_2
nickel sulfide (1173 K)		Sulfides
Transposed temperature calorimetry drop	Substance (298 K)→ transformed substance (T)	Heat of fusion, rapid phase transition, annealing, decomposition (process must be fast, <~30 min.)
Drop solution calorimetry	Solute (298 K) + solvent (T) → dilute solution (T)	Substances that would decompose to an ill-defined state at calorimetric T, hydrous phases, high P phase transitions
Gas–solid reaction	Substance (T) ±gas (T) → product (T)	Oxidation-reduction, metal-hydrogen reaction

Table 4.3. *Enthalpies of solution $2PbO \cdot B_2O_3$*
at 975 K of some compounds
in the MgO-SiO_2 system

Compound	Enthalpy of solution (kJ/mol)
MgO	4.8 ± 0.5[a]
SiO_2 (quartz)	-3.5 ± 0.2[a]
SiO_2 (glass)	-11.2 ± 0.5[a]
Mg_2SiO_4 (α)	67.1 ± 0.1[b]
Mg_2SiO_4 (β)	37.1 ± 0.7[b]
Mg_2SiO_4 (γ)	28.0 ± 0.6[b]
$MgSiO_3$ (opx)	36.3 ± 0.3[c]
$MgSiO_3$ (garnet)	0 ± 1[c]

[a] "Best" value from a large number of experiments over several years in Navrotsky's laboratory.
[b] Akaogi, Ito, and Navrotsky (1989)
[c] Akaogi, Navrotsky, Yagi, and Akimoto (1987)

The first calculation confirms the observation, known from the phase diagram, that $MgSiO_3$ pyroxene is stable relative to a mixture of Mg_2SiO_4 (olivine) and SiO_2 (quartz) at atmospheric pressure (because the ΔS of the reaction is small). The second calculation shows the modified spinel structure is indeed energetically metastable at atmospheric pressure and supports the high-pressure phase studies.

4.2 Phase equilibria

In general, the goal is to ascertain what phases with what structures and compositions are stable under various conditions of pressure, temperature, imposed chemical potential (oxygen fugacity, H_2O or CO_2 fugacity, presence of a pure metallic phase, etc.), and bulk composition. The standard methods involve annealing a sample under desired pressure and temperature conditions, quenching it to ambient, and examining the products. Two points should be (but often are not) borne in mind concerning such experiments: (1) Phase equilibrium is only proven when a phase boundary has been properly reversed. This means that one must show not only that the low-temperature assemblage transforms to the high-temperature assemblage (or low-pressure assemblage to high-pressure assemblage, or low oxygen fugacity assemblage to high oxygen fugacity assemblage) as temperature (pressure, oxygen fugacity) is raised, but that the high-temperature assemblage transforms back to the low-temperature assemblage (high pressure → low pressure, high oxygen fugacity → low oxygen fugacity) as temperature (pressure, oxygen fugacity) is lowered, and that both forward and reverse transformations take place under very similar values of temperature (pressure, oxygen fugacity). This reversal is also known as bracketing an equilibrium. An example is shown in Figure 4.3, where the reaction is fairly tightly bracketed by the growth of either the high-pressure or low-pressure assemblage from a mixture of both. (2) It is important to remember that the true equilibrium boundary for a first-order transition can lie anywhere within the brackets and need not lie in the middle of each bracket. A reaction approached from one side only produces a half bracket. The analysis of reversed and partially reversed phase boundaries by linear programming techniques to produce internally consistent sets of thermodynamic data is a well-developed specialty (Gordon 1973).

Difficulties in attaining phase equilibrium reversals are greatest at low temperature for reactions involving dry solids with no liquid phase present. Indeed, below 600 K many reactions, even on a geologic timescale, may not reach

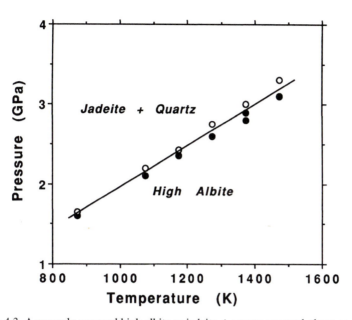

Figure 4.3. A properly reversed high albite = jadeite + quartz reversed-phase equilibrium. The reaction solid circles represent experiments in which albite grew in a mixture of albite, jadeite, and quartz. Open circles represent experiments in which jadeite + quartz grew (data from Holland 1980).

equilibrium (see also Chap. 8). Water enhances reaction rates, as does increasing temperature, the presence of even small amounts of a melt phase, and, often, high pressure. This enhancement brings out the second problem in phase studies, the inability to *quench* a high *P, T* assemblage, that is, to bring it to ambient conditions by rapid cooling and/or pressure release without change in the phases present or their compositions. This problem can be serious with melts of relatively low viscosity that produce *quench crystals*, sometimes recognizable by very fine grained or skeletal morphologies but often misleading. This problem is present in studies at very high pressure, where the quenched phase may represent the true high-pressure assemblage, a completely reverted low-pressure polymorph, an intermediate metastable phase, or even glass. Thus the design and interpretation of phase equilibrium studies is as important as performing the experiments themselves.

Table 4.4. *High-pressure apparatus*

Apparatus	Sample vol. (cm³, approx.)	Pressure range (GPa)	Temp. range (K, approx.)
Autoclaves	100–10000	0–0.1	200–600
Externally heated cold seal pressure vessels	10–500	0–0.5	300–1200
Internally heated gas vessels	5–100	0–1.5	300–1800
Piston-cylinder solid media apparatus	0.01–0.1	0–5	300–2000
Multiple anvil solid media apparatus	10^{-3}–10^{-2}	0–27	300–2000
Diamond anvil cell (laser heated)	10^{-6}–10^{-3}	0–100 (up to 400 at lower T)	200–3000
Shock apparatus (dynamic methods, pressure for very short time only)	Variable	0 to 300	300–3000

4.2.1 Phase studies at high temperature and pressure

During phase equilibrium studies, samples are heated in electrical resistance furnaces at one atmosphere total pressure (Holloway and Wood 1988). The samples are often contained in noble metal (Au, Pt, Pd, Ir) capsules or crucibles, but there is no such thing as a totally inert container. Gas mixtures provide atmospheric control (see Sec. 4.2.2). Various apparatus can be used to rapidly quench samples. Cooling rates of 10^2–10^3 K/sec are easily attainable, and rates up to 10^6 K/sec can be attained by special techniques such as splat or roller quenching (Anthony and Suzuki 1974; Coutures et al. 1978).

Table 4.4 lists the characteristics of several types of high-pressure apparatus. In general, as pressure increases, the available sample volume decreases, and at pressure greater than about 10 GPa, seldom can more than a few milligrams of sample be produced, especially at high temperature. The appropriate calibration of temperature and pressure when both are high is a subject of ongoing development and controversy. The ability to perform physical measurements (X-ray diffraction, conductivity, spectroscopy) in situ at high pressure and temperature is improving. Such experiments are crucial to understanding the

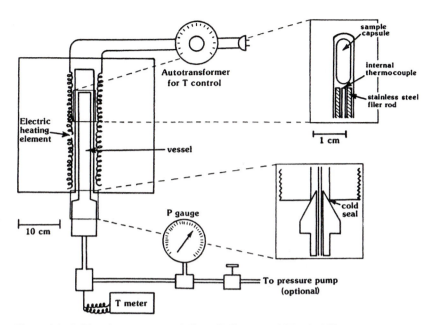

Figure 4.4. Cold seal pressure vessel (from Holloway and Wood 1988).

actual state of materials under extreme conditions, that is, in the Earth's mantle and core. At pressures below ~ 0.5 GPa, cold seal pressure vessels, or as they are called colloquially (much to the consternation of airport officials), "cold seal bombs," are cheap, relatively safe, and easy to use. They consist simply of a rod of special hard alloy (Stellite, Rene, TZM), in which a hole is drilled to hold the sample, and a sealing mechanism that remains outside the furnace (hence the term *cold seal*) (see Fig. 4.4). Cold seal vessels are pressurized by water or argon. The samples can be open to the pressurizing atmosphere or they can be sealed in gold, platinum, or silver-palladium capsules (Holloway and Wood 1988). The fugacity of H_2O, O_2, CO_2, or other volatile constituents can be controlled by various solid buffers (see following).

The diamond anvil cell is shown in Figure 4.5 (Navrotsky et al. 1992). Two gem-quality diamonds, their culets truncated into flats, are pressed together, and the sample is usually confined in a metal gasket. Pressure is controlled by the force applied to the cell and is commonly measured by a spectroscopic technique, using the pressure shift of the ruby fluorescence line, or, recently, of rare earth doped YAG. Pressures up to four million atmospheres (400 GPa) have been attained at room temperature. At temperatures above 1000 K, the

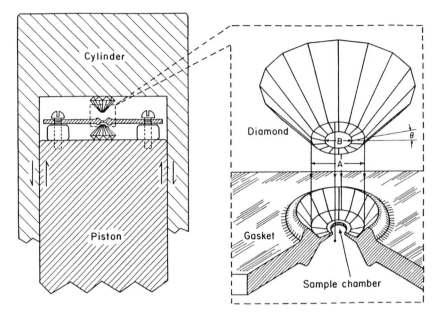

Figure 4.5. Schematic diagram of the piston-cylinder arrangement of a Mao-Bell diamond anvil cell. The left side shows two opposed gem-quality diamonds that are pressed against a sample held in a hole drilled into the gasket. The right side is an enlargement of a single diamond and the gasket. This diamond has bevels on the tip that permit higher pressures to be achieved. X-ray or laser access to the sample is provided through holes in the cylinder and piston that are aligned with the hole in the gasket (from Navrotsky et al. 1992).

highest attainable pressures are currently in the 100–200 GPa range. The interior of the cell has a sharp pressure gradient, which can sometimes be used to advantage for simultaneously studying both high- and low-pressure phases. At low to moderate temperatures, a hydrostatic environment can be achieved by filling the cell with an inert gas or a liquid organic mixture, such as methanol-ethanol. Laser heating is used to obtain temperatures to 5000 K. Both YAG laser and, more recently, CO_2 laser heating are being used. External heating of the diamond cell with a resistance heater outside the diamonds can be used to about 700 K (where the diamonds begin to convert to graphite), and internally placed resistance heaters are possible. The characterization and control of temperature gradients are areas of active development. The volume of sample produced in a diamond cell is usually a few micrograms, but larger sample volumes are actively being pursued.

Diamond cells offer relatively ready access to light of all frequencies. The

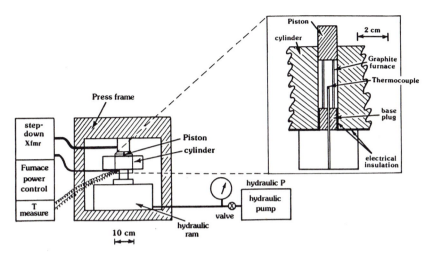

Figure 4.6. Piston-cylinder high-pressure apparatus (from Holloway and Wood 1988).

sample can be viewed optically, and phase transitions observed with the unaided eye. Infrared, Raman, visible, and ultraviolet spectroscopy can be performed. Mössbauer spectra have been obtained with a diamond cell, and recently the whole cell has been placed in an NMR spectrometer. Spectroscopic techniques in the diamond cell are presently undergoing rapid improvement.

Diamond cell geometry offers entry to X rays, using conventional, rotating anode, or synchrotron sources. The several orders of magnitude improvement in intensity available at synchrotron facilities enables single crystal refinements on very small crystals, an important achievement because many natural materials occur only as small crystals, and the crystals produced by high-pressure synthesis are usually tiny. Powder Rietveld refinements are also being done successfully using synchrotron radiation.

Larger samples are attainable in solid media pressure vessels in which the anvils are made of a hard material like tungsten carbide. The conventional piston-cylinder geometry (see Fig. 4.6) offers pressures to 5 GPa (Holloway and Wood 1988; Johannes 1973) and with the development of sintered diamond pistons, the pressure range may be extended to about 10 GPa. Belt and girdle type presses attain the 8 GPa range. Multianvil presses (see Fig. 4.7), in which, typically, six tungsten carbide anvils encroach synchronously on an octahedral sample volume, reach pressures near 23 GPa, and there is hope of extending the range to 30 GPa. Much of the development of high-pressure

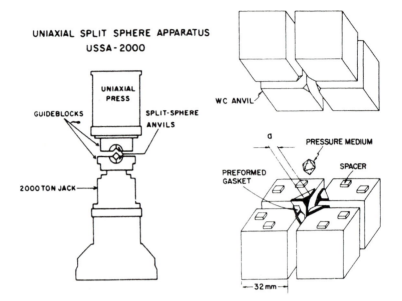

USSA - 2000

UPPER GUIDEBLOCK

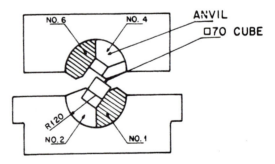

LOWER GUIDEBLOCK

Figure 4.7. Schematic diagrams of 2000-ton uniaxial split-sphere apparatus (USSA−2000) at Stony Brook High Pressure Laboratory. (Top left) overview, (top right) details of first-stage spherical anvils in guideblocks, and (bottom) second stage with ensemble of eight tungsten carbide cube anvils (MA−8) (from Navrotsky et al. 1992).

research rests on the development of new materials (such as sintered diamond in large pieces) that can perform better at high pressure and temperature.

Heating is achieved by a cylindrical resistance heater around the sample. Sometimes the heater serves as the sample container, but the sample may be further encapsulated. If the sample capsule is sealed, a solid buffer assemblage may surround the sample to fix the fugacity of oxygen, water, or carbon dioxide. This is important for controlling oxidation states, for example, Fe^{2+}/Fe^{3+} ratios, for synthesizing phases with variable stoichiometries, including oxide superconductors, and for producing phases with defined water content, such as high-pressure hydrous magnesium silicates. The measurement and control of fugacity in a high-pressure environment is a field in which earth scientists have generally had more experience than materials scientists.

The samples that can be made range from 1 mg to 100 mg, with sample size decreasing sharply with pressure. Temperatures attainable exceed 2000 K. Recent developments have extended run times from minutes to hours at 20 GPa. Because the pressure vessel is a massive heat sink, turning off the power results in cooling the sample to below 700 K in seconds and to room temperature in a minute or so. After pressure is released slowly, the high-pressure phase can often be recovered. Sometimes, however, the high-pressure phase reverts to the atmospheric pressure polymorph, a metastable crystalline phase, or an amorphous material.

In many ways, diamond cell and multianvil experiments are complementary. The former provide superior access for in situ measurements but generally larger gradients of pressure and temperature and much smaller samples. Diamond cells are extremely well suited to exploratory work because experiments can be done quickly and relatively easily and the samples are visible. Diamond cells are the only static tool for pressures above 30 GPa. Multianvil presses provide larger samples and better control of sample environment. Therefore, they are favored for synthesis and phase equilibrium studies.

4.2.2 Oxidation-reduction equilibria

For systems containing an easily reducible metal, equilibrium between the metal and an appropriate gas phase provides a useful means of measuring activities and free energies in a binary or ternary oxide (or chalcogenide) phase. Thus for the equilibrium

$$\text{``FeO''} = \text{Fe} + 1/2\ O_2 \tag{4.6}$$

$$\Delta G^\circ = -RT \ln K = -RT \ln (fO_2)^{1/2} (a_{Fe}) / (a_{\text{``FeO''}}) \tag{4.7}$$

where $\Delta G°$ is the standard free energy charge, K the equilibrium constant, R the gas constant, T the absolute temperature, and f and a refer to fugacity and activity, respectively.

For pure "FeO" in equilibrium with pure Fe, knowledge of the equilibrium value of oxygen fugacity at a given temperature gives $\Delta G°$ directly, and the temperature dependence of oxygen fugacity can be used to derive values of $\Delta H°$ and $\Delta S°$. The oxygen fugacity in equilibrium with both metal and metal oxide can be measured in two ways: by gas-mixing experiments (Biggar 1974; Holloway and Wood 1988; Huebner, 1975) or by a solid-state electrochemical cell (Sato 1971; Schmalzried and Pelton 1972). In the former, oxygen fugacity is set by mixing oxygen with an inert gas for partial pressures of oxygen, $pO_2 > 1$ to 10^{-4} atm, or to attain lower values of pO_2, by using a subsidiary gas phase equilibrium such as

$$H_2O = H_2 + \frac{1}{2}O_2, \quad pO_2 = [K(p_{H_2O})/(p_{H_2})]^2 \tag{4.8}$$

or

$$CO_2 = CO + \frac{1}{2}O_2, \quad pO_2 = [K(p_{CO_2})/(p_{CO})]^2 \tag{4.9}$$

The ratio in which the major gases (CO and CO_2 or H_2O and H_2) are mixed then determines the partial pressure (fugacity) of oxygen. This ratio is varied until a value is found for which the solid sample equilibrated with the gas mixture contains both metal and oxide. Automated mass flow controllers for gas flow make this task easier.

The electrochemical method depends on the use of a cell with a solid electrolyte (typically a CaO-stabilized zirconia) that conducts by the movement of oxide ions. For the Fe/"FeO" equilibrium, such a cell may be written as

$$Pt \mid Fe, \text{"FeO"} \parallel ZrO_2 \text{ (CaO} \parallel air \mid Pt) \tag{4.10}$$

The left side of the cell has its pO_2 fixed by the Fe/"FeO" equilibrium while the right side has $pO_2 = 0.21$ atm. The half-cell reactions may be written as

$$left \quad O^{2-} + Fe = \text{"FeO"} + 2e^- \tag{4.11a}$$

$$right \quad 1/2O_2 = 2e^- + O^{2-} \tag{4.11b}$$

and the cell reaction

$$Fe + 1/2O_2 = \text{"FeO"} \tag{4.11c}$$

with

$$\Delta G^\circ = -n F E^\circ = -2 F E^\circ \qquad (4.12)$$

where n is the number of electrons transferred, F the Faraday constant, and E° the measured cell voltage under standard state conditions.

In the preceding reactions "FeO" has been written in quotation marks because this oxide is nonstoichiometric, containing a deficiency of cations and a virtually perfect anionic sublattice in the rocksalt structure. The defect equilibria and deviations from stoichiometry are functions of pressure, temperature, and oxygen fugacity. In silicates containing iron, the deviations from stoichiometry are much smaller than in "FeO" and although the defect equilibria dominate transport and electrical properties, they are generally not major factors in the overall thermodynamics. However, in systems containing both Fe^{2+} and Fe^{3+}, for example, the iron-titanium oxides, oxidation-reduction plays a major role.

Both the gas equilibration method and the solid cell method can be used to find the free energies of ternary compounds and activities in solid solutions (Muan 1967). In general, easily reducible oxides include those of Fe, Ni, Co, Cu, and the higher oxides of manganese. Careful studies of multicomponent systems (e.g., Mg–Fe–Si–O) (Nafziger and Muan 1967) can give free-energy data for all end-member phases and activity-composition relations for their solid solutions. The temperature range of such studies is limited at low temperatures (typically ~1100 K for silicates) by sluggish equilibration and at high temperatures by melting, vaporization, reaction with the container material, and other side reactions. Thus the temperature range over which equilibrium measurements can be made is not always large enough to derive accurate entropy and enthalpy values from the temperature dependence of free energy. A combination of ΔG° from equilibrium measurements and ΔH° from solution calorimetry often gives the best estimates of ΔS°.

Noble metals such as Pt tend to absorb and alloy with transition metals under even moderately reducing conditions because these alloys show very large negative deviations from ideal solution behavior. This "iron loss problem" (Grove 1981; Merrill and Wyllie 1973) is a major nuisance in experimental petrology; it changes sample compositions and embrittles and ruins expensive Pt crucibles. This problem can be turned into an advantage by using the transition metal content of a noble metal probe wire in contact with a silicate melt as a monitor of thermodynamic activity of the transition metal (Grove 1981). Sensitive analytical techniques enable similar methods to be used to determine the activities for major oxide components (MgO, Al_2O_3, SiO_2) by measuring the concentration of Mg, Al, and Si alloyed with Pd equilibrated with a melt or mineral at fixed temperature and oxygen fugacity (Chamberlin, Beckett, and Stolper in press).

4.2.3 Vapor pressure measurements

If a constituent is relatively volatile, its vapor pressure can be used to probe its thermodynamic activity in the solid state. Measurement of vapor pressure falls into two classes: dynamic measurements in which one monitors weight loss or the amount of volatile species in a carrier gas stream as a function of time, and effusion measurements in which equilibrium in an almost closed system is attained, and the vapor is sampled through a pinhole (Paule and Margrave 1967). The detection and measurement of vapor species by mass spectrometry, spectroscopic techniques, or chemical analysis determines the sensitivity and accuracy of vapor pressure measurements.

Such measurements have proved especially useful in semiconductors, where one or both species in a binary system are often volatile and the solid phases show considerable homogeneity ranges. These methods have recently been applied quite successfully to measuring activities in aluminosilicate melts containing alkalis (Rammensee and Fraser 1982) (see also Chap. 8).

4.3 Equation of state

The relation

$$f(P,V,T) = 0 \qquad (4.13)$$

is called an equation of state, relating pressure, volume, and temperature. For an ideal gas, it is (per mole)

$$PV - RT = 0 \quad \text{or} \quad V = RT/P \qquad (4.14)$$

For a solid or liquid, one usually considers the effects of temperature and pressure separately. As a function of temperature, the volume is given by

$$V = V_{298} \left[1 + \alpha \left(T - 298 \right) \right] \qquad (4.15)$$

where α, the thermal expansivity, can itself be a function of temperature

$$\alpha = \alpha_0 + \alpha_1 T + \ldots . \qquad (4.16)$$

For pressures to about 1 GPa, the compressibility (β) and bulk modulus ($K = 1/\beta$) of a solid can be approximated as constant and

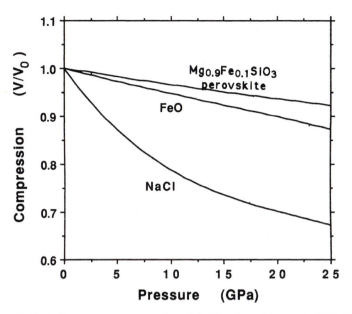

Figure 4.8. Typical compression curves for solids (data from Mao et al. 1969, Mao et al. 1991).

$$V = V_0 \left(1 - \frac{P}{K} \right)$$ (4.17)

where K is the bulk modulus.

For higher pressures (see Fig. 4.8) the compressibility decreases with pressure, that is, $K' = \partial K / \partial P > 0$. A commonly used equation of state at moderate compression is the Murnaghan equation (Anderson 1989):

$$V = V^0 \left(\frac{PK'}{K} + 1 \right)^{-1/K'}$$ (4.18)

Typical values of K are 5–10 GPa for halides and 10–20 GPa for oxides, with K' often near 4 or 5, though sometimes as large as 9 (see also Table 6.5). Above 10 GPa, additional terms are needed and finite-strain theory gives appropriate expressions (Anderson 1989).

It is generally assumed that thermal expansion and compression at pressure work independently of each other, thus

$$V(P, T) = V_{298}^0 \cdot f(P) \cdot f(T) \tag{4.19}$$

This is a reasonable first approximation but, at least at high temperature, a better approximation is that

$$\alpha K = \text{constant} \tag{4.20}$$

however, both the thermal expansivity (α) and bulk modulus, K, depend on temperature (Anderson 1989).

Thermal expansivity and compressibility of crystalline materials is best measured by in situ X-ray diffraction at high T and P. As discussed, this can be done in a diamond cell, multianvil apparatus, or in a controlled atmosphere heating stage or furnace.

Thermal expansion can also be measured by measuring the change in length of a glass or ceramic rod. The volumes of glasses can be measured by pycnometry (measuring the density by weighing the sample, first in air and then submerged in a liquid of known density). The volume of a crystal can be obtained similarly, or from its lattice parameters. At high temperature, lattice parameter measurements remain useful for crystals, but the determination of volume (or density) of a silicate by direct measurements of weight and volume become problematic. Instead, less direct methods must be used. A general method to obtain compressibilities of liquids involves the measurement of the speed of sound. It has been applied to silicate melts at atmospheric and high pressures (Rivers and Carmichael 1987). A related method is obtaining the equation of state by following the passage of a shock wave through a solid, glass, or liquid (Anderson 1989; Rigden, Ahrens, and Stolper 1984).

The density of silicate melts can be measured by an Archimedean double-bob technique, in which the weight of a bob of known density (Pt or other) is measured in air and then in a silicate liquid (Lange and Carmichael 1987). An alternative method is that of using falling spheres; the rate of descent (or ascent) of spheres of different densities can bracket a liquid's density, and more quantitative measurements give information on both density and viscosity. Methods of measuring melt density are compared by Lange and Carmichael, (1987).

4.4 References

4.4.1 General references and bibliography

Anderson, D. L. (1989). *Theory of the earth*, Oxford: Blackwell Scientific Publications.

Calvet, E., and H. Prat. (1954). *Microcalorimetrie*. Paris: Masson et Cie.

Cheetham, A. K., and P. Day, eds. (1987). *Solid state chemistry techniques*. Oxford: Oxford University Press.

References 169

Holloway, J. R., and B. J. Wood. (1988). *Simulating the Earth experimental geochemistry.* Winchester, MA: Unwin Hyman.

McCullough, J. P., and D. W. Scott, eds. (1968). *Experimental thermodynamics.* London: Butterworth.

Navrotsky, A. (1987). Thermodynamic aspects of solid state chemistry. In *Solid state chemistry,* ed. P. Day and A. Cheetham, 362–93. Oxford: Oxford University Press.

Navrotsky, A. (1991). Calorimetric studies of ceramics. In *Chemistry of electronic ceramic materials.* Proceedings of the International Conference held in Jackson, WY, August 17–22, 1990, National Institute of Standards and Technology Special Publication 804. Washington, DC: GPO.

Navrotsky, A., D. J. Weidner, R. C. Liebermann, and C. T. Prewitt. (1992). Materials science of the earth's deep interior, *MRS Bull. 17,* 19–37.

Robie, R. A., and B. S. Hemingway. (1972). *Calorimeters for heat of solution and low-temperature heat capacity measurements.* U.S. Geological Survey Professional Paper 755. Washington, DC: GPO.

Sato, M. (1971). Electrochemical measurements and control of oxygen fugacity and other gaseous species with solid electrolyte sensors. In *Research technique for high pressure and high temperature,* ed. G. C. Ulmer, 43–100. Heidelberg, Germany: Springer Verlag.

Schmalzried, H., and A. D. Pelton. (1972). Two aspects of solid-state thermodynamics at elevated temperatures: Point defects and solid-state galvanic cells. *Annual Review of Materials Science 2,* 143–80.

Schmalzreid, H., and A. Navrotsky. (1978). *Festkorperthermodynamik chemie des festen zustandes.* Berlin: Akademie-Verlag.

4.4.2 Specific references

Akaogi, M., E. Ito, and A. Navrotsky. (1989). Olivine-modified spinel-spinel transitions in the system Mg_2SiO_4–Fe_2SiO_4: Calorimetric measurements, thermochemical calculations and geophysical application. *J. Geophys. Res. 94,* 15671–85.

Akaogi, M., A. Navrotsky, T. Yagi, and S. Akimoto. (1987). Pyroxene-garnet transition: Thermochemistry and elasticity of garnet solid solutions, and application to a pyrolite mantle. In *High pressure research in mineral physics,* ed. M. H. Manghnani and Y. Syono, 251–60. Tokyo: Terra Publications.

Anthony, A. M., and T. Suzuki. (1974). Rapid quenching on the binary systems of high temperature oxides. *MRS Bull. 9,* 745–54.

Biggar, G. M. (1974). Oxygen partial pressures: Control, variation, and measurement in quench furnaces at one atm. total pressure, *Min. Mag. 39* (March), 580–6.

Callanan, J. E., and S. A. Sullivan. (1986). Development of standard operating procedures for differential scanning calorimeters. *Rev. Sci. Instrum. 57,* 2584–92.

Capobianco, C., and A. Navrotsky. (1987). Solid solution thermodynamics in $CaCO_3$–$MnCO_3$. *Amer. Mineral. 72,* 312–18.

Cemic, L., and O. J. Kleppa. (1986). High temperature calorimetry of sulfide systems. I. Thermochemistry of liquid and solid phases of Ni + S. *Geochim. Cosmochim. Acta 50,* 1633–41.

Chai, L., and A. Navrotsky. (1993). Thermochemistry of carbonate-pyroxene equilibria. *Contrib. Mineral. Petrol. 114,* 139–47.

Chamberlin, L., J. Beckett, and E. Stolper. (In press). A new experimental method for the direct determination of oxide activities in melts and minerals, *Contrib. Mineral. Petrol.*

Charlu, T. V., and O. J. Kleppa. (1973). High-temperature combustion calorimetry, 1.

Enthalpies of formation of tungsten oxides. *J. Chem. Thermodynamics 5*, 325–30.

Chatillon-Colinet, C., R. C. Newton, D. Perkins, III, and O. J. Kleppa. (1983). Thermochemistry of (Fe^{2+}, Mg)SiO_3 orthopyroxene. *Geochim. Cosmochim. Acta 47*, 1597–603.

Coutures, J. P., R. Ber Joan, G. Benezech, B. Granier, R. Renard, and M. Foex. (1978). Utilisation des fours solaires de laboratoire pour l'etude a haute temperature des proprietes physicochimiques des oxydes refractaires. *Rev. Int. Hautes Temp. Ref. 15*, 103–14.

Gordon, T. M. (1973). Determination of internally consistent thermodynamic data from phase equilibrium experiments. *J. Geol. 81*, 199–208.

Grove, T. L. (1981). Use of Fe-Pt alloys to eliminate iron loss problems in 1 atmosphere gas mixing experiments: Theoretical and practical considerations. *Contrib. Mineral. Petrol. 78*, 298–304.

Holland, T. J. B. (1980). The reaction albite = jadeite + quartz determined experimentally in the range 600–1200 °C. *Amer. Mineral. 65*, 129–34.

Huebner, J. S. (1975). Oxygen fugacity values of furnance gas mixtures. *Amer. Mineral. 60*, 815–23.

Johannes, W. V. (1973). A simplified piston-cylinder apparatus of high precision. *N. Jb. Min. Mn. 1973*, no. 7/8, 337–51.

Lange, R. A., and I. S. E. Carmichael. (1987). Densities of Na_2O-K_2O-CaO-MgO-FeO-Fe_2O_3-Al_2O_3-TiO_2-SiO_2 liquids: New measurements and derived partial molar properties. *Geochim. Cosmochim. Acta 51*, 2931–46.

Lange, R. A., J. J. DeYoreo, and A. Navrotsky. (1991). Scanning calorimetric measurement of heat capacity during incongruent melting of diopside. *Amer. Mineral. 76*, 904–12.

Mao, H. K., R. J. Hemley, Y. Fei, J. F. Shu, L. C. Chen, A. P. Jephcoat, Y. Wu, and W. A. Bassett. (1991). Effect of pressure, temperature, and composition on lattice parameters and density of (Fe, Mg)SiO_3-perovskites to 30 GPa. *J. Geophys. Res. 96*, B5, 8069–79.

Mao, H. K., T. Takashashi, W. A. Bassett, J. S. Weaver, and S. I. Akimoto. (1969). Effect of pressure and temperature on the molar volumes of wustite and of three (Fe, Mg)$_2$$SiO_4$ spinel solid solutions. *J. Geophys. Res. 74*, 1061–9.

Marucco, J., P. Gerdanian, M. Dode. (1970). Determination des grandeurs molaires partielles de melange de l'oxygene dans le protoxyde de fer a 1075 °C. *J. Chim. Phys. 67*, 906–13.

Merrill, R. B., and P. J. Wyllie. (1973). Absorption of iron by platinum capsules in high pressure rock melting experiments. *Amer. Mineral. 58*, 16–20.

Muan, A. (1967). Determination of thermodynamic properties of silicates from locations of conjugation lines in ternary systems. *Amer. Mineral. 52*, 797–804.

Nafziger, R. H., and A. Muan. (1967). Equilibrium phase compositions and thermodynamic properties of olivines and pyroxenes in the system, MgO-"FeO"-SiO_2. *Amer. Mineral. 52*, 1364–85.

Navrotsky, A. (1977). Recent progress and new directions in high temperature calorimetry. *Phys. Chem. Minerals 2*, 89–104.

Ostvold, T., and O. J. Kleppa. (1969). Thermochemistry of the liquid system lead oxide-silica at 900 °C. *Inorganic Chem. 8*, 78–82.

Parks, M. E., A. Navrotsky, K. Mocala, E. Takayama-Muromachi, A. Jacobson, and P. K. Davies. (1989). Direct calorimetric determination energetics of oxygen in $YBa_2Cu_3O_x$. *J. Solid State Chem. 79*, 53–62. *Erratum 83*, 218–19.

Paule, R. C., and J. L. Margrave. (1967). Free evaporation and effusion techniques. In *The characterization of high-temperature vapors*, ed. J. L. Margrave, 130–51. New York: Wiley.

Pawley, A. R., C. M. Graham, and A. Navrotsky. (1993). Tremolite-richterite amphiboles: Synthesis, compositional and structural characterization, and thermochemistry. *Amer. Mineral. 78*, 23–35.

Rammensee, W., and D. G. Fraser. (1982). Determination of activities in silicate melts by Knudsen cell mass spectrometry–I. The system $NaAlSi_3O_8$–$KAlSi_3O_8$. *Geochim. Cosmochim. Acta 46*, 2269–78.

Rigden, S. M., T. J. Ahrens, and E. M. Stolper. (1984). Properties of liquid silicates at high pressures. *Science 226*, 1071–4.

Rivers, M. L., and I. S. E. Carmichael. (1987). Ultrasonic studies of silicate melts. *J. Geophys. Res. 92*, B9, 9247–70.

Stebbins, J. F., D. F. Weill, and I. S. E. Carmichael. (1982). High temperature heat contents and heat capacities of liquids and glasses in the system $NaAlSi_3O_8$–$CaAl_2Si_2O_8$. *Contrib. Mineral. Petrol. 80*, 276–84.

Stebbins, J. F., I. S. E. Carmichael, and L. K. Moret. (1984). Heat capacities and entropies of silicate liquids and glasses. *Contrib. Mineral. Petrol. 86*, 131–48.

Westrum, E. F., Jr. (1984). Computerized adiabatic thermophysical calorimetry. In *Proceedings NATO Advanced Study Institute on Thermochemistry at Viana de Castello, Portugal*, ed. M. A. V. Riberio Da Silva, 745–76. New York: Reidel.

Woodfield, B. F. (1988). *I. The construction of a low-temperature adiabatic calorimeter, II. Calorimetric studies of the phase transitions in CHI_3, $K_2Mn_2(SO_4)_3$, and $K_2Mg_2(SO_4)_3$*. Master's thesis, Brigham Young University, Provo, Utah.

Ziegler, D., and A. Navrotsky. (1986). Direct measurement of the enthalpy of fusion of diopside. *Geochim. Cosmochim. Acta 50*, 2461–6.

5

Chemical bonding

5.1 Introduction

Electrons are the glue that bonds atoms together. Whether in an isolated atom, a simple molecule, or a complex solid, the electronic states are determined by solutions to the basic equation of quantum mechanics, the Schrödinger equation:

$$\mathbf{H}\Psi = E\Psi \qquad (5.1)$$

$\mathbf{H}$ is the quantum mechanical Hamiltonian operator, the sum of kinetic and potential energy expressions for the system, given by

$$\mathbf{H} = \frac{h^2}{8\pi^2 m}\left(\frac{\partial^2}{\partial x^2} + \frac{\partial^2}{\partial y^2} + \frac{\partial^2}{\partial z^2}\right) + \mathbf{V}\,(x, y, z) = \frac{h^2\,\nabla^2}{8\pi^2 m} + \mathbf{V}\,(x, y, z) \qquad (5.2)$$

with h Planck's constant and m the mass of the electron. The potential energy term, $\mathbf{V}$, includes all interactions among nuclei, among electrons, and among electrons with nuclei. One goal of quantum mechanics is to find a set of wave functions (eigenfunctions) that yield, as solutions of Eq. 5.1, eigenvalues corresponding to the energy of the system. Other observables (dipole moments, spectra, physical properties) can in principle be calculated once the wave functions are known. The difficulty of applying quantum mechanics lies not in its formulation, but in finding the solutions for a given real potential that contains numerous many-body interactions and correlations. For any system more complex than diatomic molecules of the first row elements, the problem is formidable and must be solved at various degrees of approximation. The approaches and approximations used distinguish the different schools of thought, which, though all are basically trying to solve the same problem, appear radically different to the novice.

172

Three major approaches can be identified: I call these the *physics approach*, which stresses the periodicity of the crystal as a constraint on the eigenfunctions, the *chemistry or molecular approach*, which describes bonding in molecular-sized clusters of two to perhaps twenty atoms as a model for a solid, and the *mineralogy or ionic approach*, which treats the solid as an assemblage of charged ions with classical coulombic interactions and with other attractive and repulsive terms. This mineralogy approach leads to systematization of parameters (ionic radius, polarizability, electronegativity) applicable to many systems. The terms *physics, chemistry*, and *mineralogy* are used only as familiar labels here, to indicate from which community the concepts historically emerged. Today solid-state scientists in any discipline are familiar with and use all these methodologies. All three viewpoints have been immensely successful and are now at comparable levels of sophistication. The evolution of modern computational methods blurs the boundaries between these camps, and the field of computational quantum chemistry is evolving rapidly. The goal of this chapter is to provide a guided tour of the concepts, general approach, and jargon, and to highlight some of the successes and major applications of these methods to understanding bonding in complex solids.

A few words should be said about units of energy. Thermodynamicists tend to think in terms of kilocalories or kilojoules per mole. Band theory calculations often express energies in electron volts. The conversion factor is 1 eV = 23.06 kcal/mol = 96.48 kJ/mol. Thus for a rough comparison, an electron volt corresponds to a hundred kilojoules. Molecular quantum chemistry expresses energy in hartrees (related to the energy of the ground state of the hydrogen atom). This ground state lies 43.594×10^{-19} joules per atom below the energy of isolated proton and electron. Thus one hartree corresponds to $43.594 \times 10^{-19} \times 6.023 \times 10^{23}$ joules per mol, or 2624 kJ/mol. In the figures, I have rescaled all energies to correspond to kJ/mol.

Quantum mechanics abounds with uncertainty principles. The most rigorous one is that which relates conjugate observables (momentum and position, energy and time) such that both can not be known to an accuracy that makes the product of their uncertainties smaller than Planck's constant. This is important in understanding the duality of the wave-particle description of matter and in determining spectral line widths. A second uncertainty principle is that the more sophisticated a quantum mechanical calculation, the more time-consuming and expensive it is, and the more difficult it usually is to apply the method to systems with many electrons, such as heavy atoms, especially those with incomplete d or f shells. Thus simpler systems can be studied more rigorously, but much of the systematic solid-state chemical insight has come from rather simplistic models for complex systems. A third uncertainty principle,

certainly very applicable in the 1990s, is that as soon as a user has mastered a
computational method, bigger, faster, and cheaper computers and more clever
and economical ways of doing the calculations become available. Computations are evolving rapidly in both hardware and software. Thus in computational quantum chemistry as elsewhere, you must pick your problems and tools
carefully, and keep your goals clearly in mind when applying a given method.
In my opinion, though the pursuit of ever-more-sophisticated solutions to the
Schrödinger equation is at the cutting edge of theoretical research, the modern
mineralogist and solid-state chemist continue to have much to learn and to
contribute by exploring the systematics gleaned from more approximate levels
of theory.

Figure 5.1 is a *phase diagram* that positions some compounds qualitatively
in terms of their degrees of ionic, covalent, and metallic bonding. I use these

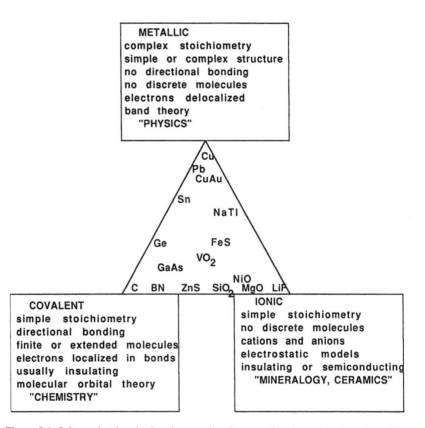

Figure 5.1. Schematic triangle showing varying degrees of ionic, covalent, and metallic
bonding in solids.

terms loosely – *ionic* to mean insulating or semiconducting, hard, generally high melting materials without obvious directional bonding; *covalent* to mean materials containing discrete molecules or strong directional bonds; and *metallic* to mean materials with high electronic conductivities and evidence of electron delocalization. Historically, the physics approach evolved largely from the desire to understand metals and semiconductors, the chemistry approach from the need to understand molecules, and the mineralogy approach from the necessity to deal with architecturally complex phases that can be considered as largely ionic. The point of Figure 5.1 is that most real materials contain some degree of each type of bonding, that this balance can change, continuously or abruptly, with pressure, temperature, and composition, and that to really understand the solid state, one must be versed in all three approaches.

5.2 Electrons and phonons in periodic solids – solid-state physics

5.2.1 Basic concepts

Consider an electron that can move freely in a solid. Its Schrödinger equation contains no potential energy term and is

$$\nabla^2\psi + \frac{8\pi^2 m}{h^2} E\psi = 0 \qquad (5.3)$$

Its energy is given by

$$E = h^2/2m\lambda^2 = h^2 k^2/8\pi^2 m \qquad (5.4)$$

where λ is the associated wavelength, and the wave vector, $\mathbf{k}$, is $2\pi/\lambda$.

The energy has a parabolic dependence on the wave vector (see Fig. 5.2a). In a real solid, the electron feels a "drag" by its interactions with the other electrons and with the vibrating atoms in the solid. Because the crystal is periodic, this potential must also be periodic. In the nearly free electron approximation, the relation of energy to wave vector is perturbed and instead of being parabolic, in general shows discontinuities as the wave vector approaches values corresponding to lattice repeats (see Fig. 5.2b).

The influence of lattice periodicity can be assessed as follows. In a periodic potential

$$\psi_k\,(\mathbf{r} + \mathbf{l}) = \psi_k\,(\mathbf{r})\,\exp\,(i\mathbf{k}\cdot\mathbf{l}) \qquad (5.5)$$

where ψ is the wave function, $\mathbf{r}$ is any arbitrary lattice point (expressed as a vector from the origin), and $\mathbf{l}$ is any lattice repeat. The periodic part of this wave function can be expressed as

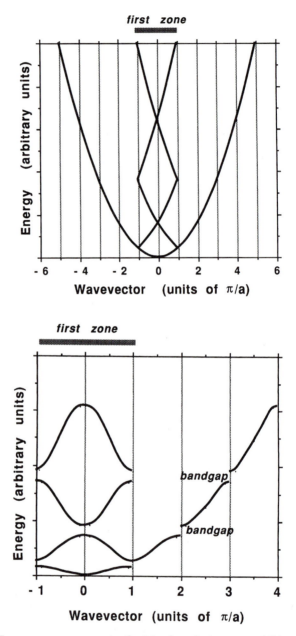

Figure 5.2. Energy versus wave vector for (a) a free electron gas and (b) a nearly free electron gas in a periodic potential, shown both in extended zone scheme (on right) and folded back into first Brillouin zone (reduced zone scheme) in center.

$$\exp(i\mathbf{k} \cdot \mathbf{l}) = \sum_n A_n \exp(in \cdot \mathbf{r}) \tag{5.6}$$

with n an integer, $i = \sqrt{-1}$, and A_n is given by a Fourier series:

$$A_n = \frac{1}{V} \int_{\text{all } \mathbf{r}} \psi(\mathbf{r}) \exp(-in \cdot \mathbf{r}) \, d\mathbf{r} \tag{5.7}$$

with V the volume of the unit cell. Wave functions of this form are called Bloch functions.

A concept of immense use to solid-state physics, lattice dynamics, and crystallography is that of the reciprocal lattice. In the real lattice (see Chap. 2), invariance occurs along any vector for which

$$\mathbf{R} = n_1\mathbf{a} + n_2\mathbf{b} + n_3\mathbf{c} \tag{5.8}$$

with n_2, n_2, n_3 integers, and $\mathbf{a}$, $\mathbf{b}$, $\mathbf{c}$ unit cell vectors. A reciprocal unit cell is defined to have unit cell vectors $\mathbf{a}^*$, $\mathbf{b}^*$, $\mathbf{c}^*$ such that

$$\mathbf{a}^* = 2\pi/V(\mathbf{b} \times \mathbf{c}), \mathbf{b}^* = 2\pi/V(\mathbf{a} \times \mathbf{c}), \mathbf{c}^* = 2\pi/V(\mathbf{a} \times \mathbf{b}) \tag{5.9}$$

where V is the volume of the unit cell and the cross indicates the vector cross product. One can then construct any lattice vector in reciprocal space such that

$$\mathbf{G} = h\mathbf{a}^* + k\mathbf{b}^* + l\mathbf{c}^* \tag{5.10}$$

with h, k, l integers. The orientation of the real and reciprocal cell is such that

$$\mathbf{a} \cdot \mathbf{a}^* = \mathbf{b} \cdot \mathbf{b}^* = \mathbf{c} \cdot \mathbf{c}^* = 2\pi \tag{5.11}$$

and

$$\mathbf{a} \cdot \mathbf{b}^* = \mathbf{c} \cdot \mathbf{b}^* = \mathbf{a} \cdot \mathbf{c}^* = \mathbf{b} \cdot \mathbf{a}^* = \mathbf{c} \cdot \mathbf{a}^* = \mathbf{c} \cdot \mathbf{b}^* = 0 \tag{5.12}$$

where the dots mean vector dot product. A wave vector in the reciprocal lattice is often called $\mathbf{k}$ and reciprocal space referred to as $\mathbf{k}$-space.

A particular unit cell, the Wigner-Seitz cell, is defined by drawing bisecting planes to the lines joining the origin to the usual unit cell vertices and choosing the smallest volume enclosed by these planes, which form a polyhedron. Doing the same in the reciprocal cell produces a polyhedron known as the first Brillouin zone, with larger cells known as second, third, or nth Brillouin zones (see Fig. 5.3).

The utility of the reciprocal lattice lies in its economy and convenience in treating phenomena of diffraction and scattering. Planes in the real lattice transform to points in the reciprocal lattice, and the mathematics becomes

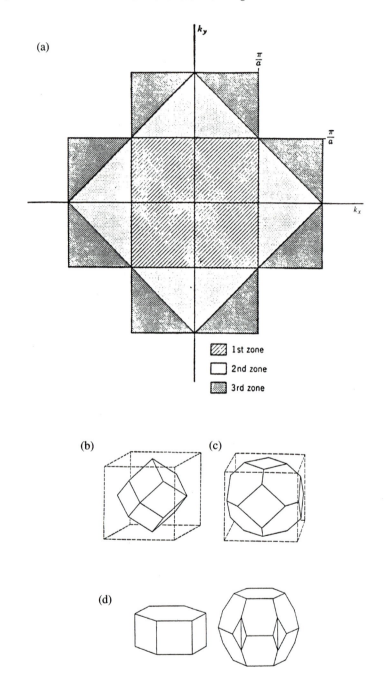

(a)

$\frac{\pi}{a}$

$\frac{\pi}{a}$

k_x

k_y

1st zone
2nd zone
3rd zone

(b)   (c)

(d)

more straightforward. It is sometimes jokingly said that chemists live in real space and crystallographers and physicists live in reciprocal space.

Returning to Figure 5.2, it is often convenient to plot all accessible energies as a function of wave vector in the first Brillouin zone only, since lattice points separated by a unit cell repeat are related by symmetry. This can be done in a "reduced zone scheme" in which the horizontal axis is folded back, literally like folding up a piece of paper, so the segments of $E(\mathbf{k})$ appear one above the other for values of $\mathbf{k}$ between $-\pi/a$ and π/a. The reduced zone scheme, which contains the same information in a more compact form, is used almost exclusively for real systems, with the extended scheme appearing mainly as a teaching device in textbooks. Because such segments of energy curves, especially for complex systems (whether one is calculating electronic or vibrational states), tend to cross each other and appear tangled, such reduced zone schemes are sometimes humorously called spaghetti diagrams.

5.2.2 Electronic energy bands in solids – qualitative concepts

Each atom in an assemblage of Avogadro's number of isolated copper atoms in the gas phase will have the same electronic energy, based on filling the atomic orbitals $1s^2$, $2s^2$, $2p^6$, $3s^2$, $3d^{10}$, $4s^1$. When the atoms are brought together in the metal, all but the outermost $4s$ electrons form core levels localized on each atom, but the outer electrons are mobile and cause copper to behave as a one-electron metal. The atomic energy levels broaden into bands of allowed energies separated by forbidden regions called *band gaps* (see Fig. 5.4).

The density of states, $g(E)$, is simply the number of states per unit energy increment expressed in arbitrary or normalized units (see Fig. 5.4). For Mg, the $3s$ and $3p$ bands overlap. The two valence electrons, which would be $3s$ in the isolated atom, have an abundance of closely spaced states available, and Mg is a metal. For MgO, there is an energy gap between the filled valence band and the unfilled conduction band, and pure silicon is an insulator.

Figure 5.3 (*Facing page*). Wigner-Seitz cells and Brillouin zones. Both are constructed by drawing the polyhedron defined by the intersection of planes that bisect the distances between lattice points: the W-S cell in real space, the BZ in reciprocal space. (a) The first, second, and third zones for a square two-dimensional lattice. (b) The first zone for a face-centered cubic lattice. This is the W-S cell for fcc and the BZ for bcc, whose reciprocal cell is fcc. (c) The first zone for a body-centered cubic lattice. This is the W-S cell for bcc and the BZ for fcc, whose reciprocal cell is bcc. (d) First and second zone for hexagonal close-packed lattice.

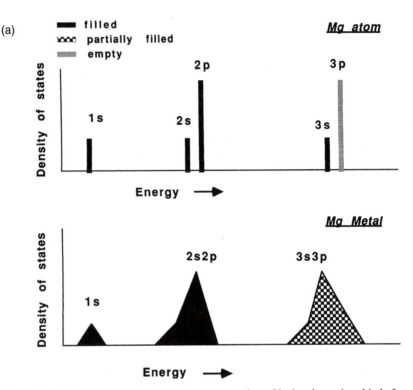

Figure 5.4. (a) Comparison of schematic representation of isolated atomic orbitals for an Mg atom (*top*) and the band developed from these orbitals for the crystalline state. The half-filled band derived from 3s and 3p orbitals is consistent with observed metallic properties. (b) Schematic representation of development of band structure in MgO: (*top*) atomic orbitals for isolated Mg atom, (*second*) atomic orbitals for isolated O atom, (*third*) molecular orbitals for isolated MgO molecule (also cf. Fig. 5.16), (*bottom*) bands in MgO crystal. The filled 2s2p band is separated from the empty 3s3p band by a large gap; MgO is an insulator.

Because electrons possess half-integral spin, they behave as *fermions* and fill the allowed (one-electron) levels successively. The average thermal occupancy of a level of energy E is given by the *Fermi-Dirac distribution function*, f_{FD}.

$$f_{FD} = \left[\exp\left(\frac{\varepsilon - \varepsilon_0}{kT}\right) + 1\right]^{-1} = \left[\exp\left(\frac{E - E_0}{RT}\right) + 1\right]^{-1} \quad (5.13)$$

where ε is energy per electron, E is energy per mole of electrons, k is Boltzmann's constant, R is the gas constant, T is absolute temperature, and ε_0 (and

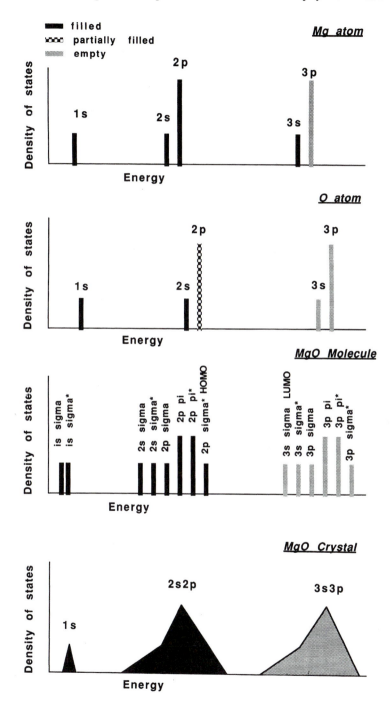

E_o) represent the Fermi energy, related to the chemical potential of the electron. The Fermi-Dirac distribution function always has value between zero and one.

At absolute zero, f_{FD} equals one for $E < E_o$ and zero for $E > E_o$. The energy E_F thus divides the occupied levels, each containing one electron, from the vacant levels at $T = 0$ (see Fig. 5.5). The Fermi energy of a metal lies within a partially filled band.

These concepts can be extended to the *Fermi surface*, which is the surface of constant energy, E_F, in **k**-space. The existence of a Fermi surface is necessary for metallic behavior. The morphology of the Fermi surface determines the electrical properties of the metal because electrical conduction reflects changes in the occupancies of states near the Fermi surface. In the free-electron approximation, the Fermi energy does not depend on crystallographic direction; it simply defines a sphere in reciprocal space of radius k_F determined by the valence electron concentration. In a real metal, the Fermi surface is more complex and is represented for aluminum in Figure 5.6.

The terms *metal, semiconductor, insulator,* and *superconductor* need further definition. In a metal – because there is an essentially continuous range of accessible states in a partially filled band – at a finite temperature some electrons are promoted to higher states, leaving "holes" below the Fermi level and

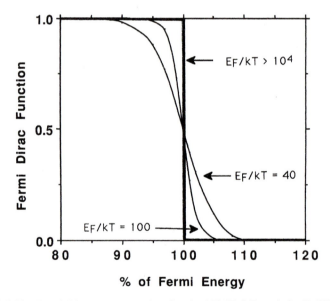

Figure 5.5. The Fermi-Dirac occupancy function for 0 K ($E_F/kT = \infty$), for $E_F/kT = 100$, and for $E_F/kT = 40$. Note "smearing out" of distribution with increasing temperature.

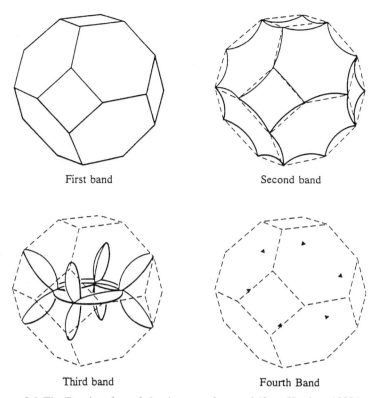

First band   Second band

Third band   Fourth Band

Figure 5.6. The Fermi surface of aluminum, an fcc metal (from Harrison 1980.)

a few electrons above. (Note: In solid-state literature the absence of an electron is called a *hole*, the absence of an atom a *vacancy*). The electrons are mobile and can readily respond to an electric field (applied voltage) by moving through the solid (current). The proportionality constant relating voltage applied and current produced is the *resistance*. The resistance is produced primarily by the interactions of the moving electrons with the atoms in the crystal, that is, with lattice vibrations or phonons. In general, impure metals and alloys show higher resistance than pure metals, and because lattice vibrations increase with temperature, resistance usually increases also.

In special cases at low temperature, the resistance drops to a very low value and the material expels any magnetic field. This phenomenon is called *superconductivity*. Until the discovery in 1986 of complex, mainly copper-containing, oxides that show superconducting transitions at temperatures up to about 120 K, T_c was limited to below 23 K. The mechanism for superconductivity appears to involve pairing of electrons and coupling of these electron pairs to the lattice vibrations (phonons) in such a manner as to minimize dissipative interactions. The mechanisms for superconductivity derive from an effective attraction between electrons, which in the normal nonsuperconducting state repel each other. This attraction comes, at least for classical superconductors, through interactions with phonons and has been described by the Bardeen-Cooper-Schrieffer (BCS) theory. It leads to pairing of electrons and a macroscopic quantum state with unusual properties. For materials with high critical temperatures for superconductivity (high T_c oxides), there is still lively discussion of the details of the mechanism for superconductivity.

If the separation between valence and conduction band is large, then thermal energy must be used to activate an electron (or hole) into the conduction band. With increasing band gap, one moves from a good semiconductor to a poor one to an insulator. Because conduction in this case is an activated process, the number of mobile species increases greatly with temperature, and the resistivity decreases exponentially with temperature. Indeed the temperature dependence can distinguish a good semiconductor from a poor metal when both have similar resistivities at room temperature. The Fermi energy of a semiconductor or insulator generally lies within the band gap, and thus does not define a unique division between occupied and vacant states. The distinction between insulator and semiconductors is one of degree – a highly insulating material can be made semiconducting by appropriate doping, that is, Fe in MgO or Mg_2SiO_4.

The energy bands and their occupancies can be critically influenced by small compositional variations, especially by impurities (dopants) introduced into a clean semiconductor. Thus, the introduction of phosphorus or arsenic into silicon adds electrons, which must occupy the conduction band, enhancing electron (n-type) conduction. Introduction of boron or aluminum depletes electrons, causing holes in the valence band and enhancing hole (p-type) conduction. Other impurities create localized states intermediate between valence and conduction band, making the activation process easier. The multibillion dollar semiconductor and electronics industry is based on properly understanding and controlling the flow of electrons and holes on increasingly more microscopic scales in devices.

Most oxide and silicate minerals are rather poor semiconductors, with the major source of conductivity being holes, namely, Fe^{3+} impurities in a predominantly Fe^{2+} matrix. This leads to complicated defect chemistry depending on P, T, and fO_2. Sulfides are better semiconductors but, again, their conductivities tend to be governed by impurities that, in natural multicomponent systems, are not readily controlled. Therefore, understanding and predicting the conductivities of minerals and rocks is a difficult task.

Pressure can affect the band structure in several ways. In general, as interatomic distances shrink, bands broaden, so a usual effect of pressure is expected to be the increase of band overlap. In cases such as FeO (Isaak et al. 1993), this broadening can lead to metallic behavior, but in other instances, such as diamond, MgO, and SiO_2 stishovite, despite broader bands, the band gap increases with pressure (Bukowinski 1980). There may also be more abrupt and specific changes as well. A pressure-induced insulator-to-metal transition may occur; this is seen in solid CsI (Aidun, Bukowinski, and Ross 1984), may occur in iron oxides, and is sought in hydrogen at megabar pressures (Mao and Hemley 1992). A pressure-induced transition to a completely different denser structure (see Chap. 6) carries with it a complete change in band structure and electronic properties. In addition, the energies of different bands may change at different rates with pressure, so that there may be crossovers that change which band represents the lower energy. Thus, for example, the role of d electrons may become more important, making metals like potassium more similar to transition metals at pressures of the Earth's core (Bukowinski 1979).

5.2.3 Band structure calculations

Why are first principles quantum mechanical calculations inherently complex? The reasons include the following: (1) The coulomb potential between electrostatic charges drops off slowly with distance, namely, as r^{-1}. Attractive and repulsive interactions must be considered together to produce well-behaved integrals and, even then, the sums or integrals converge slowly unless special "tricks" (like the Ewald sum for Madelung potentials) are employed. (2) Because electrons require quantum treatment (they have small mass and long de Broglie wavelengths compared to interatomic distances) and are indistinguishable, appropriate permutational symmetry (interchanging electrons 1 and 2 or i and j does not change the wave function) must be imposed on the solutions to the Schrödinger equation. This leads to fairly cumbersome equations. (3) The solution to the Schrödinger equation gives information about all energy levels – ground and excited states, full, partially filled, and empty bands. An accurate and detailed delineation of all these wave functions is inherently complex. (4) The n-electron "problem" is complex and *mean field approximations* require substantial correction terms. This means that each electron is influenced not just by the average distribution of all other electrons around it (the mean field), but also by the actual (fluctuating) potential generated by other electrons near it as a function of time. That is, electronic motion is correlated, and electrons avoid each other. (5) The quantities of chemical interest (e.g., energies of different structures) are small differences between large total energies, so very high accuracy is demanded.

These sources of difficulty (taken from a very readable description of ab initio calculations by Simons (1991)) apply equally to calculations for solids and for molecules. In band structure calculations for solids (as developed initially by the solid-state physics community but now used widely), the symmetry imposed on the wave functions (Eqs. 5.5–5.7) is built into the problem from the beginning, and one seeks to solve the Schrödinger equation to obtain wave functions, energies, and an electronic density of states consistent with crystal symmetry and the level of approximation used for the potentials and wave functions.

Conceptually, these approaches generally rest on several assumptions. Almost all of them use the "one-particle" approximation, namely, that the wave function for the system can be approximated as a product of one-electron wave functions, taken in properly linear combinations to reflect the fact that electrons are indistinguishable and taken to reflect crystal symmetry through Bloch's theorem (Eqs. 5.5–5.7). One of the simplest band structure methods defines the eigenstates (wave functions) of the crystal as a linear combination of plane waves. The number of such functions needed is very large because of the need to describe the rapid and complex oscillations of the wave functions near the atomic cores (the nuclei and the inner electrons). The number of plane waves can be drastically reduced if they are chosen such that those for the core states and valence states are orthogonal; this is called the orthogonal plane wave (OPW) approach (Harrison 1980). The OPW approach leads in turn to the concept of a *pseudopotential* (Harrison 1980), in which the potential $V(r)$ in the Schrödinger equation is broken down into two terms, an effective averaged term due to core states and a pseudopotential due to valence states, (Zunger 1981). One then proceeds to choose wave functions, calculate a potential, solve for wave functions using this new potential, and iterate the cycle until no significant further changes in potential occur. This procedure leads to a *self-consistent field* (SCF) solution.

An alternate approach is to choose a specific form for the potential, that is, one where it is spherical around the nuclei and essentially flat in the region of low electron density between atomic centers. Crystal symmetry is thus imposed by this "muffin tin" potential. The wave function is then chosen to be a combination of a spherical part near the nuclei and plane waves between them, and the method is thus referred to as an augmented plane wave (APW) method (Harrison 1980). A related method, linear augmented plane waves

Table 5.1. *Relations among methods of quantum mechanical calculations*

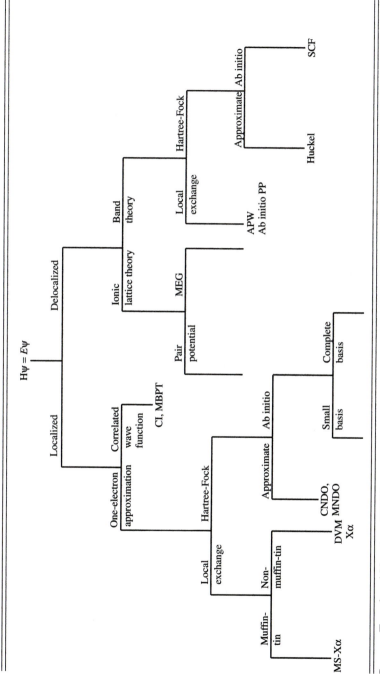

Source: Tossel and Vaughan (1992).

(LAPW), has been applied quite successfully to mineralogical systems (Cohen 1991; Isaak et al. 1993).

A family of approaches known as *density functional methods* (Callaway and March 1984; Schluter and Sham 1982) depends on the theorem that the ground state energy of a many-electron system is a unique function of its electron density distribution. This approach allows the many-electron problem to be simplified to a set of single-electron Schrödinger-like equations, where an effective potential gathers together a number of interactions. The equations are then solved by a set of iterations until self-consistency is reached. A starting point for defining the function is to use results for a uniform electron gas (EG). Refinements of this approach include modified electron gas (MEG) theories that allow for relaxation of charge densities when atoms come into contact, changing *rigid ion models* into *shell models*, where the relaxation of a *shell* of valence charge density occurs around a *core* of inner electron density and, in a physical sense, corresponds to polarization. These methods have been applied with considerable success to oxides and silicates (Cohen, Boyer, and Mehl 1987b; Post and Burnham 1986). A variation on the same theme is the potential-induced-breathing (PIB) model (Cohen 1987, 1991; Cohen et al. 1987a; Cohen and Krakauer 1990; Isaak et al. 1993).

A specific older method, which links density functional theory and the molecular orbital approach, is the $X\alpha$ method. It is basically a density functional theory that reflects electron correlation (Hess, McCarthy, and McMillan 1993). The relations among these methods and those for localized "molecular" calculations are shown in Table 5.1.

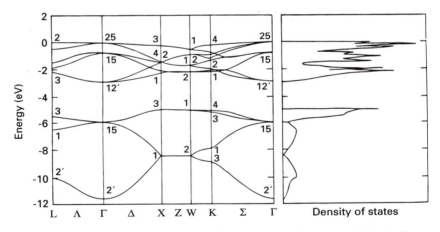

Figure 5.7. Band structure and electronic density of statics for β-cristobalite (1 eV = 96 kJ/mol) (from Harrison 1980).

Figure 5.7 shows a bond structure calculation for β-cristobalite (Harrison 1980) and the corresponding density of states of the filled bonds. A large bond gap (not shown) separates these from an empty bond, making cristobalite an insulator. The numbers and letters on the bond structure diagram refer to a standard symmetry notation. In each horizontal segment of the bond structure diagram, dispersion (variation of energy along a given direction is the Brillouin zone) is shown.

5.2.4 Some applications of band structure arguments

Figure 5.8 shows some striking deviations from linearity in lattice parameters, c/a ratio, and molar volume for hexagonal cadmium–magnesium alloys (Pearson 1972). For ideal hexagonal close packing, c/a is 1.63, so there is an increasing degree of distortion as magnesium is added to the system. The role of this

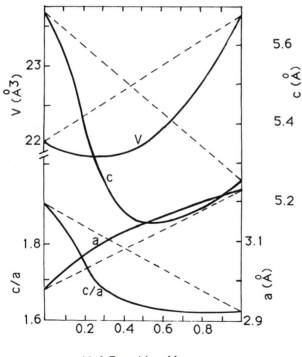

Mol Fraction Mg

Figure 5.8. Lattice parameters, c/a ratios, and molar volumes in Mg-Cd alloys (data from Pearson 1972).

Table 5.2. *Crystallographic sites in tetrahedrites*

Site	Approximate symmetry	Occupancy
12d	Tetrahedral	Cu, Fe, Zn, Hg, Cd
12e	Trigonal	Sb, As, Bi, Te, (Cu, Fe, Zn)
8c	Triangular pyramid	Sb, As, Bi, Te
24g	Tetrahedral	S
2a	Octahedral	S
24g	3-coordinate	Cu, Ag, vacancies

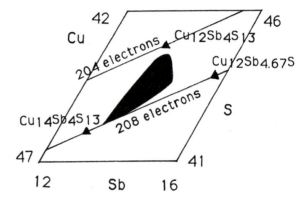

Figure 5.9. Tetrahedrite stability field (in black) from synthesis data. Horizontal axis represents mole % Sb. Oblique axis represents mole % Cu. This is a small part of the Cu–Sb–S ternary system. Lines show electron counts of 204 and 208; note that existing tetrahedrites fall within these limits. (After Johnson and Jeanloz 1983.)

distortion may be to lower the symmetry such that the valence band is split into two bands, and with the decreasing number of electrons as Mg substitutes for Cd, the remaining electrons can fill only the lower lying band to energetic advantage.

Analogous factors may provide the driving force for the existence, stability, and cation ordering in complex sulfide minerals. The tetrahedrites (Johnson and Jeanloz 1983) form a family of natural and synthetic complex sulfosalts of ideal formula $A_{10}B_2C_4D_{13}$, with A a nominally monovalent metal, Cu or Ag; B a nominally divalent metal, generally Cu, Fe, Zn, Hg, Cd, or Pb; C a large ion, Sb, As, Bi, or Te; and D generally sulfur (see Table 5.2). Synthetic materials in the Cu–Sb–S system show more complex stoichiometries, but the sum of cation charges must add to 26. Figure 5.9 shows the occurrence of tetra-

Table 5.3. *Contribution of various atoms to conduction electron count in tetrahedrites*

Element	Electrons participating in conduction band[a]	
	Number	Type
Cu^+, Cu^{2+}	1	$4s$
Ag^+	1	$5s$
Fe^{2+}	2	$4s$
Zn^{2+}	2	$4s$
Hg^{2+}	2	$6s$
Cd^{2+}	2	$5s$
Pb^{2+}	2	$6p$
As^{3+}	3	$4p$
Bi^{3+}	3	$5p$
Te^{4+}	4	$4p$
S^{2-}	6	$3p$, $4s$

[a] This assumes that structure is metallic and transition metal elements contribute $4s$ (not $3d$) electrons.

hedrites in the Cu–Sb–S system. The limits of composition in which tetrahedrites exist can not be rationalized in terms of stoichiometry, size factors, or the complete filling of certain sites. However, a consideration of the band structure lends insight. The space group I43m dictates that each Brillouin zone contain four electrons when full; thus filling the fifty-first zone involves 204 electrons total, filling the fifty-second involves 208. The stabilization as the fifty-second zone is filled is followed by a sharp rise in energy representing the band gap to the fifty-third zone. This picture would predict that tetrahedrites having a total near 208 electrons should be the most stable, and those with 209 or more should be very uncommon.

Table 5.3 lists the number of conduction electrons each element contributes. A tetrahedrite composition can then be converted into a count of valence electrons. Figure 5.10 shows the frequency of occurrence of tetrahedrites as a function of valence electron count calculated from the chemical analyses. The prediction about the stability of tetrahedrites is well substantiated. Further evidence for the band-filling mechanism is that compounds with 204 to 205 electrons have very low resistances and appear metallic, whereas a compound with an electron count of 207.8 has a much higher resistance and is probably a semiconductor. Since a filled band occurs at 208 electrons, the composition with an electron count of 207.8 probably corresponds to a filled band within the uncertainties of the microprobe analysis.

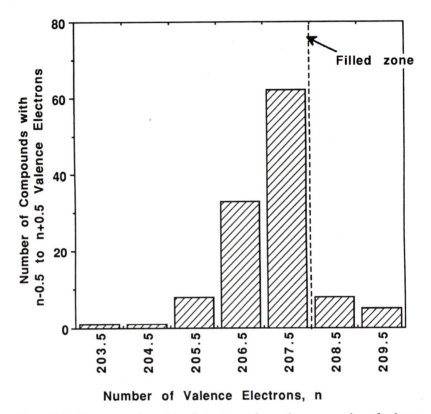

Figure 5.10. Histogram of number of structures observed versus number of valence electrons (after Johnson and Jeanloz 1983).

The preceding arguments are essentially qualitative. The point to remember is that very simple band theory arguments based on symmetry and electron counting can predict the region of stability of structures that are fiendishly complex, both crystallographically and compositionally.

Quantitative band structure calculations have of course been very successful for metals and semiconductors in solid-state physics. Applications to more "ionic" solids like MgO, $MgSiO_3$, and SiO_2 are newer, exemplified by work done by Mehl, Cohen, and Krakauer (1988) on MgO and CaO. Such calculations not only produce the electronic density of states but also provide insights into chemical bonding by predicting the spatial distribution of electron density. An example is a recent study of FeO (Issak et al. 1993), in which the driving

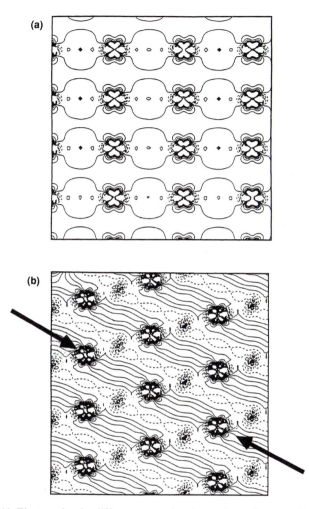

Figure 5.11. Electron density difference map showing regions where bonding increases in LAPW-calculated electron densities relative to spherical ions for FeO. (a) Undistorted cubic NaCl structure. (b) Structure strained (rhombohedrally) along (111). The latter increases Fe–Fe bonding and may be related to a transition to a metallic low-spin phase. The solid wavy curves show regions of increased bonding. The direction of increased bonding is shown by arrows. (From Isaak et al. 1993.)

force for rhombohedral distortion at high pressure can clearly be seen in terms of strong Fe–Fe interactions (see Fig. 5.11). Another example is an LAPW calculation of the electron density distribution of stishovite (see below and Fig. 5.38).

5.2.5 Phonons and lattice vibrations

The constraints imposed by periodicity and the convenience of reciprocal space can be used to advantage in analyzing atomic vibrations in a crystal. In general, the $3n$ (n = number of atoms per unit cell) vibrational degrees of freedom (also see Chap. 6) are assembled in a spectrum of vibrational energies called the vibrational density of states (VDOS), $g(E)$. Just as the allowed electronic energy levels in a molecule are broadened into electronic bands in a solid, so the discrete molecular vibrational frequencies are broadened into vibrational bands. The energy (and frequency) of a given vibration depends on the wave vector, $\mathbf{k}$, in reciprocal space. This dependence is known as a *dispersion relation*. The VDOS at $\mathbf{k} = 0$ is sampled by vibrational (infrared and Raman) spectroscopy, that at all values of $\mathbf{k}$ by inelastic neutron scattering (see Chap. 3). The lattice waves moving through a crystal can be described as individual, in a sense, particle-like, excitations known as *phonons*. These are characterized by their energy, the symmetry of the motion, and where in the (first) Brillouin zone they arise.

A set of dispersion curves for several solids is shown in Figure 5.12. Figure 5.12a presents an analysis of the linear chain. Figure 5.12b is schematic for the three-dimensional solid, emphasizing that there are three low-frequency acoustic modes that have zero frequency at the zone center ($\mathbf{k} = 0$) and an approximately sine wave dispersion, a spaghetti tangle of optical branches at intermediate energy that show complex dispersion, and, often, several high-frequency modes corresponding to localized vibrations that are not strongly dispersed. Figure 5.12c shows the real case for quartz. These will be discussed further in connection with heat capacities in Chapter 6.

A vibrating solid may be thought of as a coupled set of springs, with the spring constants given fundamentally by quantum mechanics. The description of these vibrations in terms of models of varying rigor is called *lattice dynamics*, and Born and Huang (1954) and Blakemore (1985) give good summaries. The essential difficulties are of two sorts: the proper choice of interatomic potentials and their range and directionality of interaction and the correct solution of the dynamical problem once these interactions are chosen. Some examples of lattice dynamical calculations for oxides and silicates are given in Table 5.4. Modern computational capability is revolutionizing this field. Once a lattice dynamical calculation is made, it can be used to calculate infrared and Raman spectra, the VDOS, elastic constants, sound speeds, and other observables. To maximize agreement between calculated and measured vibrational frequencies, elastic constants, or other parameters, the interatomic potentials

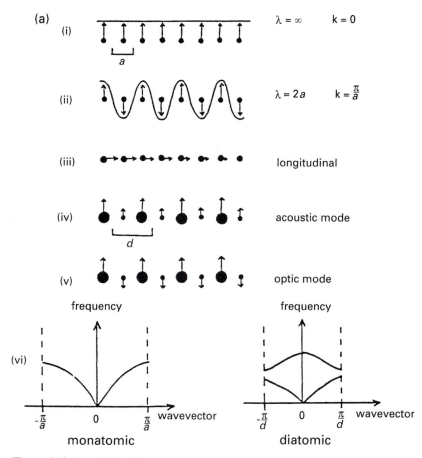

Figure 5.12. (*pp. 195–197*) (a) Origin of acoustic and optic modes and dispersion curves. Linear array of atoms: (i, ii, iii) monatomic chain sharing two transverse and one longitudinal mode; (iv, v) acoustic and optic mode for diatomic chain; (vi) frequency versus wave vector (dispersion curve) for monatomic chain and for diatomic chain. Note that optic mode has $\nu \neq 0$ at $\mathbf{k} = 0$ (after McMillan and Hofmeister 1988). (b) Phonon dispersion relations in Mg_2SiO_4 forsterite. Curves are calculated using rigid "molecular ion" model; points represent measurements by inelastic neutron scattering (from Ghose et al. 1987). (c) Phonon density of states corresponding to (b) (from Rao et al. 1987).

(b)

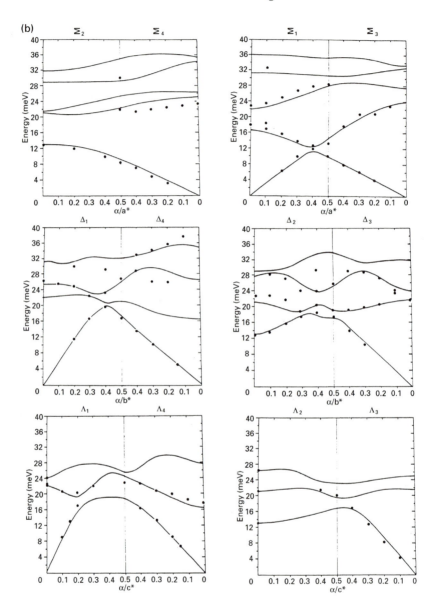

(c)

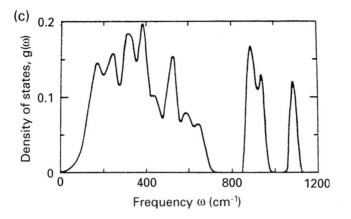

Table 5.4. *Lattice dynamical approaches*

Type of calculation	Material studied	Reference
Mg^{2+} and rigid SiO_4^{4-} groups	Mg_2SiO_4 (forsterite)	Ghose et al. (1987)
Various interatomic force models	SiO_2 (quartz)	Barron, Huang, and Pasternak (1976)
Valence force field from ab initio MO calculations	SiO_2 (quartz)	McMillan and Hess (1990)
LAPW calculations	Ferroelectricity in $BaTiO_3$ (perovskite)	Cohen and Krakauer (1990)
PIB model	$MgSiO_3$ (perovskite)	Cohen (1987)
PIB model	Alkaline earth oxides	Cohen, Boyer, and Mehl (1987a)
LAPW calculations	SiO_2 (stishovite)	Cohen (1991)
Polarizable ion model based on experimental data	$\alpha-\beta$ quartz transition	Iishi (1978b)
Rigid ion model based on experimental data	Mg_2SiO_4 (forsterite)	Iishi (1978c)
Large short-range forces in rigid or polarizable ion model, fit to experimental data	Al_2O_3 (corundum)	Iishi (1978a)

can be adjusted and the models "fine-tuned." Thus lattice dynamics can be approached as an empirical exercise, with potential parameters chosen to fit experimental data, or as a calculation based on interatomic potentials obtained from ab initio quantum mechanical computations.

Note the similarity in many respects of the phonon dispersion curves (Fig. 5.12) and of the electronic band structure diagram (Figs. 5.2, 5.4). In both cases, interactions in the condensed state cause molecular energy levels (vibrational or electronic) to be broadened into a dispersed curve or band of energies that depend on position in the reciprocal lattice unit cell. In both cases, the total number of states is fixed by the number of particles and the symmetry of the structure plays a major role in determining the distribution of states. Both electronic and vibrational energy levels are occupied with populations determined by available thermal energy. In both lattice dynamics and band theory, an appropriate choice of model is crucial.

5.3 Bonding in molecular clusters – molecular quantum chemistry

5.3.1 Rationale for a molecular approach

It is striking that local coordinations of cations in solids are often similar to those in aqueous solution, in melts, and in the glassy state. Furthermore, heats of fusion and vitrification (see Chap. 8) are only a small fraction of the total energy of the solid. Thus much of the bonding in a solid must depend largely on short-range interactions and be insensitive to specific long-range order. It therefore seems appropriate to take molecular-sized clusters as a starting point for understanding these interactions. Quantum chemists have developed the art of solving the Schrödinger equation for a cluster of atoms to a high degree of rigor, and programs exist to run on computers from a small workstation to a supercomputer; thus the software is accessible to the general user as well as the sophisticated expert. The molecular approach has been particularly useful for minerals, because it focuses naturally on the directional and covalent aspects of silicate and aluminosilicate frameworks. This directionality probably is what causes silicate tetrahedra to behave as rigid units (little deformation of T–O bond lengths and O–T–O angles within a tetrahedron) held together by loose hinges (easy changes in angles between tetrahedra (T–O–T angles) and in rotations of successive tetrahedra in a chain). This part of the chapter summarizes the formalism of molecular quantum chemistry and provides some major mineralogical applications of the molecular orbital approach to bonding in silicates.

←Rrepesentative→ ←Transition→ ←Representative→

s^1	s^2	d^1	d^2	d^3	d^4	d^5	d^6	d^7	d^8	d^9	d^{10}	p^1	p^2	p^3	p^4	p^5	p^6
H 1 1.00797																H 1 1.00797	He 2 4.0026
Li 3 6.939	Be 4 9.0122											B 5 10.811	C 6 12.01115	N 7 14.0067	O 8 15.9994	F 9 18.9984	Ne 10 20.123
Na 11 22.9898	Mg 12 24.312											Al 13 26.9815	Si 14 28.086	P 15 30.9738	S 16 32.064	Cl 17 35.453	Ar 18 39.948
K 19 39.102	Ca 20 40.08	Sc 21 44.956	Ti 22 47.90	V 23 50.942	Cr 24 51.996	Mn 25 54.9381	Fe 26 55.847	Co 27 58.9332	Ni 28 58.71	Cu 29 63.54	Zn 30 65.37	Ga 31 69.72	Ge 32 72.59	As 33 74.9216	Se 34 78.96	Br 35 79.909	Kr 36 83.80
Rb 37 85.47	Sr 38 87.62	Y 39 88.905	Zr 40 91.22	Nb 41 92.906	Mo 42 95.94	Tc 43 (99)	Ru 44 101.07	Rh 45 102.905	Pd 46 106.4	Ag 47 107.870	Cd 48 112.40	In 49 114.82	Sn 50 118.69	Sb 51 121.75	Te 52 127.60	I 53 126.9044	Xe 54 131.30
Cs 55 132.905	Ba 56 137.34	La* 57 138.91	Hf 72 178.49	Ta 73 180.948	W 74 183.85	Re 75 186.2	Os 76 190.2	Ir 77 192.2	Pt 78 195.09	Au 79 196.967	Hg 80 200.59	Tl 81 204.37	Pb 82 207.19	Bi 83 208.980	Po 84 (210)	At 85 (210)	Rn 86 (222)
Fr 87 (223)	Ra 88 (226)	Ac** 89 (227)															

*Lanthanides	Ce 58 140.12	Pr 59 140.907	Nd 60 144.24	Pm 61 (145)	Sm 62 150.35	Eu 63 151.96	Gd 64 157.25	Tb 65 158.924	Dy 66 162.50	Ho 67 164.930	Er 68 167.26	Tm 69 168.934	Yb 70 173.04	Lu 71 174.97
**Actinides	Th 90 232.038	Pa 91 (231)	U 92 238.03	Np 93 (237)	Pu 94 (242)	Am 95 (243)	Cm 96 (247)	Bk 97 (249)	Cf 98 (251)	Es 99 (254)	Fm 100 (253)	Md 101 (256)	No 102 (253)	Lw 103 (257)

Figure 5.13. The periodic table.

5.3.2 Review of electronic states in atoms

Solution of the Schrödinger equation for atoms results in the need for quantum numbers to describe atomic energy levels. The first, or principal, quantum number is n. This number refers to the filling of major shells (i.e., $n = 1$ is the K shell, $n = 2$ the L shell, etc.), is responsible for the major periodicity leading to the periodic table (see Fig. 5.13), is related to the most probable distance of the electron from the nucleus and to the size of atoms and ions, and determines the generation of X rays. Each period of the periodic table corresponds to a value of n and starts with an alkali metal with one electron in the outermost shell and ends with a noble gas with a completely filled outer shell (see Table 5.5).

The second quantum number, l, is the azimuthal or directional quantum number, related to the shape of orbitals. It can have values ranging from 0 to $n - 1$. Thus the first period (H to He) contains only electrons with $l = 0$ in spherically symmetric orbitals, the second period (Li to Ne) contains s and p electrons, the third period (Na to Ar) contains s, p, and d electrons, and the first transition series and higher rows contain s, p, d, and f electrons.

The third, or magnetic, quantum number, m, can have values from $-l$ to $+l$, giving $2l + 1$ orbitals. It thus determines that s orbitals come singly, p orbitals in sets of three, d orbitals in sets of five, and f orbitals in sets of ten. Orbitals

Table 5.5. *Atomic configurations and the buildup of the first three periods of the periodic table*

Atomic number	Element	Electronic configuration of atom	Common oxidation states	Characteristics
1	Hydrogen	$1s^1$	+1 (−1)	
2	Helium	$1s^2$	—	Noble gas, end of first period
3	Lithium	$1s^2 2s^1$	+1	Alkali
4	Beryllium	$1s^2 2s^2$	+2	Alkaline earth
5	Boron	$1s^2 2s^2 2p^1$	+3	
6	Carbon	$1s^2 2ps^2 2p^2$	+4	
7	Nitrogen	$1s^2 2s^2 2p^3$	+5, −3	
8	Oxygen	$1s^2 2s^2 2p^4$	−2	Chalcogen
9	Fluorine	$1s^2 2s^2 2p^5$	−1	Halogen
10	Neon	$1s^2 2s^2 2p^6$	—	Noble gas, end of first period
11	Sodium	$1s^2 2s^2 2p^6 3s^1$	+1	Alkali
12	Magnesium	$1s^2 2s^2 2p^6 3s^2$	+2	Alkaline earth
13	Aluminum	$1s^2 2s^2 2p^6 3s^2 3p^1$	+3	
14	Silicon	$1s^2 2s^2 2p^6 3s^2 3p^2$	+4	
15	Phosphorous	$1s^2 2s^2 2p^6 3s^2 3p^3$	+5, −3	
16	Sulfur	$1s^2 2s^2 2p^6 3s^2 3p^4$	+6, −2	Chalcogen
17	Chlorine	$1s^2 2s^2 2p^6 3s^2 3p^5$	−1	Halogen
18	Argon	$1s^2 2s^2 2p^6 3s^2 3p^6$	—	Noble gas, end of second period
19	Potassium	$1s^2 2s^2 2p^6 3s^2 3p^6 4s^1$	+1	Alkali
20	Calcium	$1s^2 2s^2 2p^6 3s^2 3p^6 4s^2$	+2	Alkaline earth
21–30	Scandium–Zinc	$1s^2 2s^2 2p^6 3s^2 3p^6 4s^2 3d^n$	variable	First transition series
21	Gallium	$1s^2 2s^2 2p^6 3s^2 3p^6 4s^2 3d^{10} 4p^1$	+3	
22	Germanium	$1s^2 2s^2 2p^6 3s^2 3p^6 4s^2 3d^{10} 4p^2$	+4	
23	Arsenic	$1s^2 2s^2 2p^6 3s^2 3p^6 4s^2 3d^{10} 4p^3$	+3, −5	
24	Selenium	$1s^2 2s^2 2p^6 3s^2 3p^6 4s^2 3d^{10} 4p^4$	+6, −2	Chalcogen
25	Bromine	$1s^2 2s^2 2p^6 3s^2 3p^6 4s^2 3d^{10} 4p^5$	−1	Halogen
26	Krypton	$1s^2 2s^2 2p^6 3s^2 3p^6 4s^2 3d^{10} 4p^6$	—	Noble gas, end of third period

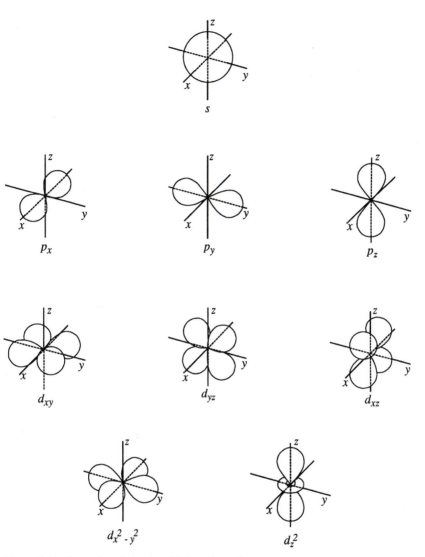

Figure 5.14. Geometry of atomic orbitals, schematic.

with different values of m are split in applied electric and magnetic fields. Thus their energies are related to "ligand field effects" (see Sec. 5.3.5) and to electron spin resonance and other spectroscopies (see Chap. 3). The geometry of s, p, and d orbitals is shown in Figure 5.14.

Finally, the spin quantum number, s, can have two values, 1/2 and $-1/2$.

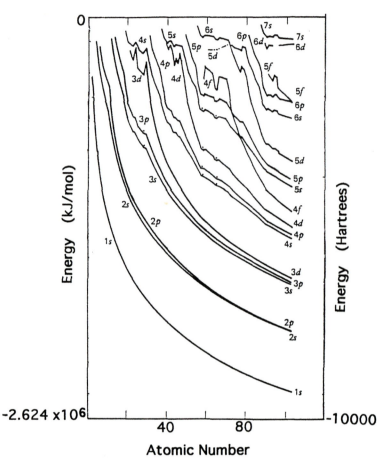

Figure 5.15. Electron energies as function of atomic number. Scale is logarithmic. (From Berry, Rice, and Ross 1980.)

Thus any orbital with the same n, l, and m values can at most contain two electrons, and these must be of opposite spin.

As the charge on the nucleus increases, inner electrons are held more tightly, and the energy of orbitals drops (relative to the unbound state as the zero of energy). This condition is shown in Figure 5.15. Note that some orbitals cross; thus the order in which they fill depends on where the element is in the periodic table. This filling scheme may be used to assign electronic configurations to atoms and ions (see Table 5.6). Elements with partially filled shells are prone to variable valence, with extra stability residing in filled, half-filled (all spins parallel), and empty shells.

Table 5.6. *Electronic configurations of some ions*

Ion	Number of electrons	Configuration
Y^{3+}	36	$Kr + 1s^2 2s^2 2p^6 3s^2 3p^6 3d^{10} 4s^2 4p^6$
La^{3+}	54	$Kr + 4d^{10} 5s^2 5p^6$
Nd^{3+}	57	$Kr + 4d^{10} 5s^2 5p^6 4f^3$
Ho^{3+}	64	$Kr + 4d^{10} 5s^2 5p^6 4f^{10}$
Ra^{2+}	86	$Kr + 4d^{10} 4f^{14} 5s^2 5p^6 5d^{10} 6s^2 6p^6$

5.3.3 Molecular orbitals

The interaction of atomic orbitals to form molecular orbitals for a symmetrical diatomic molecule (O_2) and an asymmetrical one (CO) is shown in Figure 5.16. In both cases, orbitals of s symmetry in the atoms interact to form a pair of molecular orbitals of s-like symmetry called *sigma* (σ), one lowered in energy and therefore bonding (σ), and one raised in energy and antibonding (σ^*). Atomic orbitals of p symmetry interact in a more complex fashion to form a group of molecular orbitals of both σ and π symmetry. The resulting molecular orbitals are filled with a total of sixteen electrons for O_2 and fourteen for CO; oxygen has two unpaired electrons in equivalent $2p$ π^* orbitals, whereas CO has all its electrons paired. The electron density distributions are shown schematically in Figure 5.17. In CO there is a shift of charge such that the carbon end of the molecule is slightly positive and the oxygen end slightly negative. In O_2 both ends of the molecule are identical. Even this simple case illustrates two features important to the solid state. First, the molecular orbitals begin to fan out into groups of orbitals similar in energy with larger spacings between the groups, presaging the beginning of band formation as the cluster grows larger. Second, unlike atoms carry unequal charges; this is the beginning of ionicity.

Quantitatively, one proceeds as follows. In the most straightforward approach, atomic wave functions are given simple algebraic expression. The simplest reasonable radial dependence for the wave function is a Slater-type orbital

$$\Psi = A \exp(-ar^2) \tag{5.14}$$

with A and a constants, $\mathbf{r}$ the distance from the nucleus.

But these are cumbersome in calculations. Gaussian functions

$$\Psi = A \exp(-a\mathbf{r}^2) \tag{5.15}$$

(a)

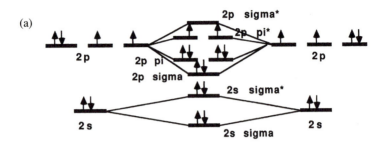

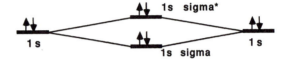

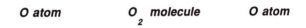

O atom   *O₂ molecule*   *O atom*

(b)

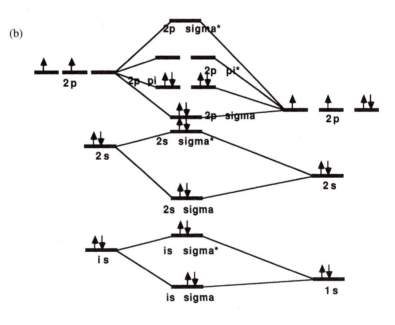

C atom   CO molecule   *O atom*

Figure 5.16. Molecular orbitals for (a) O_2 and (b) CO, schematic.

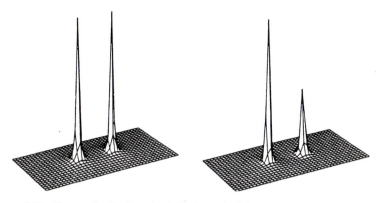

Figure 5.17. Charge distributions in (left) O_2 and (right) CO, schematic. In CO the electron density is greater near O than near C, making the molecule slightly polar. (After Berry et al. 1980.)

are more convenient because the product of two Gaussians is a Gaussian. A Slater orbital can be approximated by a sum of Gaussians, and this set of functions is taken as the basis set for calculations. The coefficients in the Gaussians have been optimized to fit atomic data and are available as part of public domain or commercial quantum chemistry packages. One then proceeds to construct molecular orbitals, for example, as linear combinations of atomic orbitals (LCAO)

$$\Psi_{mol} = \sum_i C_i \Psi_{i, \, atomic} \qquad (5.16)$$

where the sum is taken over all basis functions used. The variation theorem gives the total energy of the system as

$$E = \frac{\int \left(\sum_i C_i \Psi_i\right) \mathbf{H} \left(\sum_i C_i \Psi_i\right) d\tau}{\int \left(\sum_i C_i \Psi_i\right)^2 d\tau} \qquad (5.17)$$

where $\mathbf{H}$ is the quantum mechanical Hamiltonian (Eq. 5. 2) and the limits of integration, $d\tau$, include all available phase space (positions and momenta). Standard shorthand for the integrals is

$$\mathbf{H}_{ij} = \int \Psi_i \, \mathbf{H} \, \Psi_j \, d\tau = \int \Psi_j \, \mathbf{H} \, \Psi_i \, d\tau = \mathbf{H}_{ji} \qquad (5.18)$$

and

$$S_{ij} = \int \Psi_i \Psi_j \, d\tau = \int \Psi_j \Psi_i \, d\tau = S_{ji} \qquad (5.19)$$

As an example, suppose one wishes to construct a molecular orbital out of two atomic orbitals Ψ_1 and Ψ_2. The energy would then be

$$E = \frac{\int (C_1\Psi_1 + C_2\Psi_2) \, \mathbf{H} \, (C_1 + C_2\Psi_2) d\tau}{\int (C_1\Psi_1 + C_2\Psi_2)^2 \, d\tau} \qquad (5.20)$$

or

$$E = \qquad\qquad\qquad\qquad\qquad\qquad\qquad\qquad\qquad (5.21)$$
$$\frac{\int (C_1\Psi_1) \, \mathbf{H} \, (C_1\Psi_1) \, d\tau + 2\int (C_1\Psi_1) \, \mathbf{H} \, (C_1\Psi_2) \, d\tau + \int (C_2\Psi_2) \, \mathbf{H} \, (C_2\Psi_2) d\tau}{\int (C_1^2\Psi_1^2) \, d\tau + 2\int C_1 \, C_2 \, \Psi_1\Psi_2 \, d\tau + \int (C_2\Psi_2)^2 \, d\tau}$$

giving

$$E = \frac{H_{11} + 2H_{12} + H_{22}}{S_{11} + 2S_{12} + S_{22}} \qquad (5.22)$$

For i atomic orbitals, the corresponding expression is

$$E = \frac{\sum_i \sum_j H_{ij} \, C_i \, C_j}{\sum_i \sum_j S_{ij} \, C_i \, C_j} \qquad (5.23)$$

The problem then is to first compute the integrals H_{ij} and S_{ij} using the available basis set of atomic orbitals and, second, to minimize the energy by optimizing the coefficients C_i, C_j. The complexity of this calculation depends on the number of orbitals used (the total number of electrons in the system and the basis set for each orbital), the form of the orbitals, the form of the Hamiltonian and the number of interactions included in it, other constraints such as symmetry imposed by the positions of the nuclei, and, finally, the available computer power and cleverness in setting up the "number crunching." However, it is easy to see that the problem becomes far harder for heavier atoms and larger molecules.

In general, the crux of the calculation is the solution of a set of equations:

$$C_1 (H_{11} - S_{11}E) + C_2 (H_{21} - S_{21}E) + \ldots + C_i (H_{i1} + S_{i1}E) = 0$$
$$C_1 (H_{12} - S_{12}E) + C_2 (H_{22} - S_{22}E) + \ldots + C_i (H_{i2} + S_{i2}E) = 0 \quad (5.24)$$
$$\vdots$$
$$C_1 (H_{1i} - S_{1i}E) + C_2 (H_{2i} - S_{2i}E) + \ldots + C_i (H_{ij} + S_{ij}E) = 0$$

which can be done by setting up and solving a so-called secular determinant of the form

$$\begin{vmatrix} H_{11} - S_{11}E & H_{21} - S_{21}E \ldots H_{i1} - S_{i1}E \\ H_{12} - S_{12}E & H_{22} - S_{22}E \ldots H_{i2} - S_{i1}'E \\ \vdots \\ H_{1j} - S_{1j}E & H_{2j} - S_{2j}E \ldots H_{ij} - S_iyj_jE \end{vmatrix} = 0 \quad (5.25)$$

and of size given by the number of orbitals used. This can sometimes be done by direct matrix inversion, but often approximate methods must be introduced. The output is a set of molecular wave functions. Quantum mechanically prescribed operations on them yield not just the energy of the system but other physical observables, including bond lengths, bond angles, dipole moments, electronic and vibrational spectra, and polarizabilities.

The intrinsic difficulties encountered in Section 5.2.3 apply here also. In general, the problem can be attacked at various levels of approximation, both in terms of the form and number of wave functions used and in terms of the extent to which electronic correlations and other complicating interactions are included (Hess et al. 1993; Tossell and Vaughn 1992). The general form of the wave function for an n-electron system is written as a normalized antisymmetric product of one-electron wave functions, namely, the Slater determinant

$$\phi = \frac{1}{\sqrt{n}} \begin{vmatrix} \phi_1(1) & \phi_1(2) \ldots \phi_1(n) \\ \phi_2(1) & \phi_2(2) \ldots \phi_2(n) \\ \phi_n(1) & \phi_n(2) \ldots \phi_n(n) \end{vmatrix} \quad (5.26)$$

where the $\phi_i(i)$ terms represent electron orbitals, consisting of a spatial and a spin part, and the form of the matrix assures compliance with the Pauli exclusion principle and electron indistinguishability.

If the orbitals used are expressed as different types of functions in different parts of space, the basis set is *partitioned*; examples include those used in solid-state calculations described in Section 5.2.3. If the same functions are used throughout, the basis set is *nonpartitioned*, as is common in molecular calculations. Table 5.7 summarizes some basis sets and their nomenclature.

One then needs to write the appropriate Hamiltonian, and minimize the energy in the Schrödinger equation. Usually this is done in a series of iterations to yield a self-consistent field (SCF) solution. The two leading terms in the

Table 5.7. *A glossary of quantum mechanical and related methodology*

Method (acronym)	Brief description
Ab initio (or ab initio SCF)	A general term used to refer to calculations "from first principles," often describing Hartree-Fock calculations with no integral approximations and with self-consistent charge distributions.
Ab initio pseudopotential	Band-theoretical method replacing core electrons by pseudopotentials, using density-functional electron exchange and correlation, a plane-wave basis, and a self-consistent charge distribution.
Angular overlap model (AOM)	An approximate (non-SCF) molecular-orbital method involving extensive parameterization of integrals and with the overlap integral represented as a simple product of radial and angular terms.
Augmented plane wave (APW)	A band-theoretical method employing density-functional theory, a composite basis set, and a muffin-tin potential.
Band theory	A general term to describe a calculation in which the potential and the wave function have translational symmetry and the entire periodic solid is considered.
Basis set	The set of mathematical functions used to expand the molecular orbitals in a Hartree-Fock-Roothaan calculation.
Configuration interaction (CI)	A high-level calculation based on Hartree-Fock theory in which the HF determinants are mixed so as to describe the dynamic correlation of electrons.
Complete neglect of differential overlap (CNDO)	An approximate or semiempirical Hartree-Fock molecular-orbital method, utilizing approximate electron repulsion integrals and some Hamiltonian matrix elements to solve approximate HF equations and iterate to self-consistency.
Delta self-consistent field (ΔSCF)	Calculation of ionization potentials as differences of total energies of neutral molecules and cations.
Density-functional theory (DF)	Evaluation of total energy by focusing on the total electron density. Typically the exchange and correlation energy is expressed as a reasonably simple function of the electron density.
Discrete variational (DV or DV-Xα)	A density-functional method for molecular systems with Slater orbital bases and no (or little) potential averaging.
Extended Hückel (molecular-orbital) theory (EHMO or EHT)	An approximate Hartree-Fock molecular-orbital method involving extensive parameterization (all Hamiltonian matrix elements parameterized to fit experimental data).

Table 5.7. *Continued*

Method (acronym)	Brief description
Empirical (or approximate) methods	Methods that approximate quantum-mechanical integrals and/or incorporate experimental data to simplify computation and improve agreement with experiment.
Full-potential linearized augmented plane wave (FLAPW)	An APW method in which the energy dependence of the basis functions is linearized and no potential averaging is used.
Hartree-Fock (theory) (HF) (restricted HF, RHF for closed shells; unrestricted HF, UHF for open shells)	A method in which a many-electron wave function is written as an antisymmetric product of one-electron orbitals. The instantaneous electron-electron repulsion is replaced by interaction with the time-averaged densities of the other electrons.
Hartree-Fock-Roothaan theory (HFR)	HF theory with expansion of the MOs in a finite basis set. The equation to be solved is now a matrix equation for the expansion coefficients.
Intermediate neglect of differential overlap (INDO)	An approximate or semiempirical Hartree-Fock molecular-orbital method in which electron repulsion integrals and some Hamiltonian matrix elements are approximated and the approximate Hartree-Fock equations solved and integrated to self-consistency.
Korringa-Kohn-Rostoker (KKR) method	A band-theoretical method employing density-functional theory, a composite basis set and a muffin-tin potential.
Linear band structure	Methods in which the energy dependence of atomic basis functions is approximated through linearization.
Linear combination of atomic orbitals (LCAO)	A term used to refer to a Hartree-Fock-Roothaan calculation, often applied to calculations with small bases.
LCAO-Xα	A density-functional molecular calculation with an analytic nuclear-centered basis set (typically Gaussian).
Local-density approximation (LDA) or functional (LDF)	DF schemes in which the exchange and correlation potentials is a function of the local electron density (or spin density).
Linear muffin-tin orbital (LMTO)	Density-functional band calculation with muffin-tin orbitals as basis with their energy dependence linearized.
Many-body perturbation theory (MBPT)	A method for incorporating correlation using perturbation theory with the difference of HF and exact electron repulsion as the perturbation.
Modified electron gas (MEG)	A method using model non-self-consistent charge

Table 5.7. *Continued*

Method (acronym)	Brief description
Modified neglect of differential overlap (MNDO)	An approximate or semiempirical Hartree-Fock molecular-orbital method in which some of the Hamiltonian matrix elements are parameterized to give agreement with experiment. AMI is a current version.
Molecular-orbital theory (MO theory)	Any method representing the many-electron wave function for a molecule as an antisymmetrized product of one-electron orbitals.
Minimum basis set (MBS)	Hartree-Fock-Roothaan calculation in which each occupied atomic orbital of an atom is represented by one mathematical function with variable coefficient in the molecular orbitals.
Pseudopotential (PP)	Replacement of the orthogonality relation of core and valence electrons by an effective potential applied to the valence shell.
Potential-induced breathing (PIB)	A modification of the modified electron-gas method that incorporates the effect of the environment (described by an array of point charges) on the properties of the anion.
Self-consistent field (SCF)	Any method in which the electron density is evaluated by an iterative process.
Tight-binding method	Band-theoretical method using an LCAO expansion of the delocalized Bloch oribtal (often with integral approximations).
$X\alpha$	Density-functional scheme employing Slater's local-density functional for exchange and correlation.

Source: Tossell and Vaughan (1992)

Hamiltonian represent Coulomb and exchange terms. The former represents kinetic energy, the latter potential energy in the core Hamiltonian, that of an electron interacting with the nucleus. That electron also interacts with all other electrons; if this can be expressed as a one-electron operator (i.e., involving the coordinates of only one electron), the calculation is said to be performed at the Hartree-Fock level. The one electron then is described in the average field created by all other electrons, and iteration makes this field self-consistent (SCF).

To improve on this average field approximation, configuration interaction

(CI) can be used. In it, the effect of electron correlation is better described by using a wave function that itself is the weighted sum of a set of Slater determinants. In principle CI can provide an exact solution; in practice approximations are frequently needed.

What level of calculation is "best"? There is no simple answer. The calculated values of different properties are variously sensitive to the choice of basis set, and it is not easy to ascertain how much improvement is obtained at what cost. This is an issue under active debate.

What physical observables can be calculated? By minimizing the energy as a function of interatomic distance and bond angles, optimum geometries can be found. The variation of energy with atomic coordinates creates a potential curve or surface. The energies of different structures can be compared. Likely reaction intermediates and reaction pathways can be identified. Surface structure and reactions can be modeled. Spectra – optical, vibrational, charge transfer – can be calculated. Elastic constants and compressibilities can be computed. The calculated parameters can be starting points for other models and simulations. The nomenclature for these various methods and related jargon is summarized in Table 5.7.

5.3.4 Molecular orbital approach to silicate frameworks

The SiO_4 group is the fundamental building block in silicates. To form a neutral molecule for computations, hydrogens are attached to the terminal oxygens of a cluster, with silicic acid, H_4SiO_4, as the prototypical molecule for quantum calculations (see Fig. 5.18). Optimizing the geometry of this molecule (using any of several basis sets of increasing complexity) results in an Si–O bond length of 1.63–1.65 Å and regular tetrahedral O–Si–O angles, in good agreement with experiment (see Fig. 5.19a). Similar calculations for silicon in octahedral coordination, using the molecule $Si(OH)_4 (H_2O)_2$, give an equilibrium bond length of 1.77 Å, again in good agreement with experiment (see Fig. 5.19b). Calculations for BO_4, AlO_4, PO_4, GeO_4, and SO_4 tetrahedra and for BO_3 and CO_3 triangles also give satisfactory agreement with observed bond lengths. Histograms of observed bond lengths (see Fig. 5.19) peak near the minimum in the calculated potential curve of energy versus bond length. The steepness of these potential curves is consistent with the Si–O bond being strong and rigid, and with its generally low compressibility and thermal expansivity (see Sec. 5.5.4). The potential curve can be used to calculate Si–O stretching frequencies, with generally good agreement with observation, and the analogous curve for energy versus O–Si–O angle constrains internal bend-

(a)

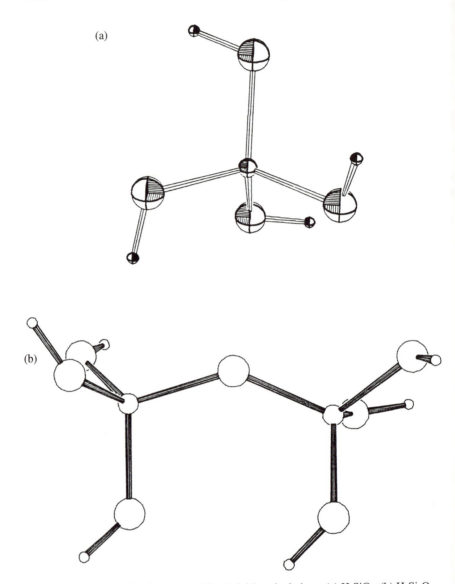

Figure 5.18. Molecular clusters used for ab initio calculations. (a) H_4SiO_4. (b) $H_6Si_2O_7$: large circles, O; medium circles, Si; small circles, H. (From Gibbs 1982.)

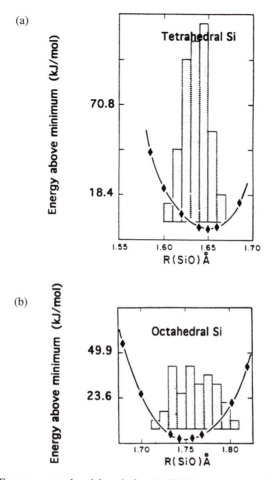

Figure 5.19. Energy versus bond length for (a) H_4SiO_4 (tetrahedral silicon) and (b) $H_4SiO_4–2H_2O$ (octahedral silicon). Minimum of potential curve is taken as zero of energy. Histogram shows observed bond lengths. (Modified from Gibbs 1982.)

ing modes. Thus it seems clear that short-range forces are adequate to describe many features of strong tetrahedral bonds in silicates.

The Si_2O_7 group can be considered the fundamental building block of silicate chains, rings, and frameworks. It is the smallest unit containing a bridging oxygen, that is, one oxygen common to two tetrahedra. The $H_6Si_2O_7$ molecule used for molecular orbital calculations is shown in Figure 5.18. The calculated bond lengths and angles show that the bridging Si–O bond (1.60 Å) is shorter than the nonbridging (1.65 Å), the tetrahedra show little internal distortion, and the Si–O–Si angle is near 143°. These results are reasonably robust under different basis sets. The potential energy surface, showing contours of equal energy relative to the most stable configuration, is displayed in Figure 5.20. It is striking that the energy needed to distort the Si–O–Si bond from its equilibrium value of 143° all the way to a linear bond (180°) is less than 10 kJ/mol, whereas, with decreasing angle, the energy only begins to rise steeply below 125° (see Fig. 5.21). This implies a very "soft" intertetrahedral angle, with a range of values between about 135° and 180° almost the same in energy, and is consistent with the large range of Si–O–Si angles seen in silicate crystals (see following) and the ease of glass formation (see Chap. 8). The bridging Si–O bond length, in contrast, increases steeply in energy when deformed. This also mirrors reality and gives theoretical backing to conceptual models that build silicates from rigid SiO_4 units with variable intertetrahedral angles.

Calculations have been performed on a variety of $H_6T_2O_7$ and $H_7T_2O_7$ molecules with T = Si, Al, P, Be, Mg, Be, C (see Table 5.8 and Fig. 5.22) (Geisinger, Gibbs, and Navrotsky 1985). All T–O bonds are rather rigid in terms of bond length variation but somewhat flexible with respect to T–O–T angles. This flexibility is greatest for Si, somewhat less for Al, and considerably less for Mg, B, Be, and P, with the optimum angle moving to lower values. This trend is accentuated because in reality the bridging oxygen in a T^{3+}–O–T^{4+} or T^{3+}–O–T^{3+} group is three coordinate, with a non-framework cation providing charge balance (see following).

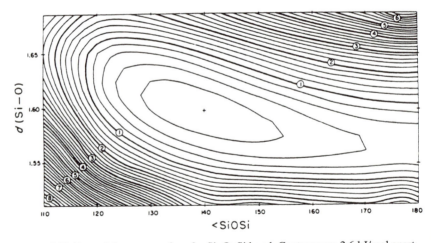

Figure 5.20. Potential energy surface for Si–O–Si bond. Contours are 2.6 kJ/mol apart. (From Gibbs 1982.)

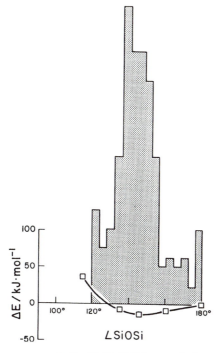

Figure 5.21. Energy versus angle for Si–O–Si linkage, relative to a 180° angle. Note the very shallow minimum and small barrier to linearity. Histogram shows observed angles. (From Gibbs et al. 1981.)

In $H_6T_2O_7$ molecules containing unlike T atoms (e.g., Si–O–Al vs. Si–O–Si and Al–O–Al), the calculations show a tendency for the weaker and longer bond to increase and the shorter and stronger bond to decrease slightly. This and other slight changes in geometry and electron distribution suggest that the Si–O–Al configuration is more favorable than the Si–O–Si and Al–O–Al linkages; that is, that the energy of the reaction

$$\text{Si–O–Si} + \text{Al–O–Al} = 2 \text{ Si–O–Al} \qquad (5.27)$$

is exothermic. This forms a theoretical basis for the Loewenstein aluminum avoidance principle, an empirical observation that, in general, Al–O–Al linkages tend to be avoided and Al–O–Si linkages preferred. However, the calculated energy is too exothermic to be consistent with Al–Si disorder observed in many minerals. Thus effects of the three-dimensional structure of the crystal, including short-range order, may play a stabilizing role for Al–O–Al linkages

Table 5.8. *Molecular orbital calculations on clusters to simulate bonding geometries in minerals*

| | | | Results | | | | | |
| | | | Calculated | | | Experimental | | |
Bonding geometry	Cluster	Level of calculation	Bridging T–O bond (Å)	Nonbridging T–O bond (Å)	T–O–T angle (°)	Bridging T–O bond (Å)	Nonbridging T–O bond (Å)	T–O–T angle (°)
Isolated SiO$_4$ tetrahedron	H$_4$SiO$_4$	HF-LCAO-SCF						
		STO-3G basis		1.655		1.650 in isolated molecules		
		6-31G basis		1.662				
		6-31G* basis	1.629			1.635 average in orthosilicates		
SiO$_6$ octahedron	H$_4$SiO$_4$ (H$_2$O)$_2$ (H$_8$SiO$_6$)	STO-3G		1.78			1.78	
Tetrahedral (T–O–T) Linkages		STO-3G						
Si–O	H$_6$Si$_2$O$_7$		1.591	1.658	144	1.61	1.65	145
Si–O	H$_6$SiAlO$_7^-$		1.575	1.671	139	1.60	1.65	138
Al–O	H$_6$SiAlO$_7^-$		1.695	1.719	139	1.70	1.72	138
Si–O	H$_6$SiBO$_7^-$		1.601	1.699	125	1.60	1.65	129
B–O	H$_6$SiBO$_7^-$		1.436	1.477	125	1.46	1.48	129

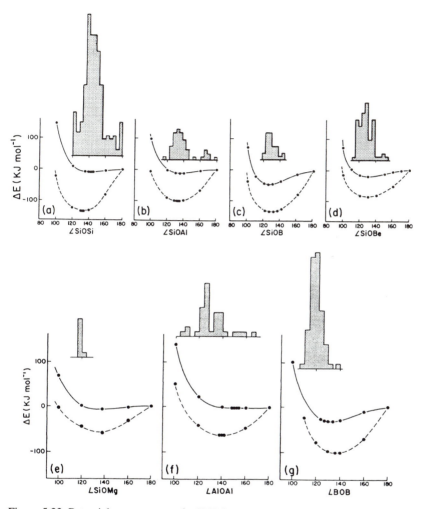

Figure 5.22. Potential energy curves for $H_6T_2O_7$ and $H_7T_2O_7$ molecules containing the T–O–T groups as a function of T–O–T angles. Each curve is referenced to $\Delta E = 0$ at 180° for ease of comparison. The solid and dashed curves were calculated for the appropriate $H_6T_2O_7$ and $H_7T_2O_7$ molecules, respectively. The histograms inserted above each pair of curves represent a frequency distribution of experimental T–O–T bridging angles observed in various solids. (From Geisinger, Gibbs, and Navrotsky 1985.)

not seen in isolated clusters. In addition, the calculated energies are sensitive to the charges on the clusters, and using four-membered cyclic clusters that keep the charges constant improves the calculated energetics substantially (Tossell, pers. comm., 1993). Aluminum avoidance can also be seen as a consequence of electrostatic interactions that favor the ordering of cations of unlike valence (Cohen and Burnham 1985).

To compensate for the negative charge on molecules of the type $(H_6T^{3+}T^{4+}O_7)^-$ and $(H_6T_2^{3+}O_7)^{2-}$, a proton may be attached to the bridging oxygen, making it three coordinate. This tends to steepen and deepen the minimum in the curve of energy versus T–O–T angle and to displace it to lower angles. Thus the barrier to linearity is higher and the range of energetically favored T–O–T angles smaller.

If the hydrogen on the bridging oxygen is replaced by another cation, the preceding effects are accentuated. Figure 5.23 shows the geometry of such a configuration, in which the approaching cation is itself coordinated by four or six oxygens. The calculated potential energy curves (energy versus T–O–T angle) are shown in Figure 5.24. The approaching cation perturbs the T–O–T linkage, sharply decreasing its angle, steepening and deepening the potential curve, and lengthening the T–O bonds slightly (with the weaker bond in an asymmetrical T–O–T linkage affected more). This effect becomes more pronounced in the series Na, Mg, Al, Si, that is, with increasing charge and closer distance of approach (Geisinger et al. 1985; Navrotsky et al. 1985). These observations agree well with crystallographic data. The thermodynamic consequences of this perturbation for glass formation are discussed in Chapter 8.

The silicon-sulfur system shows a marked contrast to the silicon-oxygen system (Geisinger and Gibbs 1981). Curves of energy versus Si–S bond length

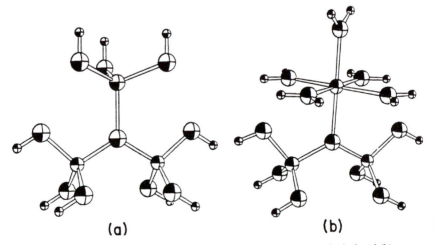

(a)   (b)

Figure 5.23. Geometry of a bridging oxygen approached by (a) tetrahedral and (b) octahedral species to form a three-coordinate arrangement. (From Geisinger et al. 1985.)

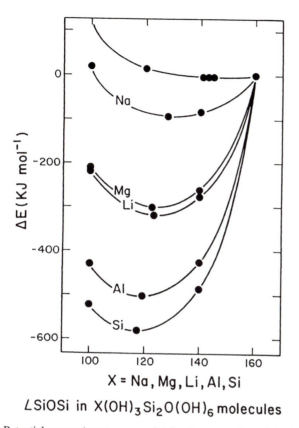

$X = Na, Mg, Li, Al, Si$

$\angle SiOSi$ in $X(OH)_3 Si_2 O(OH)_6$ molecules

Figure 5.24. Potential curves (energy vs. angle) for three-coordinate bridging oxygen. (From Geisinger et al. 1985.)

and Si–S–Si angle are shown in Figure 5.25. Clearly the bond length is much more flexible and the bond angle much more rigid than in the Si–O system. The small number of sulfosilicates that can be synthesized indeed show very different structures, and there is little or no glass formation.

The parameters calculated by molecular orbital theory are often combined with empirical concepts (ionic radii, Pauling bond strengths) to provide systematics. Together they can be used to correlate bond length and angle variations in aluminosilicates and other solids (Geisinger et al. 1985; Gibbs 1982).

A recent development has been the incorporation of SCF LCAO molecular orbital Hartree-Fock calculations in a computer program (CRYSTAL) that solves these equations under the constraint of crystal symmetry, the so-called periodic Hartree-Fock (PHF) calculations (Hess et al. 1993; Pisani, Dovesi,

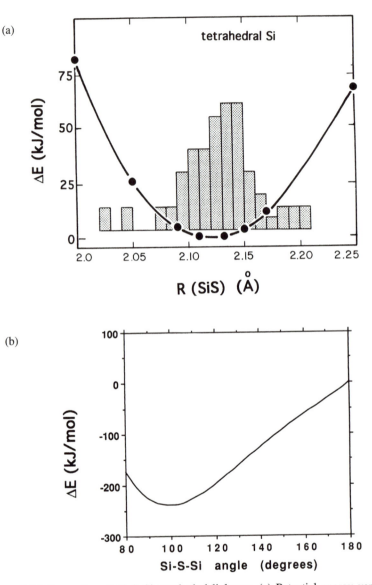

Figure 5.25. Energetics of Si–S–Si tetrahedral linkages. (a) Potential energy versus bond length. The histogram shows observed bond lengths in thiosilicates. (b) Energy versus Si–S–Si angle. Note the broader minimum in bond length but deeper minimum in angle, with a larger barrier to linearity than for Si–O–Si. (After Geisinger and Gibbs 1981.)

Table 5.9. *Oxidation states and electron configurations in transition metal and related ions*

Element	d^0	d^1	d^2	d^3	d^4	d^5	d^6	d^7	d^8	d^9	d^{10}
Calcium	Ca^{2+}										
Scandium	Sc^{3+}	(Sc^{2+})									
Titanium	Ti^{4+}	Ti^{3+}	Ti^{2+}								
Vanadium	V^{5+}	V^{4+}	V^{3+}	V^{2+}							
Chromium	Cr^{6+}	(Cr^{5+})	Cr^{4+}	Cr^{3+}	Cr^{2+}						
Manganese	Mn^{7+}	Mn^{6+}	(Mn^{5+})	Mn^{4+}	Mn^{3+}	Mn^{2+}					
Iron						Fe^{3+}	Fe^{2+}				
Cobalt						Co^{4+}	Co^{3+}	Co^{2+}			
Nickel							(Ni^{4+})	Ni^{3+}	Ni^{2+}		
Copper									(Cu^{3+})	Cu^{2+}	Cu^{+}
Zinc											Zn^{2+}
Cadmium											Cd^{2+}

and Roetti 1988). It enables calculations on systems containing relatively light atoms (Mg, Si, Al, O) but within the periodic constraints of a rather complex large unit cell, for example, in clay and zeolite structures. The method is interesting in that part of the iterative calculation is done in real space and part in reciprocal space, taking advantage of some of the strengths and simplifications of both molecular orbital and band theory approaches.

5.3.5 Bonding in transition metal and rare earth compounds

Transition metal compounds are often brightly colored. Consider the first row transition metal series, scandium through zinc. The elements and their common oxidation states are shown in Table 5.9. Note the wealth of oxidation states possible as the various d-electrons are removed. Figure 5.14 shows the spatial extent of the five d orbitals, labeled by convention d_{xy}, d_{xz}, d_{yz}, d_{z^2}, and $d_{x^2-y^2}$. The first three are referred to, from symmetry considerations, as t_{2g} orbitals and the last two as e_g. In a free atom or one surrounded by a spherically symmetric charge distribution, all five orbitals are equivalent in energy, and electrons are distributed among them according to Hund's rule to maximize spin. However, they will be different in energy in a magnetic field (Zemann splitting) or in an electric field of lower than spherical symmetry brought about by the approach of other ions in solution or in a crystal (ligand field or crystal field splitting). The latter is shown in Figure 5.26 for several symmetries. The extent of the splitting is often referred to as the crystal field splitting, Δ or 10

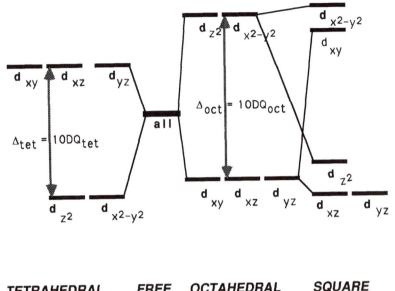

TETRAHEDRAL   **FREE**   **OCTAHEDRAL**   **SQUARE**
                    ION   **PLANAR**

Figure 5.26. Splitting of d orbital energies in fields of tetrahedral, octahedral, and square planar symmetry.

Dq. Rigorously, the term *crystal field* refers to an essentially ionic model, whereas the term *ligand field* includes covalency and sharing of electrons between cation and ligands. In common use, the terms tend to be interchangeable. The ligand field splitting is a measure of the strength of interaction between the cation and the approaching anions. It depends strongly on the bond distance; an r^{-5} dependence is a good first approximation. Thus the ligand field splitting parameter will depend strongly on factors that influence bond distances, including pressure, temperature, composition, and specific crystal structure. Δ also depends on the nature of the ligand. A series in order of increasing cation-ligand interaction (the nephelauxetic series) is shown in Table 5.10. From the point of view of mineral physics, it is noteworthy that water, oxide ion, hydroxide ion, and fluoride ion are of comparable strength, but sulfide is considerably stronger.

The splitting of d or f energy levels in a crystal has thermodynamic consequences. Consider the case of Co^{3+} or Fe^{2+}, with the d^6 configuration in an octahedral environment (see Fig. 5.27). The electrons can occupy the lower-

Table 5.10. *Ligand field splittings, Δ = 10 Dq (kJ/mol) for ions in octahedral coordination*

Ion	Δ pyroxene	Δ corundum	Δ periclase	Δ $M(H_2O)_6$ (aq)
Ti^{3+}	221	228	~135	227
V^{3+}	200	209	224	229
V^{2+}			151	
Cr^{3+}	185	217	194	208
Cr^{2+}	107			166
Mn^{3+}	177	232		251
Mn^{2+}	99		117	101
Fe^{3+}	138	183	160	164
Fe^{2+}	109	138	129	108
Co^{2+}	100		111	87
Ni^{2+}	101	128	102	101
Cu^{2+}	153			156

Note: Pyroxene is diopside (M1 site) when available, otherwise another pyroxene structure.
Source: Data from Burns (1993).

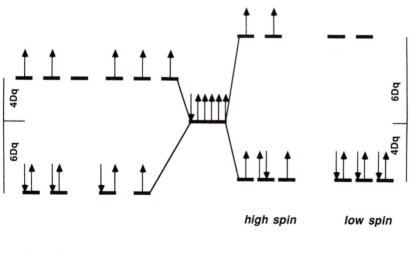

Figure 5.27. Occupancy of d orbitals for a d^6 ion in (e.g., Co^{3+}, Fe^{2+}) octahedral and tetrahedral environments for high-spin and low-spin configurations.

lying orbitals as shown, leading to a net stabilization of $2(0.6\Delta) - 4(0.4\Delta) = -0.4\Delta$. If the energy gained by putting another electron in a paired configuration in the lower-lying t_{2g} orbitals, more than compensates the spin-pairing energy (see Fig. 5.27), then the spin-paired configuration will be stable. Because the ligand field splitting, Δ, increases with charge on the cation, Co^{3+} at ambient conditions is in the spin-paired state, while Fe^{2+} is not. However, Co^{3+} can undergo a spin unpairing transition with increasing temperature (the unpaired configuration has higher energy and higher entropy). Because in a solid the d-orbitals are spread into a band, this transition occurs over a range of temperatures (see Fig. 5.28) (Mocala, Navrotsky, and Sherman 1992). In addition, the spin-paired state exhibits a shorter Co–O bond distance than the unpaired state. In Co_3O_4, the spin-unpairing transition is therefore accompanied by an anomalous increase of lattice parameter (see Fig. 5.28), though no change in symmetry of the cubic spinel occurs. The smaller volume of the low spin state implies that pressure may be a driving force for spin pairing in iron compounds in the Earth. The possible existence of low-spin Fe^{2+} in the lower mantle and its consequences are still being debated. For example, Isaak et al. (1993) used LAPW calculations for the rocksalt, distorted rocksalt (rhombohedral), and hypothetical cesium chloride forms of stoichiometric FeO at different unit cell volumes to simulate the effects of pressure. They predict a change from a high-spin magnetic state to a low-spin metallic state with increasing pressure (see Fig. 5.11) and consider their results in qualitative agreement with experimental observations of phase transitions in diamond cell experiments.

Because the ligand field splitting, Δ, is greater in sulfides than in oxides, one might expect more frequent occurrence of low-spin states in sulfides. This is indeed observed.

For high-spin tetrahedral coordination (as would apply to Fe^{2+}), the occupancy of orbitals is shown in Figure 5.27. The overall stabilization relative to the free ion is $3(0.4\Delta) - 3(0.6\Delta) = -0.6\Delta$. The difference in stabilization between octahedral and tetrahedral coordination is $-0.4\Delta - (-0.6\Delta) = 0.2\Delta$. This would imply that Fe^{2+} might slightly prefer tetrahedral coordination. For the low-spin tetrahedral case, as applied to Co^{3+} under ambient conditions (see Fig. 5.27), the difference from the free ion is $2(0.4\Delta) - 4(0.6\Delta) = -1.6\Delta$. The low-spin octahedral case is stabilized relative to the free ion by $6(-0.4\Delta) = -2.4\Delta$. The difference between octahedral and tetrahedral low-spin states is $-2.4\Delta + 1.6\Delta + E_{\text{spin pair}}$, since the low-spin octahedral case has three sets of spin-paired electrons, and the high-spin tetrahedral case has two. Thus one would predict that low-spin Co^{3+} would prefer octahedral coordina-

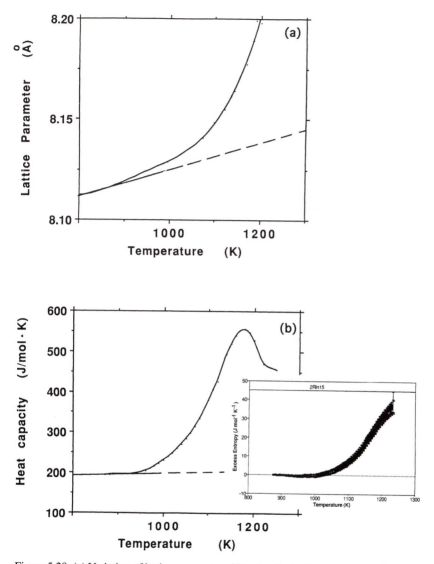

Figure 5.28. (a) Variation of lattice parameter of Co_3O_4 spinel with temperature. Dashed line shows extrapolation of data at 400–800 K (from Liu and Prewitt 1990). (b) Heat capacity of Co_3O_4. The peak indicates an excess heat capacity relative to the vibrational contribution shown as a dashed line. The insert shows the calculated excess entropy, with the value of $2R\ln15$ representing the configurational term arising from complete spin unpairing. The shaded area indicates the uncertainty in the calculated excess entropy. (Data from Mocala, Navrotsky, and Sherman 1992.)

Table 5.11. *Octahedral site preference energies (kJ/mol) in oxides*

Ion	From ligand field spectra	In spinels, from cations distributions	In binary oxides, from solid solubilities
Cr^{3+}	-157	-88	
Mn^{3+}	-95	-59	
Mn^{2+}	0	17	-11
Fe^{2+}	17	-4	
Co^{2+}	12	$+3$	-8
Ni^{2+}	-87	-50	-38
Cu^{2+}	-64	-38	

Source: Data from Navrotsky and Kleppa (1967).

tion, as is indeed observed. In reality, the values of Δ and the spin-pairing energy are somewhat different for octahedral and tetrahedral coordination (reflecting different bond lengths), so the argument becomes more complicated. In addition, for solids the spreading of orbitals into bands adds complexity.

This difference in crystal field stabilization energy (CFSE) between octahedral and tetrahedral coordination is often referred to as the octahedral site preference energy. Values obtained from spectroscopic considerations and from other experiments are listed in Table 5.11. This ligand field related site preference energy is certainly one of the factors determining crystal structures, coordination of ions in aqueous solution, glasses, and melts, and other structural trends. However, early enthusiasm for using it as the main predictor for structure and thermodynamics has waned. The ligand field term is but one of several factors in the lattice energy, and may not always be dominant. Nevertheless, it can safely be concluded that ions with strong crystal field preference for octahedral coordination (Co^{3+}, Cr^{3+}) rarely occupy tetrahedral coordination, and those for which crystal field effects favor tetrahedral coordination (Fe^{2+}, Co^{2+}) frequently do occur in tetrahedral sites. This point is discussed further in connection with spinel cation distributions (see Sec. 7.3).

Figure 5.29 shows the variation of ionic radii (derived from bond lengths as discussed in Sec. 5.4) for ions containing d and f electrons. The double-humped curve for transition metals reflects the effect of the unequal occupation of orbitals (Fig. 5.29a). Put simply, the ligand is drawn closer to the transi-

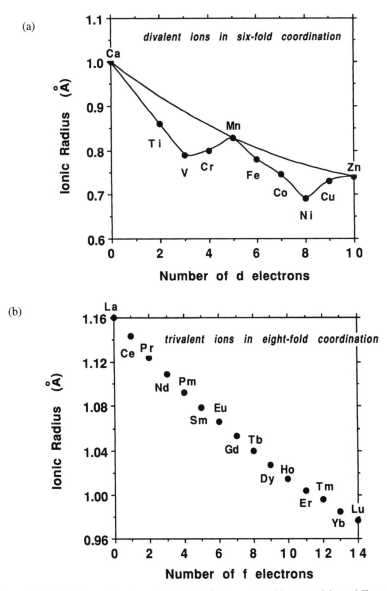

Figure 5.29. Variation of ionic radius for (a) first row transition metal ions, M^{2+}, and (b) lanthanides, Ln^{3+}, with number of d or f electrons.

tion metal when, for a given coordination, the d electrons can be concentrated in orbitals whose spatial extent minimizes the repulsion between the cation's and ligand's electron clouds. A similar but much smaller effect can be seen in the rare earths. Because such effects are already included in common tables of ionic radii and in calculations of lattice energies and other properties based on them, one must be careful to avoid, in effect, double counting these ligand field effects in further computations or considerations of systematics.

Further lowering of symmetry and splitting of energy levels can occur, commonly by distorting an octahedral coordination site. Figure 5.26 shows the splittings resulting in elongating an octahedron along one direction, taken as z, to produce square planar coordination. The latter is quite common in Cu^{2+}, and is related to the Jahn Teller effect (Cotton 1963). This square planer coordination is responsible for the characteristic turquoise blue color of many copper minerals and of Cu^{2+} in aqueous solution.

5.4 Ionic models of crystals

5.4.1 Introduction

The mineralogical community has found it extremely fruitful to view crystals as packings of spheres of relatively well defined sizes and charges. The idea of charge stems directly from that of chemical valence, giving a formal charge to an ion equal to its valence, with the tacit assumption of complete charge transfer to form cations and anions, that is, completely ionic bonding. The concept of radius comes from two considerations. The first is the empirical observation that the distance between a cation and its nearest neighbor anions in a series of different crystals containing the same cation and anion is nearly the same, defining a nearly constant bond length that can be resolved (though not uniquely) into the sum of cation and anion radii. The second is that in order to define an equilibrium distance between two atoms, whether in a diatomic molecule or a crystal, the electrostatic forces between point charges (cations and anions) must be balanced by repulsive forces stemming from the unfavorable interpenetration of electron clouds. This balance defines an interatomic potential similar in shape to curves shown in Figure 5.19, but based, in the ionic model, on more empirical parameterization. Thus one linkage between the ionic point of view and molecular quantum mechanics is through the forms and parameters in potentials used (see the following discussion).

What is amazing about the ionic model is how much mileage we get out of it in describing the systematics of crystal chemistry. The approach rationalizes crystal structures and their distortions, rationalizes which structures occur at what compositions, and systematizes the variation of crystallographic parameters with composition, temperature, and pressure. The semiempirical potentials are used to calculate the energies and equilibrium configurations of crystals and, through molecular dynamics, dynamic parameters as well. The strength of the ionic model lies in its simplicity, chemical intuition, and in the fact that its parameters can be readily compared for different atoms. We are beginning to see the convergence of the ionic model, molecular quantum mechanics, and electronic structure calculations in crystals. I consider it very important to remember that the ionic model, though naive in many ways, retains concepts and tools that our brains can grasp and play with, even when a thunderstorm knocks out the computer. To creatively meld these intuitive concepts with the power of modern computational techniques is a major challenge for the future.

5.4.2 Lattice energy

The energy of a crystal relative to isolated ions at infinite separation is called the *lattice energy*. It may be broken up, at least conceptually, into a number of contributions. The energy consists of a static contribution and one resulting from atomic vibrations that, even at absolute zero, contribute in zero point energy:

$$E_{\text{lattice}} = E_{\text{static}} + E_{\text{vibrational}} \qquad (5.28)$$

The static contribution may be viewed as coming from several sources:

$$E_{\text{static}} = E_{\text{electrostatic}} + E_{\text{repulsion}} + E_{\text{polarization}} + E_{\text{crystal field}} + E_{\text{other}} \qquad (5.29)$$

At equilibrium (rigorously at a low temperature where entropy effects are negligible) the energy is a minimum with respect to any variation with interatomic distance, r:

$$dE_{\text{total}}/dr = 0 \qquad (5.30)$$

The leading term arises from the attractions and repulsions among charged ions and has the form

$$E_{\text{electrostatic}} = \sum_i Nn_i z^+ z^- e^2/r_i \qquad (5.31)$$

where N is the number of particles, generally taken as Avogadro's number, r_i is the distance from a reference atom to its neighbor, n_i is the number of neighbors at that distance, z^+ and z^- are the charges of cation and anion, respectively, and e is the charge on the electron. For example, for a crystal $A^{2+}B^{2-}$ in the NaCl structure,

$$E_{\text{electrostatic}} = -4Ne^2 \left(6/r_{\text{AB}} - 12/\sqrt{2}\, r_{\text{AB}} + 8/\sqrt{3}\, r_{\text{AB}}\right) \qquad (5.32)$$
$$- 6/\sqrt{2}\, r_{\text{AB}} + 24/\sqrt{3}\, r_{\text{AB}} \ldots$$

This reflects the geometry: Each A^{2+} has 6 B^{2-} neighbors at a distance of r_{AB}, 12 A^{2+} next-nearest neighbors at a distance of $\sqrt{2}\, r_{\text{AB}}$, and so on. This sum may be rewritten as

$$E_{\text{electrostatic}} = NMz^+z^-e^2/r_{\text{AB}} \qquad (5.33)$$

where M, the Madelung constant for the sodium chloride structure, is given by

$$M_{\text{NaCl}} = 6 - 12/\sqrt{2} + 8/\sqrt{3} - 6/\sqrt{2} + 24/\sqrt{3} \ldots = 1.748 \quad (5.34)$$

For the zincblende structure with tetrahedral coordination, $M = 1.638$ and for the cesium chloride structure with eight-fold coordination, $M = 1.763$. Thus one would be tempted to say that higher coordination numbers are favored on electrostatic grounds, but one must bear in mind that the nearest interatomic distances increase with increasing coordination number, offsetting the larger Madelung constant and, in many cases, several structure types are quite comparable in energy.

In complex crystals one can not factor out a simple Madelung constant. Rather, the electrostatic sum (Eq. 5.31) converges slowly because it contains an alternation of negative and positive terms arising from successive anion and cation shells around a reference cation. General ways of dealing with this summation in reciprocal space (the so-called Ewald sum (Born and Huang 1954)) ease the computation.

There are several commonly used forms for the repulsive term. These include the simple Born exponent (B, n constants):

$$E_{\text{repulsion}} = B/r^n \qquad (5.35)$$

or an exponential form (B', B'' constants)

$$E_{\text{repulsion}} = B'e^{-rB''} \qquad (5.36)$$

the Buckingham potential (B, B', ρ constants)

$$E_{\text{repulsion}} = B \exp(-r/\rho) - B'r^{-6} \qquad (5.37)$$

and several others. In each case the form of the equation assures that repulsion increases sharply as the atoms come close together but becomes negligible at larger interatomic separations. Thus the repulsion term is a short-range interaction, whereas the electrostatic term is long-range.

The repulsion coefficients can be fitted empirically from experimental data,

Table 5.12. *Values of the Born exponent,* n,
for various solids

Compound	Born exponent, n
LiF	5.9
NaCl	9.1
NaBr	9.2
MgO	6.1

commonly from the compressibilities. When this is done in the simple Born model (Eq. 5.35), values of the Born exponent, n, commonly range between six and nine for oxides and halides (see Table 5.12). The repulsion terms can also be used as adjustable parameters to optimize empirical potentials for calculating various physical properties.

In general, the electrostatic term accounts for 75–90% of the lattice energy, the repulsion energy is 10–20%, and the other terms in Eq. 5.29 are less than 10%. Thus much insight can be gained by considering only the first two terms and, especially because the repulsion term may scale similarly for a number of solids of similar structure, by considering the electrostatic energy alone.

5.4.3 The Born-Haber cycle and the determination of lattice energies

The lattice energy, E_{lattice}, of a binary crystal A^+B^- is the energy of the reaction

$$A^+(\text{gas}) + B^-(\text{gas}) = A^+B^-(\text{crystal}) \tag{5.38}$$

It may be analyzed using a thermodynamic cycle involving the sum of the following reactions, the Born-Haber cycle:

$$A^+(\text{gas}) + e^- = A(\text{gas}) \tag{5.39}$$
$$\Delta E = - \text{ ionization potential of A} = -IP_A$$

$$B^-(\text{gas}) = B(\text{gas}) + e^- \tag{5.40}$$
$$\Delta E = - \text{ electron affinity of B} = -EA_B$$

$$A(\text{gas}) = A(\text{xl}) \tag{5.41}$$
$$\Delta E = - \text{ sublimation energy of A} = -E_{\text{sub,A}}$$

$$B(\text{gas}) = 1/2\, B_2(\text{gas}) \tag{5.42}$$
$$\Delta E = - 1/2 \text{ dissociation energy of } B_2$$
$$= 1/2 \text{ energy of the B–B bond} = 1/2\, E_{\text{B–B}}$$

Table 5.13. *Values of terms in Born Haber cycle calculation for a number of AX compounds (kJ/mol)*

Compound	$E_{sub,A}$	IP_A	$1/2\,E_{BB}$	$-EA_B$	$-\Delta H_{f,AB}$	$-E_{lattice}$
LiF	155	526	79	344	612	1029
NaF	109	502	79	344	569	915
KF	90	425	79	344	563	813
RbF	86	409	79	344	549	779
LiCl	155	526	121	362	409	849
NaCl	109	502	121	362	411	781
KCl	90	425	121	362	436	710
RbCl	86	409	121	362	431	685
LiBr	155	526	95	338	350	804
NaBr	109	502	95	338	360	743
KBr	90	425	95	338	392	680
RbBr	86	409	95	338	389	656
LiI	155	526	76	306	271	753
NaI	109	502	76	306	288	699
KI	90	425	76	306	328	643
RbI	86	409	76	306	328	624
AgF	277	728	79	344	204	943
AgCl	277	728	121	362	127	890
AgBr	277	728	95	338	100	877
AgI	277	728	76	306	63	867

$$A(xl) + 1/2\,B_2(gas) = AB(xl)$$

ΔE = standard energy ($\cong$ enthalpy) of formation   (5.43)
of AB from the elements = $\Delta H_{f,AB}$

Thus

$$E_{lattice} = -IP_A - EA_B - E_{sub,A} + E_{B-B} + \Delta H_{f,AB} \qquad (5.44)$$

The point of this cycle is that it enables the lattice energy to be calculated from directly observable physical properties. The lattice energy calculated by the Born-Haber cycle thus does not depend on assumptions made about the form of repulsive and other terms but gives the net effect of all these in the real crystal. This experimental lattice energy may then be usefully compared to values calculated using different models to assess the adequacy of their approximations.

Table 5.13 gives values of terms in the Born-Haber cycle for a number of compounds. Table 5.14 compares experimental (Born-Haber cycle) and theoretical lattice energies. Several general features stand out. First, lattice energies

Table 5.14. *Comparison of lattice energies of crystals calculated using Born-Haber cycle and lattice energy calculations*

Compound	Structure	M-X dist. (Å)	Lattice energy (kJ/mol)		Melting point (K)
			Calculated	Born-Haber	
LiF	NaCl	2.01	1021	1021	1143
LiCl	NaCl	2.57	837	845	886
LiBr	NaCl	2.75	782	803	820
LiI	NaCl	3.00	736	753	718
NaCl	NaCl	2.81	552	774	1074
KCl	NaCl	3.14	695	702	1049
RbCl	NaCl	3.27	665	686	988
CsCl	NaCl	3.56	640	661	915
MgO	NaCl	2.10	3933		3073
CaO	NaCl	2.40	3523		2887
SrO	NaCl	2.57	3310		2703
BaO	NaCl	2.76	3125		2191
MgS	NaCl	2.60	3255		
BaS	NaCl	3.18	2745		
MgF_2	Rutile	2.02	2883		1648
CaF_2	Fluorite	2.36	2581		1573
SrF_2	Fluorite	2.50	2427		1443
BaF_2	Fluorite	2.68	2289		1553

are very large in magnitude, typically 1000–5000 kJ/mol of AB. These values reflect the great stability of the condensed phase relative to the gaseous ions and are at least two orders of magnitude larger than the energies associated with typical solid state reactions (see Chap. 6). Thus gaining insight into the stability of phases from lattice energies requires the uncertainties in the lattice energies to be far less than 1%, because tiny differences between huge numbers are being observed when comparing different possible structures. This is a very tall order. Even though quantitative predictions of thermodynamic parameters are rarely attainable, considerable qualitative insight, especially as to the order of stability of possible polymorphs, can be obtained.

The second outstanding feature is that the lattice energies of compounds $A^{2+}B^{2-}$ are roughly four times those of compounds A^+B^-, confirming their ionic nature and the use of ionic charges in Eq. 5.31. Within a given structure type, the lattice energy decreases in magnitude (becomes less stable) as the interatomic distance increases, that is, as one goes down the periodic table to heavier and larger ions. The melting points generally follow these trends; the compounds with highest melting points are those with largest (in magnitude) lattice energies.

Third, the "experimental" and "theoretical" lattice energies agree rather well for the alkali halides and alkaline earth oxides having the sodium chloride structure, but less so for more complex structures, especially for bromides, iodides, and sulfides showing layered structures. This confirms intuition that the latter groups of materials are less "ionic" in character.

5.4.4 Ionic radii

There have been numerous attempts to derive internally consistent radii from bond lengths. Van der Waals radii derive from the closest approach of rare gas atoms and of other weakly interacting molecules; they represent the distance at which repulsions and very weak Van der Waals attractions are balanced. Nonbonded radii can be derived from the closest approach of two anions in a structure and represent the onset of strong repulsions. Metallic radii can be derived for alloys and covalent radii for three-dimensional molecular solids such as diamond. The early work of Pauling (1927) sets the stage for all present discussions of radii, and is clear and inspired. By far of greatest importance to mineral physics and ceramics are ionic radii, the sizes of ions in solids. A number of such sets exist. Those of Pauling (1927), Ahrens (1952), Goldschmidt (1926), Shannon and Prewitt (1969), and Whittaker and Muntus (1970) are of general applicability; sets have also been optimized for best fit to specific structure types, for example, spinel (O'Neill and Navrotsky 1983).

In general, the process of finding a set of radii is similar in all cases. The empirical observation is of constancy or near constancy of a given bond length for a fixed coordination number and cation charge (e.g., the Si–O bond length for tetrahedral coordination). The task is to optimize these bond lengths to a best average value and then divide up the contribution of cation and anion. Because this last step has essentially no theoretical basis and no unique solution (see following), this is the point at which different sets of radii differ the most. Usually it is the oxygen (O^{2-}) radius that is fixed at some value between 1.36 and 1.40 Å, based on ideas about the contact distances of oxygens in simple structures, and other radii depend on this choice. For applications involving bond lengths or differences in ionic radii, these differences cancel out, but when arguments invoke radius ratios, they become sensitive to which set of radii is used.

Here I discuss the Shannon and Prewitt (1969) radii because they are most commonly used. A selected set of ionic radii is shown in Table 5.15. Figures 5.30, 5.31, and 5.32 show some trends. The cation radius increases strongly as

Table 5.15. *Shannon and Prewitt ionic radii in oxides and fluorides*

Ion	2	4	6	8	10	12
Li^+		0.590	0.76	0.92		
Na^+		0.99	1.02	1.18		1.39
K^+		1.37	1.38	1.51	1.59	1.64
Rb^+		1.52	1.56	1.61	1.66	1.72
Cs^+			1.67	1.74	1.81	1.88
Be^{2+}		0.27	0.45			
Mg^{2+}		0.57	0.72	0.89		
Ca^{2+}			1.00	1.12	1.23	1.34
Sr^{2+}			1.18	1.26	1.36	1.44
Ba^{2+}			1.35	1.42	1.52	1.61
B^{3+}		0.11	0.27			
Al^{3+}		0.39	0.535			
Ga^{3+}		0.47	0.620			
La^{3+}			1.032	1.16	1.270	1.36
C^{4+}		0.15	0.16			
Si^{4+}		0.26	0.400			
Ge^{4+}		0.390	0.530			
Sn^{4+}		0.55	0.690	0.81		
Ti^{4+}		0.42	0.605	0.74		
Mn^{2+}		0.66	0.67	0.96		
Mn^{3+}			0.645 high spin			
			0.58 low spin			
Mn^{4+}		0.39	0.530			
Mn^{6+}		0.255				
Mn^{7+}		0.25	0.46			
Zn^{2+}		0.60	0.74	0.90		
O^{2-}	1.35	1.38	1.40			
S^{2-}			1.84			
F^-	1.285	1.31	1.33			
Cl^-			1.81			
Br^-			1.96			
OH^-	1.32	1.35	1.37			

we move down a group in the periodic table, for example, H^+, Li^+, Na^+, K^+, Rb^+, Cs^+ (Fig. 5.30). Anion radii increase likewise, for example, H^-, F^-, Cl^-, Br^-, I^-. Ionic radii shrink with increasing cation charge, both in moving across a period in the periodic table (Na^+, Mg^{2+}, Al^{3+}, Si^{4+}, P^{5+}) (Fig. 5.32b), and in taking electrons away from a given cation (uranium and chromium shown in Fig. 5.32a). The cation radius increases with increasing coordination number (Mg^{2+} in 4-, 6-, 8-coordination; Ca^{2+} and La^{2+} in 6-, 8-, 10-, 12-coordination). Variation in anion radii with charge and coordination are smaller (compare F^- and O^{2-}, and O^{2-} surrounded by two, four, and six cations). The behavior of

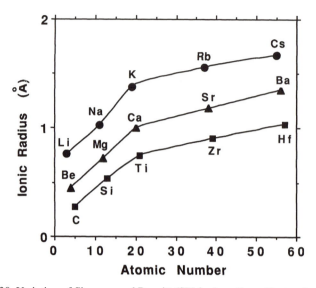

Figure 5.30. Variation of Shannon and Prewitt (SP) ionic radius with atomic number.

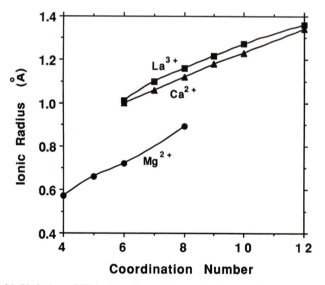

Figure 5.31. Variation of SP ionic radius with coordination number.

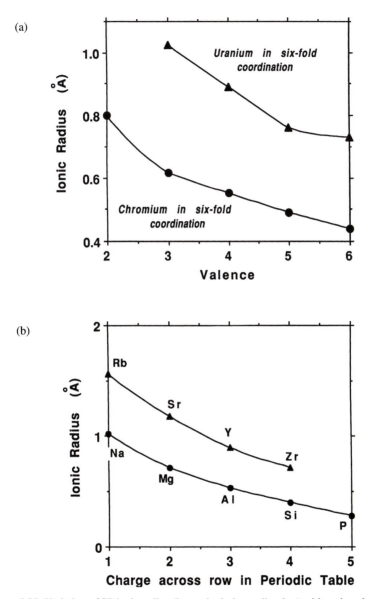

Figure 5.32. Variation of SP ionic radius (in octahedral coordination) with cation charge (a) for chromium and uranium and (b) across row in periodic table.

transition metal and rare earth ions, particularly the double-humped curve of radius versus number of d or f electrons, has already been discussed (see Sec. 5.3.5). These trends generally hold (though numerical values will change) if other sets of radii are chosen; because anion radii vary little with coordination number in the Shannon and Prewitt set of radii, very similar trends are seen in bond lengths.

Ionic radii can be used to correlate, interpolate, and extrapolate many physical properties. Certainly they are successful in predicting bond lengths and lattice parameters of complex oxides and halides and, to a more limited extent, sulfides. Figure 5.33 shows correlations between volume and $(r_A + r_X)^3$ for a number of AX compounds in the rocksalt, zincblende or wurtzite, and nickel arsenide structures. A good straight line is obtained for the rocksalt structure; a straight line, but with more scatter, is seen for the tetrahedral structures; while the NiAs type, populated mainly by sulfides, selenides, and tellurides, shows definite curvature. This pattern clearly reflects the increasing role of covalent and metallic bonding in the last two cases. Shannon (1976) tried to correct cation radii for use in sulfides to more accurately predict the bond shortening brought about by decreased ionicity, but it is not clear that this set of radii has the same utility as those for halides and oxides. Figure 5.34 shows

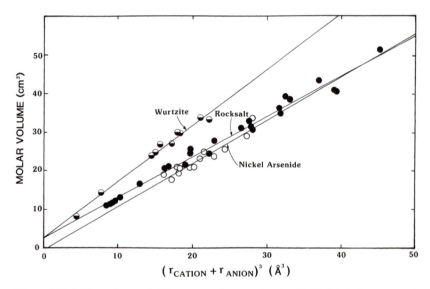

Figure 5.33. Molar volume of AX compounds versus $(r_A + r_X)^3$. (Solid circles represent rocksalt structure, open circles represent wurtzite and zincblende structures, and half-filled circles represent nickel arsenide structure.)

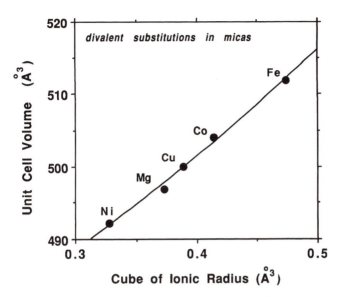

Figure 5.34. Unit cell volume versus cube of ionic radius for micas.

analogous correlations of unit cell volume versus the cube of the cation radius for a series of micas. Again, useful trends are seen.

It is tempting to use geometric arguments based on ionic radii to predict the existence and stability of structures. The "radius ratio" argument is a simple example. Figure 5.35 shows its geometric basis. Consider the rocksalt structure. If the cation is too small, it will "rattle" in the cage made by anion-anion contacts: This will occur at a ratio of $R_c{:}R_a = 0.414$. If it is too big, anion-anion contacts will be spread apart. The cesium chloride structure has precisely correct cation-anion and anion-anion contacts at $R_c{:}R_a = 0.732$. This argument suggests that the NaCl structure should be stable for a ratio between 0.414 and 0.732, the zincblende or wurtzite structure for smaller radius ratios, and the cesium chloride for larger. Table 5.16 shows the experimental evidence. Certainly tetrahedral structures occur for small cations and cesium chloride for large ones, but the changes in structure do not occur clearly at the predicted ratios. This reflects both the difficulty of uniquely defining ionic radius and the fact that the notion of ions as hard spheres is too simplistic.

In reality, arguments based on sums or differences of ionic radii (bond lengths) are much more sound than reasoning based on radius ratios because the former do not change if all radii change by an additive constant (as happens when one redefines the radius of oxygen, for example), whereas the latter are

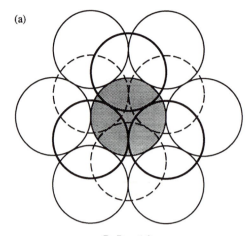

(a)

$$R_c{:}R_a = 1.0$$
$$C.N. = 12$$

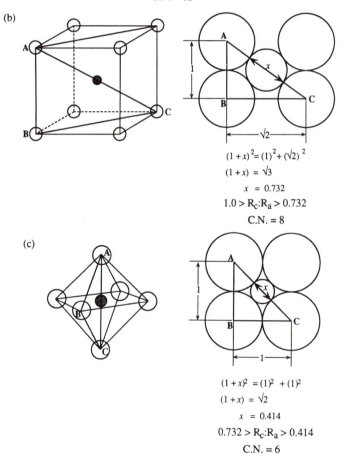

(b)

$$(1+x)^2 = (1)^2 + (\sqrt{2})^2$$
$$(1+x) = \sqrt{3}$$
$$x = 0.732$$
$$1.0 > R_c{:}R_a > 0.732$$
$$C.N. = 8$$

(c)

$$(1+x)^2 = (1)^2 + (1)^2$$
$$(1+x) = \sqrt{2}$$
$$x = 0.414$$
$$0.732 > R_c{:}R_a > 0.414$$
$$C.N. = 6$$

Figure 5.35. Coordination of cations by anions – restrictions imposed by cation:anion ratio ($R_c{:}R_a$) criteria. (a) 12-coordination occurs when cations and anions are the same size ($R_c{:}R_a = 1$). These are close-packed arrangements (hcp and ccp) and commonly occur in simple metals like Cu and Mg. (b) Cubic coordination (C.N. = 8) occurs when the cations are slightly smaller than anions. The geometry of the limiting conditions

(d)

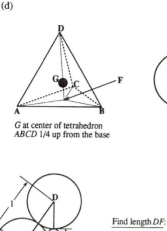

G at center of tetrahedron
ABCD 1/4 up from the base

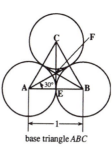

base triangle ABC

Find length AF:

$\cos 30° = AE/AF$

$\therefore \quad AF = AE/\cos 30°$

$AF = 0.5/\cos 30°$

$AF = 1.0/\sqrt{3}$

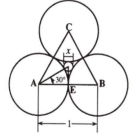

vertical triangle ADF

Find length DF:

$DF = \sqrt{AD^2 - AF^2}$

$DF = \sqrt{(1)^2 - (1/\sqrt{3})^2}$

$DF = \sqrt{(2/3)} = 0.81649$

Consider length DG:

$DG = 0.75\, DF$ since G is
0.25 up from the base ABC

And, $DG = 0.5 + 0.5x$ so:

$0.5 + 0.5x = 0.75(0.81649) = 0.6124$

$0.5x = 0.6124 - 0.5 = 0.1124$

$x = 0.225$

$0.414 > R_c:R_a > 0.225$

C.N. = 4

(e)

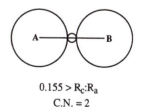

(f)

$0.155 > R_c:R_a$

C.N. = 2

$\cos 30° = 0.5/(0.5 + 0.5x)$

$0.5 + 0.5x = 0.5/\cos 30° = 0.5774$

$0.5x = 0.5774 - 0.5 = 0.0774$

$x = 0.155$

$0.225 > R_c:R_a > 0.155$

C.N. = 3

are shown which indicate a minimum $R_c:R_a$ of 0.732. (c) Octahedral coordination (C.N. = 6) occurs when $R_c:R_a$ is less than 0.732. The minimum for $R_c:R_a$ in this configuration is shown to be 0.414. (d) Tetrahedral coordination (C.N. = 4) occurs for $R_c:R_a$ ratios between 0.414 and 0.225. The geometry of the limiting conditions is illustrated. (e) Planar triangular coordination (C.N. = 3) occurs for even smaller cations, with the minimum for $R_c:R_a$ being 0.155. (f) For $R_c:R_a$ less than 0.155, linear coordination by two anions (C.N. = 2) may occur.

Table 5.16. *Radius ratio, $R = r_{cation}/r_{anion}$ and structure types in AX compounds*

	R	Structure
BeO	0.33	Wurtzite
MgO	0.51	Rocksalt
CaO	0.71	Rocksalt
ZnO	0.52	Wurtzite[a]
NaCl	0.55	Rocksalt
CsCl	0.92	Cesium chloride, sodium chloride[a]
AgI	0.45	Wurtzite[a]
NaI	0.45	Rocksalt
SiO_2	0.19, 0.29	Quartz, rutile at high P [a]
GeO_2	0.28, 0.39	Quartz, rutile[a]
TiO_2	0.44	Rutile
CaF_2	0.85	Fluorite
ZnS	0.32	Wurtzite
MnS	0.36	Rocksalt[a]

Notes: Calculated for Shannon and Prewitt radii for observed coordination. Prediction: $R < 0.414$, tetrahedral; $0.414 < R < 0.732$, octahedral; $R > 0.732$, eight-coordinated.
[a] This structure violates the predictions.

sensitive to such shifts. Indeed, one can agree that the classic radius ratio argument has poor success at predicting structures, is very vulnerable to how the system of radii used is defined (Burdett 1981, 1982a, b), and should be relegated to the same obscurity in mineralogical history as blowpipe analysis.

The tolerance factor in perovskites (see Eq. 2.11) illustrates another use of the ionic model in predicting structural stability. Figure 5.36 shows the enthalpy of formation of a perovskite from its component oxides:

$$AO_2 + BO = ABO_3 \qquad \Delta H_{f,ox} \qquad (5.45)$$

as a function of the absolute value of $(1 - t)$. Those perovskites with t close to unity have the most exothermic enthalpies of formation. This is an obvious indication of stability of the structure. Lattice energy calculations confirm this and show further that the electrostatic energy becomes more favorable with decreasing unit cell volume (and decreasing t), but that the repulsive energy increases significantly as t decreases (Takayama-Muromachi and Navrotsky 1988). This increased repulsion is what limits the structure as the tolerance factor decreases. These calculations support the ideas of packing of ions of reasonably well defined sizes implicit in the ionic model. The entropy of formation becomes more positive with decreasing t. This slightly counteracts the

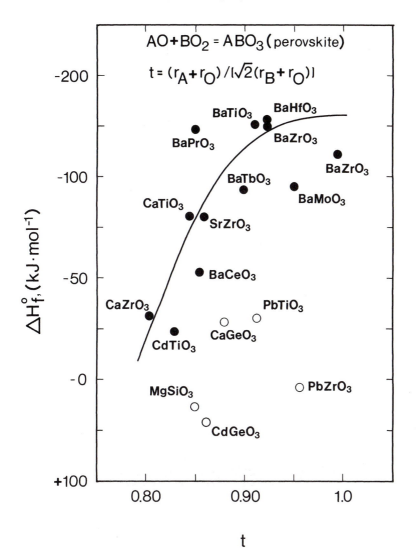

Figure 5.36. Enthalpy of formation of $A^{4+}B^{2+}O_3$ perovskites versus absolute value for $(1-t)$, with t the tolerance factor defined by Eq. 2.11. Bond length is calculated using the Shannon and Prewitt radius appropriate to the observed coordination number.

energetic destabilization; nevertheless, it is clear that perovskite stability (especially with respect to other structures) drops off sharply for t below about 0.85. The increasing entropy probably reflects greater vibrational freedom for a small cation in the A-site (see Sec. 6.2).

The distance-least-squares (DLS) method in crystallography (Dempsey and Strens 1976; Meier and Villiger 1969) is a direct application of the assumption of constant bond lengths. If one knows the space group of an unknown structure but does not have enough data to refine positional parameters, one can estimate atomic positions by requiring that each bond length be as close as possible to the relevant sum of ionic radii. The structure can then be refined by minimizing the difference between observed and predicted bond lengths, hence the name DLS. Constraints can also be placed on other refinements to weed out unreasonable trial structures; an otherwise possible structure might contain some absurdly long or short bond lengths, making it physically unlikely. Obviously DLS constraints are more meaningful for the more rigid bonds (e.g., Si–O, Al–O, P–O) than for the more compliant ones (e.g., Na–O, K–O). Certainly a structure with a tetrahedral Si–O bond length below 1.55 or above 1.68 Å is suspect, but so is one with an Na–O bond length below 0.9 Å, or an O–O distance (in the absence of a peroxide bond) of less than 2.6 Å. Questioning structures with anomalous bond lengths is just crystallographic common sense, yet common sense is as uncommon a commodity in crystallography as in other human endeavors, and the literature has numerous examples of structures with dubious bond lengths.

What can be said about the sizes of cations and anions from electron density maps of a structure obtained either directly from diffraction data or calculated from theory? One might be tempted to place the demarcation line between cation and anion at the point of minimum electron density between them. Figures 5.37 and 5.38 show such maps for coesite. Although charge densities can be computed with fair accuracy, it is difficult (and not always unique) to determine how much of the charge belongs to the cation and how much to the anion. Thus degree of ionicity and ionic radius, though very useful concepts, remain somewhat "soft" when it comes to unique quantitative definition. These ambiguities are brought home in a recent paper by Gibbs, Spackman, and Boisen (1992). They compare conventional atomic and ionic radii with two other sets of radii: bonded radii, which divide the bond length between cation and anion at the point of minimum electron density in the experimental or ab initio calculated electron density, and promolecule radii, which are similar to bonded radii in both concept and magnitude but refer to spherical electron distributions. The bonded radii fall systematically between values of ionic and of atomic radii. Thus the bonded radii for cations are larger than ionic radii and those for anions smaller. Bonded radii for anions depend more strongly on coordination number than do ionic radii of anions, but bonded radii of cations have a weaker coordination number dependence than corresponding ionic radii. The bonded

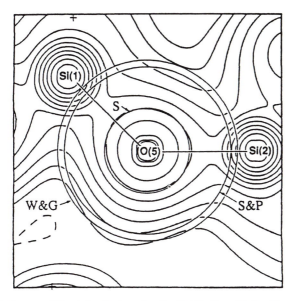

Figure 5.37. A map of the static electron density distribution obtained in a pseudoatom model refinement of coesite, calculated through the plane of the Si1-O5-Si2 group. The contour interval is 0.0625, 0.125, 0.250, . . . , e/Å^3. The circle centered on O5 and labeled S (Slater) defines the outermost limit of the O atom as defined by its atomic radius, and those labeled $S\&P$ (Shannon and Prewitt) and $W\&G$ (Wasastjerna and Goldschmidt) define the outermost limits of the oxide ion as defined by its crystal and ionic radii, respectively. (From Gibbs, Spackman, and Boison 1992.)

radii are relatively close to a set of radii derived by Fumi and Tosi (1964) by fitting a repulsive potential to experimental compressibility and thermal expansion data.

The classical ionic radii describe cations as being small and anions as large. This stems from the idea that adding electrons to an atom should expand its size and taking electrons away should make the remaining ones adhere more tightly and decrease the size. Physical pictures of crystals as cations nestled in protective polyhedra of anions and of oxide melts and glasses as viscous puddings of anions with cations distributed like raisins arise from this mindset. These ideas have proven very useful but are not a unique description. Indeed, O'Keeffe and Hyde (1981) argued that in many solids, nonbonded repulsions are important in limiting stability and that much is to be learned by focusing on polyhedra that are centered on oxygen, that is, anion polyhedra. This formulation discovers analogies between atomic packings in alloys and anion

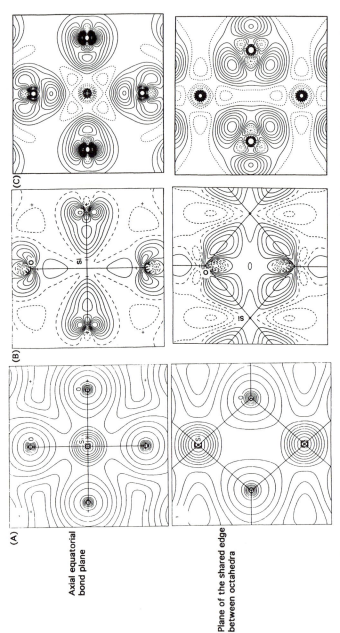

Axial equatorial bond plane

Plane of the shared edge between octahedra

Figure 5.38. (A) Total electron density for stishovite mapped from X-ray diffraction study of Spackman, Hill, and Gibbs (1987) using the pseudoatom model described in the text. Top is the axial equatorial bond plane. Bottom is the plane of the shared-edge between SiO_6 octahedra. The plane in (A) is 5 Å by 5 Å and in (B) 4 Å by 4 Å. Contours are at 0.0625, 0.125, 0.25, 0.5, ..., $eÅ^{-3}$; inner contours are noncircular due to poor numerical interpolation near the nuclei. Crosses indicate atomic positions in or projected onto the plane. (B) Deformation density, showing deviations from a spherically symmetric electron distribution around each atom, from X-ray refinement of Spackman et al. (1987). Same geometry as (A). (C) Calculated deformation density using LAPW method (from Cohen 1991).

arrangements in ternary oxide structures. It also shows that consistent and interesting insights can be had by assuming that anions are small and cations large. It underscores the arbitrariness of conventional radii.

5.4.5 Pauling's rules

The following rules, formulated by Linus Pauling over sixty years ago (Pauling 1927), are guidelines for constructing stable structures.

First rule: A coordination polyhedron of anions is formed about each cation, the bond length is the sum of ionic radii, and the coordination number of the cation is determined by the radius ratio. This embodies the heart of the classical ionic model and is generally valid. Gross exceptions indicate a large deviation from ionic bonding. Thus, for example, compounds with peroxide groups, sulfur-sulfur bonds, or polynuclear metallic clusters disobey this rule. The radius ratio rule, as discussed, is only a very approximate criterion for coordination number.

Second rule: The total strength of the bonds that reach an anion from all neighboring cations is equal to the charge of the anion. This is the electrostatic valency principle that requires that local charge balance (electroneutrality) be maintained. The Pauling bond strength, p_i, is then defined as the cation charge divided by the number of bonds it makes, that is, by the coordination number. This rule fails most often in disordered systems, and how much local order is maintained to make electroneutrality hold locally, if not on the average, is still controversial. Structures in which the Pauling bond strength sum and the anion charge are unbalanced by more than two charge units exist but are often of marginal stability. This rule has been generalized into a numerical classification of structures and bonding geometries, for example, in recent work by Ellison and Navrotsky (1991). It also forms the basis of empirical correlations of bond strengths and bond lengths and is a starting point for the bond valence formalism of I. D. Brown (1987) (see Sec. 5.5.3). Local charge balance is an important constraint in melts and glasses as well (see Chap. 8).

Third rule: Shared edges and, especially, shared faces of coordination polyhedra destabilize a structure. This effect is strongest for cations of high charge and small coordination number, especially if near the lower limit of the radius ratio for that coordination. This rule stems from both anion and cation repulsions. Shared edges tend to be shortened and shared edges and faces lead to distorted polyhedra.

Fourth rule: Cations of high valency and small coordination number tend not to share polyhedral elements. This is a corollary of the third rule. Thus silicate, borate, and phosphate tetrahedra link only by corners, and carbonate, sulfate, and nitrate groups are isolated.

Fifth rule: The number of essentially different kinds of constituents in a crystal structure tends to be small. This means that a structure is built up of a relatively small number of polyhedral elements – tetrahedra, octahedra, larger polyhedra. Though these may be split by crystal symmetry into many inequivalent sites (e.g., up to five M sites in amphiboles), these sites are often similar in local environment. This may also account for why, despite a seemingly infinite number of possible structures, the same ones recur over and over as pressure, temperature, and composition vary. This rule is probably a consequence of maximizing the electrostatic contribution to the lattice energy while minimizing repulsions by keeping local environments fairly symmetric and anions as far apart from each other as possible.

These rules are very useful for assessing whether proposed structures are reasonable. When exceptions occur, they offer insight into special cases where specific factors stabilize an otherwise unfavorable structure.

5.4.6 Lattice energy minimizations, optimized potentials, and other calculations within the ionic model

Techniques to calculate the lattice energy can be used to find the minimum in lattice energy with respect to variations in lattice parameters and positional parameters. Table 5.17 lists several examples. In many such methods the coefficients in the repulsion terms can be used as adjustable parameters to force the minimum to occur at the observed structure. In some programs the charges on the ions can also be varied. This appears desirable if one wishes to match the energies and elastic constants of different structures; generally using full ionic charges overestimates energy differences. These partial charge models fall into two classes. The first contains those in which the charges are all allowed to "float" and, though electroneutrality is maintained for the given stoichiometry, the potentials are not transferable to a different structure or stoichiometry. Each structure generates its own partial charges and using the charges from one stoichiometry (e.g., AB_2O_4) to describe another (e.g., ABO_3) would result in macroscopic charge imbalance. The second class is that of transferable potentials, where ionic charges remain proportional to the formal charges and where an effort is made to find parameters that do a reasonable job describing several structures and, in some cases, several coordination numbers. The former class is sometimes more successful at specific applications; the latter is more general. In both cases, it is unclear whether the partial charges have any real physical significance.

Table 5.17. *Lattice energy minimization methods*

Name of method	Potential	Applications	Reference
WMIN and related	*Full ionic charges* Coulomb, repulsion, other short-range forces		Busing (1970, 1981)
		Hydrous minerals	Abbott, Post, and Burnham (1989)
		Feldspars	Post and Burnham (1987)
	Coulombic and repulsion terms optimized to structural and thermodynamic data	Spinels	Ottonello (1986)
PLUTO	Coulomb, repulsion, and short-range forces	Pyroxene structures	Catlow et al. (1982)
		Zeolites	Jackson and Catlow (1988)
	Partial ionic charges Central forces, electrostatic plus repulsion, charges optimized to fit elasticity	$MgO-SiO_2$ high-pressure phases	Matsui, Akaogi, and Matsumoto (1987)

Table 5.17. *Continued*

Name of method	Potential	Applications	Reference
	Similar to above	MgO–SiO$_2$ high-pressure phases	Price and Parker (1984)
	Central forces, Coulomb, Born, and Van der Waals terms, charges vary but keep same ratios to full ionic charges	MgO–SiO$_2$ phases with both octahedral and tetrahedral silicon	Leinenweber and Navrotsky (1988)
	Noncentral potentials		
	Rigid ions plus three-body and polarization terms	Feldspars	Purton and Catlow (1990)
		Silica	Sanders, Leslie, and Catlow (1984)
	Covalent potential–bond stretching, bending, and O–O repulsion	SiO$_2$	Stixrude and Bukowinski (1988)
	Partial charges and many body forces	Rocksalt oxides, and olivines	Weidner and Price (1988)

Energy minimization techniques have been useful in applications (see Table 5.17). Particularly in the programs developed by the English school (Catlow, Price, and others), the parameterization of empirical potentials appears fairly robust, and the methods have been applied widely to ceramics and minerals, with considerable success in predicting structural features and some success in predicting energetics. Yet, because of the empirical nature of the potentials used and the number of adjustable parameters, it is not clear to me what physical insight these computations give about bonding.

The potentials derived from quantum chemistry are generally calculated point by point for different interatomic distances. Recently there has been considerable excitement in taking these potentials and parameterizing them in terms of simple functions (Lasaga and Gibbs 1987). These can then be used in calculations like lattice energy minimization and molecular dynamics (see following).

There have been numerous attempts to incorporate many-body forces, angle-bending forces, and other forms of covalency into ionic models to make them better describe silicates with polymerized tetrahedra. Though such grafts can improve models for specific applications, they lack generality. In my opinion, the beauty of the ionic approach lies in its simplicity. The phenomena it does not describe well should be attacked by other means. The power of rigorous computational techniques is constantly improving, and the case for "souped up" ionic models is becoming weaker.

5.4.7 Molecular dynamics simulations

Consider an arbitrary assemblage of n atoms or ions confined to a specific region of space. If the particles have a set of initial positions, they will interact with each other according to their interatomic potentials and redistribute their positions and energies in accord with whatever constraints (constant volume, constant pressure, constant energy, constant temperature, etc.) are imposed. This problem can be posed purely classically (despite the fact that the interactions arise from quantum mechanics) at high temperature, and if one has the computational power and patience to follow the trajectories of all the individual particles through many short time steps, one will see the system relax into an equilibrium state. The techniques of doing so are called *molecular dynamics* (MD) and they yield information on equilibrium configurations of crystals and liquids, on melting and crystallization, on phase transitions, and on diffusion, conductivity, and other transport properties. Though a detailed discussion of the method is beyond the scope of this book (and is presented well in Soules

1979), here I give a brief synopsis and illustrate some of the successful applications of molecular dynamics.

The number of atoms one can deal with is inversely related to the complexity of the potentials used. Generally several hundred to several thousand atoms are used. If an atom's motion carries it outside the confines of the "box" used, commonly used cyclic boundary conditions make it come back in on the other side. This behavior conserves mass and makes effects related to the finite size of the box less important. However, it tends to keep the density in the box fairly uniform, so for phenomena involving liquids, especially near critical points, cyclic boundary conditions may underestimate fluctuations. The easiest MD calculations are done at constant volume and constant energy, in which case the entropy, temperature, and pressure of the system evolve with time, but formalisms for constant pressure and constant temperature conditions are now available. The time steps taken must be of the order of interatomic vibrations and a computation may run from several thousand to several million steps (following the system for 10^{-9} to 10^{-6} sec). For this time duration to be long enough to reach equilibrium, temperature must be relatively high; many MD simulations refer to temperatures between 1000 and 6000 K. An MD simulation produces a sequence of "snapshots" of the system evolving with time (see Fig. 5.39 for an example of the crystallization of MgO) that show the positions of all the particles. Computer graphics that trace trajectories and label specific particles are very useful. Indeed the user of MD is swamped with data, and part of the growing impact of the field arises from improved visualization techniques.

The following accomplishments are some of the successes of MD. Early simulations strongly suggested that diffusion in tetrahedral framework liquids such as BeF_2 and SiO_2 involves five-coordinate Si or Be as a transient intermediate (Brawer 1981). After years of searching, these have been confirmed by IR and NMR (Stebbins and McMillan 1989; Williams and Jeanloz 1988). The simulations also suggest that pressure should enhance the concentrations of five- and six-coordinate aluminum and silicon in oxide melts. This has also been confirmed, and the search for coordination number changes in silicate liquids has led to very fertile ideas for both structure-property relations and magma evolution. Molecular dynamics also provides a good picture of the glass transition. It gives atomic scale insight into the mechanisms of melting and crystallization (see Chap. 8).

Molecular dynamics has suggested that perovskite structures could have rapid anion conduction at high temperature (Miyamoto 1988). This has probably been seen in fluorides (O'Keeffe and Bovin 1979) and remains to be tested for $MgSiO_3$ perovskite.

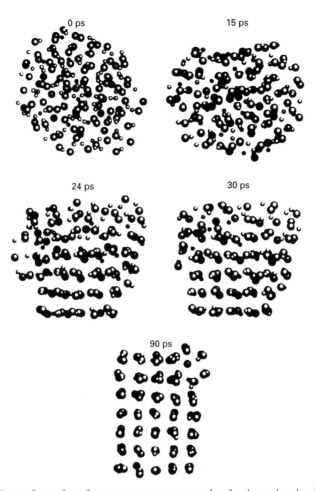

Figure 5.39. A set of snapshots from a constant energy molecular dynamics simulation using the PIB model for MgO. The large atoms are O and the smaller ones are Mg. A finite cluster of MgO liquid was cooled below the melting point and then time was stepped forward by integrating Newton's laws until the cluster crystallized. The crystal nucleates on the surface and then grows, and melt and crystal coexist during the process. At the end most defects anneal out except for a few surface defects. The temperature is about 2200 K at the beginning and there are 256 atoms in this simulation. (Figure provided courtesy of Ron Cohen.)

Ab initio and MD simulations of surface structure and of the reaction of silicate surfaces with water leads to insight about corrosion of ceramics and dissolution of minerals (Lasaga and Gibbs 1990) (see Fig. 5.40). Indeed, MD is an excellent probe of surface chemistry and reactions at surfaces.

The quality of MD simulations depends on the number of particles, the number of time steps, and the quality of the interatomic potentials used. The latter

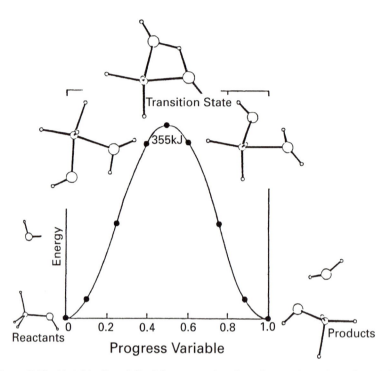

Figure 5.40. Ab initio "movie" of the approach, adsorption, and reaction of an H_2O molecular to hydrolyze an Si–O bond. This series of configurations was obtained by method of linear synchronous transit to define the reaction coordinate and activated complex. This approach uses ab initio (STO3G) molecular orbital calculations to produce the reaction pathway. It provides a linkage between ab initio potentials and dynamics. (After Lasaga and Gibbs 1990.)

can be made more rigorous by incorporating the potentials obtained from ab initio quantum calculations (generally in parameterized form) into the MD simulation. This approach has proven quite fruitful in simulating crystal structures, melting, and amorphization at high pressure and temperature in Mg_2SiO_4 and $MgSiO_3$ (Kubicki and Lasaga 1992).

5.5 Systematic crystal chemistry

5.5.1 General comments

The goal of understanding bonding is to be able to predict structure and properties. The insights gained by ab initio computations and by simulations, combined with experimental data, have led to a series of semiempirical trends and correlations that work toward this goal. They are semiempirical because their rationale lies in reasonably well founded theory but their parameters come from experiment. This section describes several such correlations.

5.5.2 Sorting diagrams for stability of structures

The basic question here is, Given a composition, what structure will it have? Bond length arguments are a first attempt at an answer. Intuition dictates that in addition to cation size, parameters related to the nature of chemical bonding should influence structure, and there are a number of two-dimensional "sorting diagrams" that plot the occurrence of structure types in terms of two parameters – one related to size and one related to bond type. A simple example is the Mooser-Pearson diagrams (Pearson 1972) for metallic alloys (Fig. 5.41) in which the structure type is displayed on a graph of period in the periodic table

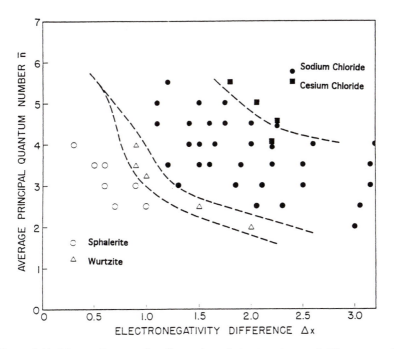

Figure 5.41. Mooser-Pearson plot: Separation of structure types of AB compounds according to electronegativity and average principal quantum number (from Pearson 1972).

Li	Be											B	C	N	O	F
1.0	1.5											2.0	2.5	3.0	3.5	4.0
Na	Mg											Al	Si	P	S	Cl
0.9	1.2											1.5	1.8	2.1	2.5	3.0
K	Ca	Sc	Ti	V	Cr	Mn	Fe	Co	Ni	Cu	Zn	Ga	Ge	As	Se	Br
0.8	1.0	1.3	1.5	1.6	1.6	1.5	1.8	1.8	1.8	1.9	1.6	1.6	1.8	2.0	2.4	2.8
Rb	Sr	Y	Zr	Nb	Mo	Tc	Ru	Rh	Pd	Ag	Cd	In	Sn	Sb	Te	I
0.8	1.0	1.2	1.4	1.6	1.8	1.9	2.2	2.2	2.2	1.9	1.7	1.7	1.8	1.9	2.1	2.5
Cs	Ba	La-Lu	Hf	Ta	W	Re	Os	Ir	Pt	Au	Hg	Tl	Pb	Bi	Po	At
0.7	0.9	1.1-1.2	1.3	1.5	1.7	1.9	2.2	2.2	2.2	2.4	1.9	1.8	1.8	1.9	2.0	2.2
Fr	Ra	Ac	Th	Pa	U	Np-No										
0.7	0.9	1.1	1.3	1.5	1.7	1.3										

Figure 5.42. Pauling electronegativity (data from Pauling 1960).

(essentially related to radius) versus the difference in Pauling electronegativity. The Pauling electronegativity (Pauling 1960) is an empirical parameter (see Fig. 5.42) indicating ease of electron donation. Indeed Pauling related the degree of ionic character of a bond to the differences in electronegativity. A reasonably good separation of structures is seen, with compounds falling near the boundaries of the regions often being dimorphic.

Phillips (1970, 1974) devised an ionicity scale based on the following argument from semiconductor physics. In each row of the periodic table, there is a homopolar tetrahedral semiconductor (diamond, silicon, germanium). Because all its atoms are the same, it is by definition covalent. Its band gap, E_H, is the homopolar bandgap. Group III–V semiconductors (e.g., AlP, GaAs, InSb) are derived by substituting a group III and a group V atom for the group IV atom, II–VI semiconductors (e.g., BeO, MgS, CaSe, ZnSe, CdTe) by substituting a group II and group VI element, and I–VII semiconductors (e.g., LiF, NaCl, KBr) by substituting a group I and group VII element. These materials may be of rocksalt or of tetrahedral (zincblende or wurtzite) type. In all cases their bandgap, E_{gap}, is greater than that in the homopolar material in the same period. Phillips proposed the following relation:

$$(E_{gap})^2 = (E_H)^2 + C^2 \qquad (5.46)$$

where C is a charge transfer energy. The ionicity, f_i, is then defined as

$$f_i = C^2/(E_{gap})^2 \qquad (5.47)$$

The observed structures, when plotted on a graph of covalent (E_H) versus ionic (C) energy gaps obtained from spectroscopic measurements, cleanly separate into fields of octahedral and tetrahedral structures at a value of $f_i = 0.785$ (see Fig. 5.43). The Pauling ionicity and Phillips ionicity scales appear to be linearly related.

Numerous attempts have been made to extend these sorting diagrams to more complex materials, notably by Bloch and Schatteman (1981), Miedema (1973), Burdett (1981), Zunger (1980, 1981), Chelikowsky (1982), Mooser (1983), and Andreoni (1985). Each defines two generalized *radii* (orbital radii, pseudopotential radii), one of which usually relates to size and one to electron transfer tendencies. These methods also try to include nonoctet electron configurations, that is, transition metals. Figure 5.43 shows two typical sorting diagrams. The original references should be consulted for further details.

(a)

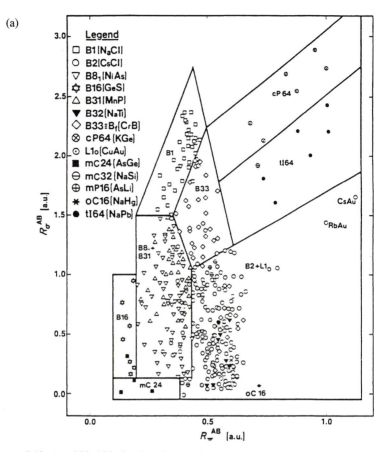

Figure 5.43. (*pp. 258–259*) Sorting diagram for structures. (a) Separation plot for 356 binary compounds using density functional orbital radii (from Zunger 1981; Burdett 1981). (b) Sorting of spinel cation distributions using Zunger's orbital radii (Burdett 1982a).

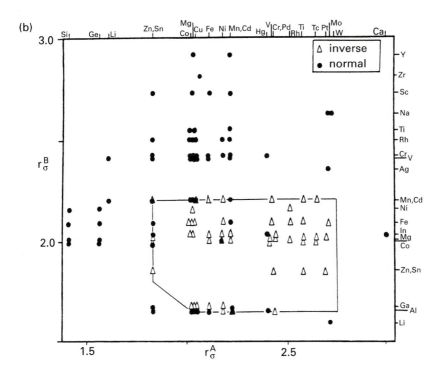

5.5.3 *Bond strength–bond length correlations and the bond valence model*

Although the *average* bond length in a coordination polyhedron is usually near that given by a sum of ionic radii, *individual* bond lengths vary, especially if the polyhedron is strongly distorted. It is intuitive to surmise that the longer bonds are weaker and that the electron density is distributed unequally among the different bonds. A number of empirical correlations between individual bond lengths and local bonding factors have been proposed. For linked tetrahedra (SiO_4, AlO_4), the work of Gibbs's group (Gibbs 1982) suggests that three factors are of primary importance. The first is the sum of the Pauling bond strengths to the bridging oxygen, p_o (defined previously). Thus p_o accounts for whether the oxygen is surrounded by two, three, or more cations. The second factor is the fractional s-character of the bridging oxygen, $f_s(O)$, which can be related to the T–O–T angle, θ_{TOT}, by quantum mechanical arguments that give

$$f_s(O) = (1 - \sec \theta_{TOT})^{-1} \tag{5.48}$$

The third factor is the fractional s-character of the T atom, given analogously as

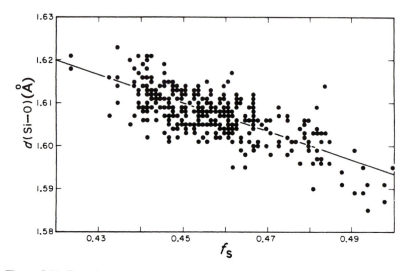

Figure 5.44. Experimental Si–O bond length in silica polymorphic versus selectron character of oxygen, $fs(O)$ (Eq. 5.48) (after Gibbs et al. 1981; Gibbs 1982). (Figures provided courtesy of G. V. Gibbs.)

$$f_s(T) = (1 - \sec \theta_{OTO})^{-1} \qquad (5.49)$$

Together, $f_s(O)$ and $f_s(T)$ give a measure of hybridization in the T–O bond. Multiple linear regression analysis using equations of the form (Gibbs 1982)

$$r_{TO} = a + bp_O = cf_s(O) + df_s(T) \qquad (5.50)$$

account well for the observed variation in r_{TO} (see Fig. 5.44) both in experimental crystallographic data and as calculated by molecular orbital methods for various clusters. Many other empirical equations have been proposed to fit tetrahedral bond length variations in silicates, phosphates, and vanadates (Baur 1981).

For nonframework cations, bond lengths are in general more variable than for tetrahedral Si–O and Al–O. In cases where deviations from Pauling's second rule are pronounced, bond length variations within a polyhedron seem unusually large, suggesting that the bond lengths are changing to compensate the charge imbalances. This suggests a relation between bond length and Pauling bond strength of the form (Baur 1981)

$$r_{AX} = r_{AX,av} + b\,(\Delta p) \qquad (5.51)$$

where Δp is the difference between Pauling bond strength of that particular bond and the average Pauling bond strength in the polyhedron. This expression has been fit to more than five thousand individual bond lengths involving twenty-nine cation-anion pairs and is generally quite satisfactory except for cations with low formal charge and high coordination numbers (the alkalis).

A more general formulation of Pauling's second rule is the inspiration for the bond valence method (Brown 1981, 1987). The bond valence of any bond in a crystal can be determined empirically, subject to the constraint that the sum of bond valences around any atom is equal to its atomic valence, by constructing a set of curves that correlate bond valence and bond length. These curves can be approximated as

$$s = (r/r_0)^{-N} \quad \text{or} \quad s = \exp[(r - r_0)/B] \tag{5.52}$$

with s the bond valence, r the bond length, r_0 the length of a bond of unit valence (really a fitting constant), and B and N constants. The bond valence approach can be used to predict bond lengths and correlate acid–base character and other structure–property relations.

5.5.4 The Duffy optical basicity concept

It is generally recognized that the most stable compounds form when there is a complete or nearly complete transfer of electrons from donor to acceptor or, in terms of an ionic crystal, from A to X to form a cation and an anion. Thus the most stable binary compounds generally form between elements with very different electronegatives (however defined). Since a Lewis acid is an electron pair acceptor and a Lewis base is an electron pair donor, concepts of electronegativity are naturally tied to an acid–base scale. But for the study of multi-component crystals, glasses, and melts, one is interested in small differences in electronegativity and relatively minor adjustments in the electron distributions related to compound formation. Thus a relative acidity scale, easily applied to a variety of ions in a number of environments, is desirable. The concept of optical basicity developed by Duffy and coworkers (Baucke and Duffy 1991; Binks and Duffy 1990; Duffy 1989; Duffy and Ingram 1971a,b, 1976) offers a very useful approach. Initially applied to metallurgical slags and glasses, it is now finding much wider use.

When a cation is coordinated by a Lewis base (simple or complex anion), its outer orbitals are profoundly affected. In the case of d electrons, changes in the ligand field spectra are related to the nature of the ligand and are a measure

Table 5.18. *Optical basicities of oxides*

Oxide	Optical basicity
Cs_2O	1.7
K_2O	1.4
Na_2O	1.15
Li_2O	1.0
BaO	1.15
SrO	1.1
CaO	1.00
MgO	0.78
ZnO	0.95
MnO	1.0
FeO	1.0
Al_2O_3	0.60
SiO_2	0.48
B_2O_3	0.42
P_2O_5	0.33

Source: Data from Duffy and Ingram (1971); Duffy (1989).

of the spatial extent of orbitals and the degree of covalency. The spectral parameter most sensitive to these changes is the Racah B parameter. But d-d transitions are weak, and spectra hard to measure, so a better probe ion turns out to be a $d^{10}s^2$ ion such as Tl^+ or Pb^{2+}, whose strong ultraviolet $^1S_0 \rightarrow {}^3P_1$ band is equally sensitive to changes in the ionic/covalent nature of bonding. With greater covalency, the UV band is shifted to lower energies (red shifted). The frequency of the absorption of the probe ion in the sample of interest, v, is compared to that in the gas phase (where no electron donation is possible), v_ι, and in a very ionic oxide such as CaO, $v_{O^{2-}}$. The optical basicity, Λ, is defined as

$$\Lambda = \frac{v_\iota - v}{v_\iota - v_{O^{2-}}} \tag{5.53}$$

Thus the measurement of basicity is transformed into a fairly straightforward optical spectroscopic measurement on a sample doped with trace amounts of Pb^{2+} or Tl^+. The optical basicities of a number of oxides are shown in Table 5.18.

5.5.5 Variation of structure with temperature and pressure

The response of crystal structures to variations in P and T may be grouped into three categories: the changes in bond lengths and bond angles within what remains essentially the same structure (and space group), systematic changes in the degree of distortion of a structure leading eventually to a new phase of a higher or lower symmetry, and phase transformations to totally unrelated structures. Since Chapter 6 covers the systematics of the last two cases, here I focus on the variation of bond lengths and angles within a given structure type. If one views structures as linkages of cation-centered polyhedra, they may be placed in three groups. The first is structures in which the polyhedra are so closely linked by edge and face sharing that it is impossible to change one bond length without affecting other polyhedra. Examples are the pseudobrookite and ilmenite structures. Their thermal expansivities and compressibilities are in general quite anisotropic (reflecting the anisotropy of edge sharing) and hard to predict. The second class is structures with large open cavities that can accommodate changes in volume by changing the geometry of the cavities, generally by rotating tetrahedra. Cordierite and zeolites are examples. Their responses to P and T are hard to systematize, but such open structures often have very small (or even negative) thermal expansivities because the expansion is compensated by movement into the cages. The third class is "normal" materials consisting of corner-shared polyhedral units. Here the polyhedra can respond relatively independently to changes in P and T, and the response of the crystal is some appropriately weighted sum of these almost-additive individual adjustments. These are the materials for which a comparative systematic approach based on individual polyhedra is the most useful.

Figure 5.45 shows the variation of average Mg–O distances with temperature and with pressure in a number of different compounds. Although there are considerable variations in the Mg–O distance at ambient conditions, the lines show roughly parallel slopes, suggesting that constant bond expansivity and compressibility are useful approximations. If this is the case, then it is useful to think of the volumetric expansivity and compressibility of a coordination polyhedron, for example, of an MgO_6 octahedron. Such polyhedral expansivities and compressibilities are shown in Figure 5.46.

The effect of pressure is to contract the bond. One expects weak bonds to contract more than strong ones, and the Si–O bond is nearly incompressible to pressures of about 10 GPa, above which distortions in the tetrahedra begin to play a role. For isostructural crystals, the bulk modulus, K, times the molar volume, V, is a constant (Anderson and Anderson 1970).

$$KV = \text{constant} \tag{5.54}$$

This can be justified on theoretical grounds and is also found to apply to individual cation polyhedra such that

$$r^3 K/z = \text{constant} \tag{5.55}$$

where r is the bond length, K the polyhedral bulk modulus, and z the cation charge (Hazen and Finger 1982).

For crystals in which the polyhedra are not strongly constrained by edge and/or face sharing, the effects of pressure and temperature often scale in the sense that a large thermal expansivity and a large compressibility go together. On a microscopic level, the thermal expansivity probes the anharmonicity of the interatomic potential, whereas the compressibility probes the entire potential in a region where repulsion begins to dominate. Thus the relations between compression and expansion should be complex, but the correlations mentioned here show that empirical concepts and systematics remain very useful.

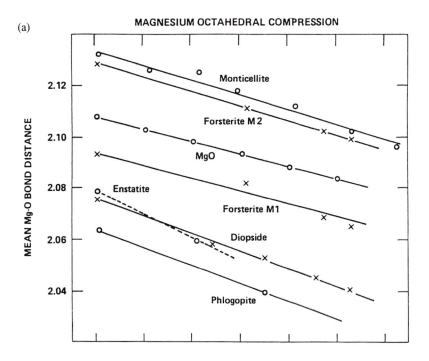

(a)

(b)

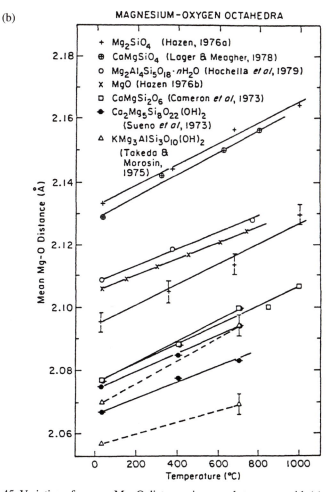

Figure 5.45. Variation of average Mg–O distances in several structures with (a) pressure and (b) temperature. (From Hazen 1988.)

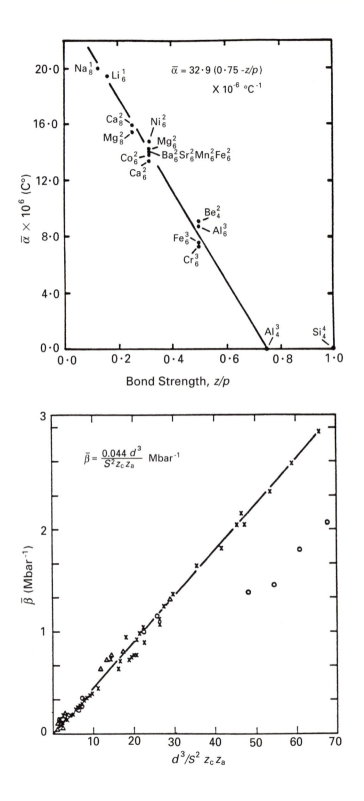

5.6 References

5.6.1 General references and bibliography

Adams, D. M. (1974). *Inorganic solids*. New York: Wiley.

Berry, R. S., S. A. Rice, and J. Ross. (1980). *Physical chemistry*. New York: Wiley.

Blakemore, J. S. (1985). *Solid state physics*, 2nd ed. Cambridge: Cambridge University Press.

Born, M., and K. Huang. (1954). *Dynamical theory of crystal lattices*. Oxford: Clarendon Press.

Callaway, J., and N. H. March. (1984). Density functional methods: Theory and applications. *Solid State Phys.* 38, 135–221.

Cotton, F. A. (1963). *Chemical applications of group theory*. New York: Wiley Interscience.

Cox, P. A. (1987). *The electronic structure and chemistry of solids*. Oxford: Oxford Science Publications, Oxford University Press.

Gibbs, G. V. (1982). Molecules as models for bonding in silicates. *Amer. Mineral. 67*, 421–50.

Harrison, W. A. (1980). *Electronic structure and the properties of solids*. San Francisco: W. H. Freeman.

Hess, A. C., M. I. McCarthy, and P. F. McMillan. (1993). Ab initio methods in geochemistry and mineralogy. In *Advances in electronic structure theory*, 2nd ed. T. H. Dunning, Jr. New York: JAI Press.

Hume-Rothery, W., and G. V. Raynor. (1944). *The structure of metals and alloy*. 2nd ed. London: Institute of Metals.

Kieffer, S. W., and A. Navrotsky, eds. (1985). *Microscopic to macroscopic – atomic environments to mineral thermodynamics*. Reviews in Mineralogy, vol. 13. Washington, DC: Mineralogical Society of America.

Kittel, C. (1960). *Introduction to solid state physics*, 2nd ed. New York: Wiley.

Lawley, K. P., ed. (1987). *Ab initio methods in quantum chemistry*. New York: Wiley.

McWeeny, R. (1978). *Methods of molecular quantum mechanics*. New York: Academic Press.

O'Keeffe, M., and A. Navrotsky, eds. (1981). *Structure and bonding in crystals*, Vol. I and II. New York: Academic Press.

Phillips, J. C. (1973). *Bonds and bands in semiconductors*. New York: Academic Press.

Schaefer, H. F. (1977). *Methods of electronic structure theory*. New York: Plenum Press.

Figure 5.46. (*Facing page*) (Top) Mean coefficients of thermal expansion for cation-oxygen bonds in cation polyhedra versus Pauling bond strength. Cations are indicated in the form R_p^z where z is the cation valence and p is the coordination number. (Bottom) Mean polyhedral compressibility versus mean compressibility for cation-anion bonds in cation polyhedra versus $d^3/S^2 z_e z_a$, where d is cation-anion mean distance. S^2 is a factor that is constant for each type of anion (0.50 for oxygen, 0.75 for all halides, 0.40 for all chalcogenides, 0.25 for phosphides and nitrides, and 0.20 for carbides), and Z_e and Z_a are cation and anion valences, respectively. Triangles, crosses, and circles are for four-, six-, and eight-coordinated cations, respectively. (From Hazen and Prewitt 1977.)

Schlüter, M., and L. J. Sham. (1982). Density functional theory. *Phys. Today 35*, 36–43.

Simons, J. (1991). An experimental chemist's guide to ab initio quantum chemistry. *J. Phys. Chem. 95*, 1017–29.

Szabo, A., and N. S. Ostlund. (1989). *Modern quantum chemistry, introduction to advanced electronic structure theory,* rev. ed. New York: McGraw-Hill.

Tossell, J. A., and D. J. Vaughan. (1992). *Theoretical geochemistry: Application of quantum mechanics in the earth and mineral sciences.* New York: Oxford University Press.

5.6.2 Specific references

Abbott, R. N., Jr., J. E. Post, and C. W. Burnham. (1989). Treatment of the hydroxyl in structure-energy calculations. *Amer. Mineral. 74*, 141–50.

Ahrens, L. H. (1952). The use of ionization potentials – 1. Ionic radii of the elements. *Geochim. Cosmochim. Acta 2*, 155–69.

Aidun, J., M. S. T. Bukowinski, and M. Ross. (1984). Equation of state and metalization of CsI. *Phys. Rev. B 29*, 2611–21.

Anderson, D. L., and O. L. Anderson. (1970). The bulk modulus-volume relationship for oxides. *J. Geophys. Res.* 3494–500.

Andreoni, W. (1985). On the structural classification of double-octet AB_2 compounds. *Helv. Phys. Acta 58*, 226–33.

Baucke, F. G. K., and J. A. Duffy. (1991). The effect of basicity on redox equilibria in molten glasses. *Phys. Chem. Glasses 32*, 211–19.

Barron, T. H. K., C. C. Huang, and A. Pasternak. (1976). Interatomic forces and lattice dynamics of α-quartz. *J. Physics C, Solid State Phys. 9*, 3925–40.

Baur, W. H. (1981). Interatomic distance predictions for computer simulation of crystal structures. In *Structure and bonding in crystals*, ed. M. O'Keeffe and A. Navrotsky, vol. 2, 31–52. New York: Academic Press.

Binks, J. H., and J. A. Duffy. (1990). Chemical bonding in rocksalt structured transition metal oxides. *J. Solid State Chem. 87*, 195–201.

Bloch, A. N., and G. Schatteman. (1981). Quantum-defect orbital radii and the structural chemistry of simple solids. In *Structure and bonding in crystals,* ed. M. O'Keeffe and A. Navrotsky, vol. 1, 49–72. New York: Academic Press.

Brawer, S. A. (1981). Defects and fluorine diffusion in sodium fluoroberyllate glass: A molecular dynamics study. *J. Chem. Phys. 75*, 3516–21.

Brown, I. D. (1981). The bond-valence method: An empirical approach to chemical structure and bonding. In *Structure and bonding in crystals*, ed. M. O'Keeffe and A. Navrotsky, vol. 2, 1–30. New York: Academic Press.

Brown, I. D. (1987). Recent developments in the bond valence model of inorganic bonding. *Phys. Chem. Minerals 15*, 30–4.

Bukowinski, M. S. T. (1979). Compressed potassium: A siderophile element. *High Pressure Science and Technology 2*, 237–44.

Bukowinski, M. S. T. (1980). Effect of pressure on bonding in MgO. *J. Geophys. Res. 85*, 285–92.

Burdett, J. K. (1981). Factors influencing solid-state structure – An analysis using pseudopotential radii structural maps. *Phys. Rev. B 24*, 2903–12.

Burdett, J. K. (1982b). Predictions of the structure of complex solids. *Adv. Chem. Phys. 49*, 47–113.

Burdett, J. K. (1982a). New ways to look at solids. *Acc. Chem. Res. 15*, 34–9.

Burns, R. G. (1993). *Mineralogical applications of crystal field theory.* Cambridge: Cambridge University Press.

Busing, W. R. (1970). An interpretation of the structures of alkaline earth chlorides in terms of interionic forces. *Trans. Am. Crystallogr. Assoc. 6,* 57–72.

Busing, W. R. (1981). *WMIN, A computer program to model molecules and crystals in terms of potential energy functions.* National Technical Information Service ORNL-5747. Springfield, VA: National Technical Information Service.

Catlow, C. R. A., J. M. Thomas, S. C. Parker, and D. A. Jefferson. (1982). Simulating silicate structures and the structural chemistry of pyroxenoids. *Nature 295,* 658–62.

Chang, K. J., and M. L. Cohen. (1984). High-pressure behavior of MgO: Structural and electronic properties. *Phys. Rev. B 30,* 4774–81.

Chelikowsky, J. R. (1982). Diagrammatic separation scheme for transition-metal binary compounds. *Phys. Rev. B26,* 3433–5.

Cohen, R. E. (1987). Elasticity and equation of state of $MgSiO_3$ perovskite. *Geophys. Res. Lett. 14,* 1053–6.

Cohen, R. E. (1991). Bonding and elasticity of stishovite SiO_2 at high pressure: Linearized augmented plane wave calculations. *Amer. Mineral. 76,* 733–42.

Cohen, R. E., L. L. Boyer, and M. J. Mehl. (1987a). Lattice dynamics of the potential-induced breathing model: Phonon dispersion in the alkaline-earth oxides. *Phys. Rev. B 35,* 5749–60.

Cohen, R. E., L. L. Boyer, and M. J. Mehl. (1987b). Theoretical studies of charge relaxation effects on the statics and dynamics of oxides. *Phys. Chem. Minerals 14,* 294–302.

Cohen, R. E., and C. W. Burnham. (1985). Energetics of ordering in aluminous pyroxenes. *Amer. Mineral. 70,* 559–67.

Cohen, R. E., and H. Krakauer. (1990). Lattice dynamics and origin of ferroelectricity in $BaTiO_3$: Linearized-augmented-plane-wave total-energy calculations. *Phys. Rev. B 42,* 6416–23.

Dempsey, M. J., and R. G. J. Strens. (1976). Modelling crystal structures. In *The physics and chemistry of minerals and rocks,* ed. R. G. J. Strens, 443–58. New York: Wiley.

Duffy, J. A. (1989). A common optical basicity scale for oxide and fluoride glasses. *J. Non-cryst. Solids 109,* 35–9.

Duffy, J. A., and M. D. Ingram. (1971a). Establishment of an optical scale for Lewis basicity in inorganic oxyacids, molten salts and glasses. *J. Amer. Chem. Soc. 93,* 6448–54.

Duffy, J. A., and M. D. Ingram. (1971b). A new correlation between s-p spectra and the nephelauxetic ratio: Applications in molten salt and glass chemistry. *J. Chem. Phys. 54,* 443–4.

Duffy, J. A., and M. D. Ingram. (1976). An interpretation of glass chemistry in terms of the optical basicity concept. *J. Non-cryst. Solids 21,* 373–410.

Ellison, A. J. G., and A. Navrotsky. (1991). Stoichiometry and local atomic arrangements in crystals. *J. Solid State Chem. 94,* 130–48.

Fumi, F. G., and M. P. Tosi. (1964). Ionic sizes and Born repulsion parameters in the NaCl-type alkali halides (I) Huggins-Mayer and Pauling forms. *J. Phys. Chem. Solids 25,* 31–43.

Geisinger, K. L., and G. V. Gibbs. (1981). SiSSi and SiOSi bonds in molecules and solids: A comparison. *Phys. Chem. Minerals 7,* 204–10.

Geisinger, K. L., G. V. Gibbs, and A. Navrotsky. (1985). A molecular orbital study of bond length and angle variation in framework silicates. *Phys. Chem. Minerals 11,* 266–83.

Ghose, S., J. M. Hastings, L. M. Corliss, K. R. Rao, S. L. Chaplot, and N. Choudbury. (1987). Study of phonon dispersion relations in forsterite, Mg_2SiO_4 by inelastic neutron scattering. *Solid State Comm. 63*, 1045–50.

Gibbs, G. V., E. P. Meagher, M. D. Newton, and D. K. Swanson. (1981). A comparison of experimental and theoretical bond length and angle variations for minerals, inorganic solids, and molecules. In *Structure and bonding in crystals*, ed. M. O'Keeffe and A. Navrotsky, vol. 1, 195–225. New York: Academic Press.

Gibbs, G. V., M. A. Spackman, and M. B. Boisen, Jr. (1992). Bonded and promolecule radii for molecules and crystals. *Amer. Mineral. 77*, 741–50.

Goldschmidt, V. M., T. Barth, G. Lunde, and W. H. Zachariasen. (1926). Geochemical Distribution Law of the Elements, Skrifer Norske-Videnskaps. *Akademi 2*, 1–117.

Hazen, R. M. (1988). A useful fiction: Polyhedral modeling of mineral properties. *Amer. J. Science 288-A*, 242–69.

Hazen, R. M., and L. W. Finger. (1981). Module structure variation with temperature, pressure, and composition: A key to the stability of modular structures? In *Structure and bonding in crystals*, ed. M. O'Keeffe and A. Navrotsky, vol. 2, 109–16. New York: Academic Press.

Hazen, R. M., and L. W. Finger. (1982). *Comparative crystal chemistry*. New York: Wiley-Interscience.

Iishi, K. (1978a). Lattice dynamics of corundum. *Phys. Chem. Minerals 3*, 1–10.

Iishi, K. (1978b). Lattice dynamical study of the α-β quartz phase transition. *Amer. Mineral. 63*, 1190–7.

Iishi, K. (1978c). Lattice dynamics of forsterite. *Amer. Mineral. 63*, 1198–208.

Isaak, D. G., R. E. Cohen, M. J. Mehl, and D. J. Singh. (1993). Phase stability of wustite at high pressure from first principles LAPW calculations. *Phys. Rev. B47*, 7720–31.

Jackson, R. A., and C. R. A. Catlow. (1988). Computer simulation studies of zeolite structure. *Molecular Simulation 1*, 207–24.

Johnson, M. L., and R. Jeanloz. (1983). A brillouin-zone model for compositional variation in tetrahedrite. *Amer. Mineral. 68*, 220–6.

Kubicki, J. D., and A. C. Lasaga. (1992). Ab initio molecular dynamics simulations of melting in forsterite and $MgSiO_3$ perovskite. *Amer. J. Science 292*, 53–83.

Lasaga, A. C., and G. V. Gibbs. (1987). Applications of quantum mechanical potential surfaces to mineral physics calculations. *Phys. Chem. Minerals 14*, 107–17.

Lasaga, A. C., and G. V. Gibbs. (1990). Ab-initio quantum mechanical calculations of water-rock interactions: Adsorption and hydrolysis reactions. *Amer. J. Science 290*, 263–95.

Leinenweber, K., and A. Navrotsky. (1988). A transferable interatomic potential for crystalline phases in the system $MgO–SiO_2$. *Phys. Chem. Minerals 15*, 588–96.

Liu, X., and C. T. Prewitt. (1990). High-temperature X-ray diffraction study of Co_3O_4: Transition from normal to disordered spinel. *Phys. Chem. Minerals 17*, 168–72.

Mao, H. K., and R. J. Hemley. (1992). Hydrogen at high pressure. *American Scientist 80*, 234–47.

Matsui, M., M. Akaogi, and T. Matsumoto. (1987). Computional model of the structural and elastic properties of the ilmenite and perovskite phases of $MgSiO_3$. *Phys. Chem. Minerals 14*, 101–6.

McMillan, P. F., and A. C. Hess. (1990). Ab initio valence force field calculations for quartz. *Phys. Chem. Minerals 17*, 97–107.

McMillan, P. F., and A. M. Hofmeister. (1988). *Infrared and raman spectroscopy*. Reviews in Mineralogy, vol. 18, 99–159. Washington, DC: Mineralogical Society of America.

Mehl, M. J., R. E. Cohen, and H. Krakauer. (1988). Linearized augmented plane wave electronic structure calculations for MgO and CaO. *J. Geophys. Res. 93*, B7, 8009–22.

Meier, W. M., and H. Villiger. (1969). Die methode der abstandsvergeinerung zur bestimmung der atomkoordinaten idealisierter geruststrukturen. *Zeitschrift für Kristallographie 129*, 411–23.

Miedema, A. R. (1973). The electronegativity parameter for transition metals: Heat of formation and charge transfer in alloys. *J. Less-Common Metals 32*, 116–36.

Miyamoto, M. (1988). Ion migration in $MgSiO_3$-perovskite and olivine by molecular dynamics calculations. *Phys. Chem. Minerals 15*, 601–4.

Mocala, K., A. Navrotsky, and D. M. Sherman. (1992). High temperature heat capacity of Co_3O_4 spinel: Thermally induced spin unpairing transition. *Phys. Chem. Minerals 19*, 88–95.

Mooser, E. (1983). Crystal chemistry and classification of multinary semiconductors. *Il Nuovo Cimento 2D*, 1613–27.

Navrotsky, A., K. L. Geisinger, P. McMillan, and G. V. Gibbs. (1985). The tetrahedral framework in glasses and melts – Inferences from molecular orbital calculations and implications for structure, thermodynamics, and physical properties. *Phys. Chem. Minerals 11*, 284–98.

Navrotsky, A., and O. J. Kleppa. (1967). The thermodynamics of cation distributions in simple spinels. *J. Inorg. Nucl. Chem. 29*, 2701–14.

O'Keeffe, M., and J.-O. Bovin. (1979). Solid electrolyte behavior of $NaMgF_3$: Geophysical implications. *Science 206*, 599–600.

O'Keeffe, M., and B. G. Hyde. (1981). The role of nonbonded forces in crystals. In *Structure and bonding in crystals*, ed. M. O'Keeffe and A. Navrotsky, vol. 1, 227–54. New York: Academic Press.

O'Neill, H. St. C., and A. Navrotsky. (1983). Simple spinels: Crystallographic parameters, cation radii, lattice energies, and cation distribution. *Amer. Mineral. 68*, 181–94.

Ottonello, G. (1986). Energetics of multiple oxides with spinel structure. *Phys. Chem. Minerals 13*, 79–90.

Pauling, L. (1927). The sizes of ions and the structure of ionic crystals. *J. Amer. Chem. Soc. 49*, 765–90.

Pauling, L. (1960). *The nature of the chemical bond.* 3rd ed. Ithaca, NY: Cornell University Press.

Pearson, W. B. (1972). *The crystal chemistry and physics of metals and alloys.* New York: Wiley-Interscience.

Phillips, J. C. (1970). Ionicity of the chemical bond in crystals. *The Reviews of Modern Physics 42*, 317–56.

Phillips, J. C. (1974). Chemical bonds in solids. In *Treatise on solid state chemistry*, ed. N. B. Hannay, 1–41. New York: Plenum Press.

Pisani, C., R. Dovesi, and C. Roetti. (1988). *Hartree-Fock ab initio treatment of crystalline systems.* New York: Springer-Verlag.

Post, J. E., and C. W. Burnham. (1986). Ionic modeling of mineral structures and energies in the electron gas approximation: TiO_2 polymorphs, quartz, forsterite, diopside. *Amer. Mineral. 71*, 142–50.

Post, J. E., and C. W. Burnham. (1987). Structure-energy calculations on low and high albite. *Amer. Mineral. 72*, 507–14.

Price, G. D., and S. C. Parker. (1984). Computer simulations of the structural and physical properties of the olivine and spinel polymorphs of Mg_2SiO_4. *Phys. Chem. Minerals 10*, 209–16.

Purton, J., and C. R. A. Catlow. (1990). Computer simulation of feldspar structures. *Amer. Mineral. 75*, 1268–73.

Rao, K. R., S. L. Chaplot, N. Choudbury, S. Ghose, and D. L. Price. (1987). Phonon density of states and specific heat of forsterite, Mg_2SiO_4. *Science 236*, 64–5.

Sanders, M. J., M. Leslie, and C. R. A. Catlow. (1984). Interatomic potentials for SiO_2. *J. Chem. Soc., Chem. Commun.*, No. 19, 1271–3.

Shannon, R. D. (1976). Revised effective ionic radii and systematic studies of interatomic distances in halides and chalcogenides. *Acta Cryst. A32*, 751–67.

Shannon, R. D., and C. T. Prewitt. (1969). Effective ionic radii in oxides and fluorides. *Acta Cryst. B25*, 925–46.

Soules, T. F. (1979). A molecular dynamics calculation of the structure of sodium silicate glasses. *J. Chem. Phys. 71*, 4570–8.

Spackman, M. A., R. J. Hill, and G. V. Gibbs. (1987). Exploration of structure and bonding in stishovite with fourier and pseudoatom refinement methods using single crystal and powder X-ray diffraction data. *Phys. Chem. Minerals 14*, 139–50.

Stebbins, J. F., and P. McMillan. (1989). Five- and six-coordinated Si in $K_2Si_4O_9$ glass quenched from 1.9 GPa and 1200°C. *Amer. Mineral. 74*, 965–8.

Stixrude, L., and M. S. T. Bukowinski. (1988). Simple covalent potential models of tetrahedral SiO_2: Applications to α-quartz and coesite at pressure. *Phys. Chem. Minerals 16*, 199–206.

Takayama-Muromachi, T., and A. Navrotsky. (1988). Energetics of compounds $(A^{2+}B^{4+}O_3)$ with the perovskite structure. *J. Solid State Chem. 72*, 244–56.

Weidner, D. J., and G. D. Price. (1988). The effect of many-body forces on the elastic properties of simple oxides and olivine. *Phys. Chem. Minerals 16*, 42–50.

Whittaker, E. J. W., and R. Muntus. (1970). Ionic radii for use in geochemistry. *Geochim. Cosochim. Acta 34*, 945–56.

Williams, Q., and R. Jeanloz. (1988). Spectroscopic evidence for pressure-induced coordination changes in silicate glasses and melts. *Science 239*, 902–5.

Zunger, A. (1980). Systematization of the stable crystal structure of all AB-type binary compounds: A pseudopotential orbital-radii approach. *Phys. Rev. B22*, 5839–72.

Zunger, A. (1981). A pseudopotential viewpoint of the electronic and structural properties of crystals. In *Structure and bonding in crystals*, ed. M. O'Keeffe and A. Navrotsky, vol. 1, 73–135. New York: Academic Press.

6

Mineral thermodynamics

6.1 Linking macroscopic and microscopic

Thermodynamic parameters (enthalpies, entropies, free energies) are useful in at least two ways. First, they are essential to understanding the pressure, temperature, and composition ranges over which a phase is stable. In earth sciences, geothermometers and geobarometers based on thermodynamic formalisms (and real thermodynamic data or a spectrum of sound to shaky empirical calibrations) serve to decipher geologic history (pressure, temperature, oxygen fugacity, compositional constraints as a function of time) from phases present in a rock. In materials science, synthesis and processing, compatibility, and device degradation are governed by thermodynamic (and kinetic) parameters. In both geologic and materials processes, heat balance, energy input/output of a reacting system, and rates of heating and cooling are governed by heat capacities, heats of chemical reaction and phase change, and thermal conductivity. Thus macroscopic thermochemical parameters have important practical uses.

Second, thermochemical parameters, especially when analyzed systematically for a group of materials with the same structure but different compositions or for a set of polymorphs, give insight into structure and bonding, that is, into the microscopic (atomistic) realm. Such systematic considerations in turn let one make rational estimates of thermodynamic properties that have not yet been measured and to predict under what conditions a phase transition is likely or a new structure might form. Such predictions are especially useful for extreme conditions of pressure and temperature like those found in the interior of the Earth and other planets. The more they are based on sound theory (including the proper forms of equations for long extrapolations) and extensive systematics, the more reliable such predictions are likely to be.

It is well to remember that the macroscopic parameters of thermodynamics are linked by statistical mechanics to microscopic interatomic interactions described by quantum mechanics. The thermodynamic parameters are the result of averages over many particles of the occupations of available energy levels. Thus, whereas the solution (or approximation to the solution) of the Schrödinger equation (see Chap. 5) gives the energy level spacings, statistical mechanics define their occupancy at a given temperature and pressure (also requiring fairly severe approximations to make the problem tractable) and define the macroscopic properties. A rigorous or even semiquantitative presentation of how this is done is well beyond the scope of the book; some useful texts are in the general references and bibliography list. Here I attempt some qualitative "feel" for certain fundamental concepts and relations.

At a given temperature, the occupation of energy levels relative to the ground state is given by the Boltzmann factor

$$N_i/N_o = \exp\left(-\Delta\varepsilon_i/kT\right) = \exp\left(-\Delta E_i/RT\right) \tag{6.1}$$

where N_i is the number of particles in the ith excited state, N_o is the number in the ground state, k is Boltzmann's constant, $\Delta\varepsilon_i$ the energy difference per particle between the excited state and ground state, and T the absolute temperature (or ΔE is the energy difference per mole of particles and R the gas constant). This means that the larger the energy difference, the fewer particles can be excited. As T approaches absolute zero, only the ground state is populated.

The sum over all states of their Boltzmann factors is the partition function or Zustandsumme, variously given the symbol q or z.

$$q = \sum_i e^{-\Delta\varepsilon_i/kT} \tag{6.2}$$

This fundamental quantity can be related to macroscopic thermodynamic parameters as shown in Table 6.1.

Consider a system with a spacing of energy levels; two examples are given in Figure 6.1. Figure 6.1a is a simple *two-level system* with one ground state and one excited state, as might be approximated by a transition to a low-lying excited electronic state. Figure 6.1b shows an *infinite ladder* of equally spaced states, as would be seen for a harmonic oscillator (see Sec. 6.2).

In the two-level system, the partition function, energy, entropy, and heat capacity can be calculated very straightforwardly, since

$$q = 1 + e^{-\varepsilon/kT} \tag{6.3}$$

As temperature approaches zero, q approaches unity and E, S, and C_V approach zero. As $\Delta\varepsilon/kT$ approaches zero (kT is high relative to the energy difference), q approaches 2, the energy approaches $\Delta\varepsilon/2$, the entropy (per particle) approaches $k\ln 2$, and the particles are distributed equally between ground and

Table 6.1. *Relation of thermodynamic functions to partition function*

Partition function

$$q = \sum_i \exp\left(-\varepsilon_i/kT\right)$$

q = sum over states (Zustandsumme)
  of Boltzmann factors
q = some number ≥ 1

Helmholtz free energy
  $A = E - TS = -NkT \ln q = -RT \ln q$
  where N = Avogadro's number, k = Boltzmann's constant, R = gas constant

Pressure
  $P = -(\partial A/\partial V)_T = NkT\,(\partial \ln q/\partial V)_T$

Internal energy
  $E = NkT^2\,(\partial \ln q/\partial T)_V$

Heat capacity
  $C_V = 2NkT\,(\partial \ln q/\partial T)_V + NkT^2\,(\partial^2 \ln q/\partial T^2)_V$

Entropy
  $S = Nk \ln W = Nk \ln q + NkT\,(\partial \ln q/\partial T)_V$
  where W = number of accessible independent configurations

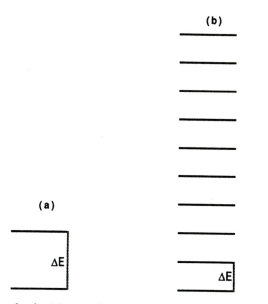

Figure 6.1. Energy levels: (a) a two-level system, (b) an infinite ladder of equally spaced states.

excited state (see Fig. 6.2). The heat capacity passes through a maximum at intermediate temperatures and diminishes to zero at high temperature, because once both states are equally populated, there is no further mechanism to raise the energy of the system. In other words, the two levels are equally populated and essentially equivalent when the available thermal energy (kT) is high relative to their spacing.

The system with an infinite ladder of levels behaves somewhat differently (see Fig. 6.3). The partition function, with $\varepsilon = h\nu$, given by

$$q = \sum_{n=0}^{\infty} \exp\left(-n\varepsilon/kT\right) = \frac{\exp(-\varepsilon/kT)}{[1 - \exp(-\varepsilon/kT)]} \tag{6.4}$$

continues to increase as temperature increases, because there are always higher levels to populate. Therefore the energy and entropy continue to increase, but at high temperature the heat capacity reaches a constant value of k per particle (R per mole). This is the classical equipartition value for one vibrational degree of freedom. Systems with more complex energy level spacings show more complex behavior, but the qualitative essence of the two-level system is main-

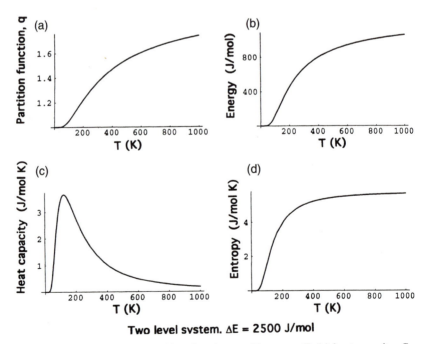

Two level system. ΔE = 2500 J/mol

Figure 6.2. Variation of (a) partition function, q; (b) energy, E; (c) heat capacity, C_V; and (d) entropy, S, with temperature, T, in two-level system.

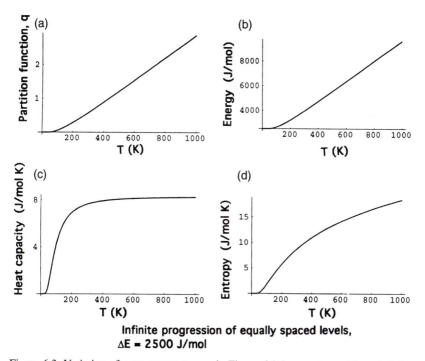

Infinite progression of equally spaced levels,
ΔE = 2500 J/mol

Figure 6.3. Variation of same parameters as in Figure 6.2 for a system with an infinite
ladder of equally spaced levels.

tained for a finite set of levels that saturate at high temperature, whereas that
of the infinite ladder is kept even when the rungs are unevenly spaced.

At equilibrium the two-level system can not have more than half its particles
in the higher state. Similarly, the infinite ladder always has a decreasing population of particles with increasing energy. However, nonequilibrium population
inversion can be attained if the exciting radiation is absorbed more rapidly than
thermal equilibrium can be maintained, that is, if the system is driven or
pumped. This situation is important in optical and NMR spectroscopy, and the
spontaneous and coherent decay from such an inversion is the basis of the laser
(see Sec. 3.6).

Increasing temperature drives any system to states of higher energy and
higher entropy. If a number of physically distinguishable states are energetically equivalent or close to equivalent, they will have comparable populations.
This is the basis of order–disorder, be it of d-electron populations, magnetic
moments, or site occupancies. In general, a disordered distribution of particles
will produce a configurational entropy. A special case pertains to distinguish-

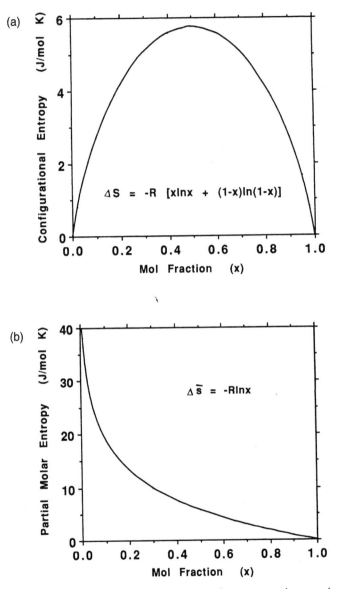

Figure 6.4. (a) Configurational entropy in a system where two species are mixed randomly over one mole of sites. (b) Partial molar entropy of mixing of components corresponding to mole fraction, X.

able particles (e.g., different atomic species such as Si and Al) occupying crystallographically and energetically equivalent sites. Writing the partition function for such a system generally leads to a configurational entropy contribution per mole of sites of the form

$$S_{conf} = -R \sum_i X_i \ln X_i \qquad (6.5)$$

where X_i is the mole fraction of a given species on that site. For the case where two species are mixed over one mole of sites, the configurational entropy is shown in Figure 6.4a. Note the vertical slope of the entropy curve at either end of the composition range. This implies that the partial molar entropy of each component becomes infinite (see Fig. 6.4b) as that component approaches an infinitely dilute state. Proper calculation of the entropy of mixing (and of partial molar entropies) is essentially a statistical exercise that takes into account the numbers of distinguishable particles and the types and numbers of sites. I stress here that the average user of thermodynamics makes this choice unwittingly, for example, by picking and using an ideal mixing equation. Though hidden, the choice has been made, and if it does not fit the physical nature of the system, the thermodynamic parameters one gets are often unduly large, with cumbersome activity coefficients trying to compensate for a wrong choice of configurational entropy (see Chapter 7). The proper link between microscopic structure (site occupancies, number of sites, speciation in melts) and macroscopic thermodynamics is essential yet often very difficult to attain.

6.2 Lattice vibrations, heat capacity, and entropy

6.2.1 Basic relations and magnitudes

The heat capacity of a substance is defined under conditions of constant pressure or constant volume:

$$C_V = (\partial E/\partial T)_V \qquad (6.6)$$

$$C_P = (\partial H/\partial T)_P \qquad (6.7)$$

where C_V and C_P are the heat capacities, E and H are internal energy and enthalpy, respectively, T is temperature, P is pressure, and V is volume. Because work must be done to expand a material during heating at constant pressure, C_P is larger than C_V, and

$$C_P - C_V = TV\alpha^2/\beta \qquad (6.8)$$

where α and β are thermal expansivity and compressibility, respectively. Thus

$$\alpha = V^{-1} (\partial V/\partial T)_P \qquad (6.9)$$

$$\beta = -V^{-1} (\partial V/\partial P)_T \qquad (6.10)$$

The reciprocal of the compressibility is the bulk modulus, K, used in the physics and geophysics literature.

For an ideal gas, $C_P - C_V$ is equal to the gas constant, R (8.314 JK^{-1}), α is $1/T$ (3.3 $\times$ 10^{-3} K^{-1} at room temperature), and β is $1/P$ (1 bar^{-1} at atmospheric pressure). For typical solids (also see Table 6.3), α is of the order of 10^{-5} K^{-1}, β is of the order of 10^{-6} bar^{-1}, and $C_P - C_V$ is only a few percent of C_V, and that only at high temperatures (> 1000 K).

For a crystalline material, the thermal expansivity and compressibility reflect the response of bond lengths and bond angles to changes in T and P. Although single parameters, the volume expansion and volume compressibility, can be defined as above, much insight can be gained by looking at the response along individual crystallographic axes or of individual bond lengths. Thus, for example, the Mg–O bond in silicates such as Mg_2SiO_4 (olivine) expands with increasing temperature and contracts with increasing pressure to a much greater extent than does the Si–O bond, (see Chap. 5).

6.2.2 Degrees of freedom and equipartition at high temperature

Consider an assemblage of N (Avogadro's number $= 6.024 \times 10^{23}$) noninteracting atoms in a low-density gas phase, for example, Ar gas at 1000 K and 0.01 atm. Each atom can move independently of the others in each of three directions (x,y,z). Classical kinetic theory of gases tells us that the average kinetic energy is 0.5 kT per degree of freedom. Thus the kinetic energy is 1.5 kT per atom or 1.5 RT per mole, and the heat capacity at constant volume, C_V, is 1.5 R (12.5 J/mol·K) per mole independent of temperature and pressure. The heat capacity at constant pressure, $C_P = C_V + R$, is 20.8 J/mol·K. Now suppose the same atoms are tied up in molecules, each containing n atoms (diatomic, triatomic, or polyatomic). The whole system still has $3N$ degrees of freedom. Each molecule can move in three directions, thus there are $3(N/n)$ translational degrees of freedom per mole, instead of $3N$ in isolated atoms. The remaining degrees of freedom, $3N(1-1/n)$, are modes internal to the molecules, rotations, and vibrations. Focusing now on each molecule, each has $3n-3$ internal modes. If the molecule is linear (e.g., any diatomic, a linear triatomic like HCN, or a linear tetratomic like acetylene, C_2H_2), it has two rotational degrees of freedom perpendicular to the axis of the molecule (a rotational about the

Table 6.2. *Heat capacities for molecules in the gas phase at high temperature and 1 atm*

Molecule	C_p, calc (J/mol·K)	C_p, obs (J/K·mol)		
		298 K	1000 K	1500 K
Ar	20.8[a]	20.8	20.8	20.8
H_2	37.4[b]	28.8	30.2	32.3[f]
O_2	37.4[b]	29.4	34.9	36.5[f]
H_2O	58.8[c]	33.6	41.3	47.1[f]
CO_2	62.4[d]	37.1	54.3	58.4[f]
CH_4	108.1[e]	35.6	71.8	86.6[f]

[a] $C_V = 3/2\ R$, $C_p = 5/2\ R = 20.8$ J/mol·K
[b] $C_V = 3/2\ R + R + R$, $C_p = 37.4$ J/mol·K
[c] $C_V = 3/2\ R + 3/2\ R + 3\ R$, $C_p = 58.2$ J/mol·K, nonlinear molecule
[d] $C_V = 3/2\ R + R + 4\ R$, $C_p = 62.4$ J/mol·K, linear molecule
[e] $C_V = 3/2\ R + 3/2\ R + 9\ R$, $C_p = 108.1$ J/mol·K, nonlinear molecule
[f] The calculated C_p values represent limits the experimental values approach as vibrations become fully excited.

molecular axis does not displace any mass). The remaining $3n-2$ degrees of freedom are vibrational. In the diatomic molecule, this leaves only one vibrational degree of freedom, namely, the bond stretching. In the linear triatomic, there are four degrees of freedom left, in the linear tetratomic, seven. In a nonlinear molecule, there are three translations, three rotations, and $3n-6$ vibrations. Classical theory assigns $0.5R$ to the heat capacity per translational mode, $0.5R$ per rotational mode, but R per vibrational mode. The reasons for this really stem from the high-temperature limit of statistical mechanical calculations, but the larger contribution per vibration can be rationalized by noting that a vibrating system is constantly interconverting kinetic and potential energy, whereas rotation and translation involve only kinetic energy. Table 6.2 summarizes the contributions to heat capacity for a variety of molecules based on this prescription and compares the calculated value of C_p to those measured experimentally at high temperature. This approach gives a useful way of estimating heat capacities of ideal gases. I stress that this approximation is not valid once intermolecular interactions become important when the gas is no longer ideal at low temperature or high pressure.

What if the $3N$ atoms are tied together in one giant molecule, that is, a crystal or glass, in which their movement past each other (diffusion) is very slow? The probability that all Avogadro's number of atoms move together (that a 56 gram chunk of iron metal will spontaneously start twirling or walk off the lab bench) is vanishingly small. Thus all $3N$ degrees of freedom are vibrational, and the heat capacity is $C_V = 3R = 24.9$ J/mol·K per gram atom of solid, or $3nR$ per mole, where a mole contains n atoms (e.g., for Mg_2SiO_4, $n = 7$). This limit of C_V is the basis of the law of Dulong and Petit, which states that the heat capacities of all solids approach a constant value of $3nR$ at high temperature.

6.2.3 Low-temperature behavior

The most important contribution of quantum mechanics to understanding thermodynamic behavior lies in the realization that because all energy levels in a bound system are quantized, one needs specific quanta of energy to excite systems into higher energy states. If these quanta are small relative to thermal energy, the system behaves classically. If they are large, only the first few energy levels are thermally accessible, and if the available thermal energy is totally insufficient to promote a significant number of molecules to excited states, the system remains in its ground state with no means of absorbing the thermal energy that is available. Put in anthropomorphic terms, an excitation can not be bought by credit card and paid off a little at a time; it must be cash (energy) up front. An immediate consequence is that as temperature approaches absolute zero, the heat capacity must diminish and approach zero. This is necessary for proper behavior of entropy and the third law of thermodynamics, but the consideration of quantized energy levels gives the macroscopic thermodynamic behavior microscopic meaning.

Figure 6.5 shows measured heat capacities of olivines M_2SiO_4 (M = Mg, Mn, Fe, Co). Apart from the spikes seen in the transition metal silicates, the heat capacity at a given temperature is smallest for Mg_2SiO_4 but fairly similar for the transition metal silicates. Within a given structure type, lattice vibrations are usually most easily excited (C_P is largest, the Debye temperature, Sec. 6.2.4, is lowest) for the material with the largest unit cell and with the longest and weakest M^{2+}–O bonds. This same trend is generally seen in S_{298}° (see Table 6.3). The spikes in the heat capacity are due to magnetic transitions. At high temperature, the spins are generally completely disordered, so that there is an added entropy of $2R \ln n$, where n is the number of electrons involved. However, this full entropy is not achieved until well above the actual transition temperature (Neel point) as Figure 6.5b shows. This implies that a significant amount of residual short-range order initially exists in the high-temperature state, disappearing rather gradually with increasing temperature. Thus heat capacity measurements, especially when combined with structural and magnetic studies, are very useful in elucidating the magnetic ordering of solids.

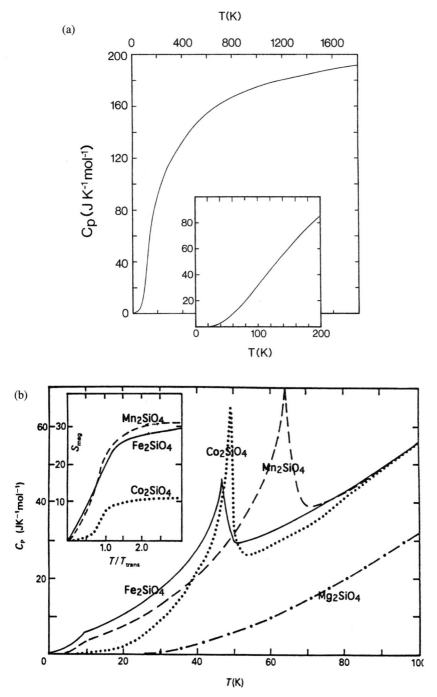

Figure 6.5. Heat capacities of some orthosilicates: (a) Mg_2SiO_4, (b) transition metal silicates at low temperature (Navrotsky 1987b, 1988b).

Table 6.3. *Heat capacities and entropies of minerals (J/(K·mol))*

Mineral	298 K		1000 K		1500 K	
	C_p	$S°$	C_p	$S°$	C_p	$S°$
MgO (periclase)	37.8	26.9	51.2	82.2	53.1	103.5
Al_2O_3 (corundum)	79.0	50.9	124.9	180.2	132.1	232.3
"FeO" (wüstite)	48.12	57.6	55.8	121.4	63.6	145.3
Fe_2O_3 (hematite)	103.9	87.4	148.5	252.7	144.6	310.5
Fe_3O_4 (magnetite)	150.8	146.1	206.0	390.2	201.0	471.5
TiO_2 (rutile)	55.1	50.3	73.2	129.2	79.5	160.1
$FeTiO_3$ (ilmenite)	99.5	105.9	133.7	249.3	155.0	307.4
Fe_2TiO_4 (titanomagnetite)	142.3	168.9	197.5	375.1	243.2	463.4
$MgAl_2O_4$ (spinel)	115.9	80.6	178.3	264.5	191.3	339.5
Mg_2SiO_4 (forsterite)	117.9	95.2	175.3	277.2	187.7	350.8
$MgSiO_3$ (enstatite)	82.1	67.9	121.3	192.9	127.6	243.5
$NaAlSi_3O_8$ (low albite)	205.1	207.4	312.3	530.1		
$KAlSi_3O_8$ (microcline)	202.4	214.2	310.3	533.8		
$Mg_3Al_2Si_3O_{12}$ (pyrope)	325.5	222.0	474.0	730.8		
$Ca_3Al_2Si_3O_{12}$ (grossular)	330.1	255.5	491.7	773.0		
$CaSiO_3$ (wollastonite)	85.3	82.0	123.4	213.4		
$CaSiO_3$ (pseudowollastonite)	86.5	87.5	122.3	217.6	132.3	269.1
$CaMgSi_2O_6$ (diopside)	166.5	143.0	248.9	401.7	269.7	506.3
$Mg_2Al_2Si_5O_{18}$ (cordierite)	452.3	407.2	698.3	1126.6	753.6	1420.9
$CaCO_3$ (calcite)	83.5	91.7	124.5	220.2		
$MgCO_3$ (magnesite)	76.1	65.1	131.5	190.5		
$CaMg(CO_3)_2$ (dolomite)	157.5	155.2	253.1	406.0		

Source: Data from Robie, Heminway, and Fisher (1978) and Berman (1988).

6.2.4 The interpretation of lattice heat capacities at intermediate temperatures

A crystal containing n atoms per formula unit has $3n$ vibrational degrees of freedom. If one assumes that all these vibrations are independent and occur at the same frequency, the Einstein model of lattice vibrations results (see Fig. 6.6). Although it correctly predicts the low- and high-temperature limits of C_V, this model does not properly describe the behavior at low T, predicting C_V is proportional to T^2 rather than T^3, and at intermediate temperatures it does not allow a good fit to experimental data.

The next level of complexity in approximating vibrational properties is the *Debye model*. Here one assumes that the density of vibrational states varies with the square of the vibrational frequency, with a maximum frequency or cutoff occurring when all degrees of freedom are accounted for (see Fig. 6.6).

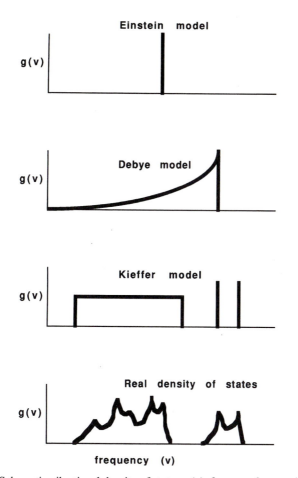

Figure 6.6. Schematic vibrational density of states, $g(\nu)$, for several approximations.

This maximum frequency is the Debye frequency, ν_D, and can be related to a Debye temperature, θ_D, by

$$\theta_D = h\nu_D/k \qquad (6.11)$$

with h Planck's constant and k Boltzmann's constant. The heat capacity at constant volume is given by

$$C_V = 9nNkT^3/\theta_D^3 \int_0^T e^x x^4 (e^x - 1)^2 \, dx = 3nNk\, D(\theta/T) \qquad (6.12)$$

where $x = h\nu_D/kT = \theta_D/T$. Thus the thermodynamic properties are a function of the ratio of the Debye temperature to the actual temperature, and the integral $D(\theta/T)$, known as the Debye function, is available in tabulations. The proper high-temperature and low-temperature limits obtain, and at low T

$$C_V = \frac{12}{5} nR\pi^4 \left(\frac{T}{\theta}\right)^3 \tag{6.13}$$

In general θ_D increases with the strength of bonding in the crystal and decreases with increasing molar volume for a given structure.

A test of how well the Debye model describes complex solids can be made by calculating the apparent Debye temperature from the heat capacity at various temperatures using Eq. 6.12. If the Debye approximation were completely adequate, θ_D would be constant. As an example, data for NaCl and MgO (rocksalt), Mg_2SiO_4 (olivine), and SiO_2 (quartz) are shown in Figure 6.7. The Debye approximation works fairly well for solids of high symmetry containing only one kind of atom (Pb, C) or two atoms of similar masses and bonding

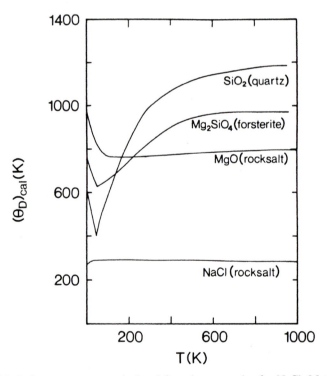

Figure 6.7. Debye temperature calculated from heat capacity for NaCl, MgO, SiO_2 (quartz), and Mg_2SiO_4 (olivine) (from Navrotsky 1988).

characteristics (NaF, NaCl, KCl) but becomes increasingly inadequate as the bond type, coordination, and mass of different atoms become more different (quartz, silicates). This is because vibrational frequencies are not uniformly distributed, and specific covalently bonded groups (e.g., SiO_4) retain some of their individual vibrational identity in crystals. For a few crystals the vibrational density of states and its dispersion across the Brillouin zone are well characterized by both infrared and Raman spectroscopy and by inelastic neutron scattering, and a rigorous calculation of heat capacity can be made. For most materials this is not the case. Nevertheless, there are important features, not predictable from a Debye model, that need explanation. For example, heat capacities (and entropies) may "cross over." Thus pyrope garnet $(Mg_3Al_2Si_3O_{12})$ has a higher heat capacity than grossular $(Ca_3Al_2Si_3O_{12})$ at low temperature, but the reverse is true above 400 K. This leads to an anomalously high entropy for pyrope, which has been linked to the low frequency vibrations of Mg in the eight-coordinated site. Similarly, a number of high-pressure phases (e.g., the perovskite form of $CdTiO_3$, $CdGeO_3$, and $MgSiO_3$) have high vibrational entropies.

From an empirical point of view, the high-temperature heat capacity, C_p, is often expressed by polynomials. To give the proper "knee" in C_p near the Debye temperature, various equations have been proposed. A form commonly used is

$$C_p = a + bT + cT^{-2} + dT^{-1/2} \tag{6.14}$$

The difficulty is finding functional forms that fit the knee in C_p well and, at the same time, extrapolate to reasonable high-temperature behavior. Because a number of polynomials in the literature fit the available data but actually show decreases in C_p above 1200 K, it is advisable to graph any calculated heat capacity, especially if it is an extrapolation, and test if it is physically reasonable before blindly using it in further computations.

6.2.5 Lattice dynamics and reasonable simplifications, particularly the Kieffer model

In order to calculate thermodynamic properties of minerals, Kieffer (1985) describes a generally applicable model that is consistent with lattice dynamical theory and that tries to remedy some of the inadequacies of the Debye model for complex silicates. The goal of this model is to estimate heat capacities from relatively readily available data for crystals (crystallographic parameters, sound speeds, elastic constants, vibrational spectra, compressibility, and thermal expansion) when a complete vibrational density of states (VDOS), such as that obtained by inelastic neutron scattering, is not available. This model tries to specifically address the following points. (1) Since many minerals show a high degree of acoustic anisotropy, three different acoustic modes, rather than a mean sound velocity, must be used. (2) Dispersion, that is, the change in frequency of a lattice mode across the Brillouin zone, can be complex. Approximations to dispersion relations are still a weak point of modeling. (3) The VDOS may show additional modes at both low and high frequencies relative to those accounted for in the Debye spectrum, which in a sense averages about a mean frequency. The excess high-frequency modes frequently represent local vibrations of discrete chemical species (SiO_4, CO_3, OH, etc.) typically above 900 cm^{-1}. The excess low-frequency modes may represent cooperative distortions and motions of the structure over relatively long distances or the vibrations of very loosely bound species, perhaps low-charged cations in large sites.

The VDOS in the Kieffer model (see Fig. 6.6) consists of three parts. First, there are three acoustic branches, whose maximum (cutoff) frequency is obtained from sound speeds or elastic constants and whose dispersion is given by a simple sinusoidal relation. Second, high-frequency optic modes are assigned to specific localized vibrations (Si–O stretches in a tetrahedron, C–O stretches in a planar CO_3 group, O–H stretches) and treated as a set of Einstein oscillators (each at a single frequency). The number of such oscillators is constrained by considering the symmetry and bonding geometry of the crystal. Third, the remaining modes are lumped into an *optic continuum*, a constant density of states starting at a lower frequency, ν_L, and stopping at a higher frequency, ν_H. These frequencies are estimated from vibrational frequencies seen in infrared and Raman spectra. Finally, some hopefully sensible approximations to dispersion relations in the optic modes are made. Sometimes, because in a real solid some modes disperse to higher and others to lower frequencies (see Fig. 6.8), the reasonable approximation is to ignore the effects of dispersion on the average VDOS in the optic continuum.

This synthetic VDOS is then taken through a proper lattice dynamical formalism to calculate the thermodynamic properties. Programs to do this run on personal computers or workstations. C_V is calculated and then converted to C_P by Eq. 6.8. The enthalpy and entropy are given by

$$H_T^\circ - H_0^\circ = \int_0^T C_P dT \qquad (6.15a)$$

or

$$H_T^\circ - H_{298}^\circ = \int_{298}^T C_P dT \qquad (6.15b)$$

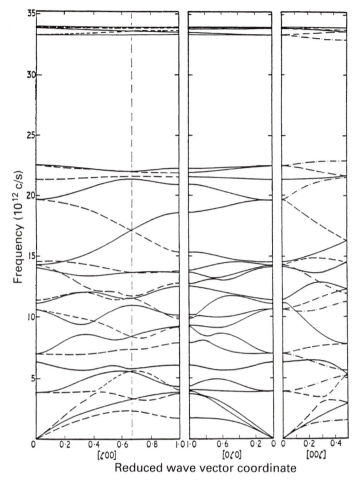

Figure 6.8. Dispersion relations for SiO_2 (quartz) (from Elcombe 1967).

and

$$S_T^\circ = \int_0^T (C_P/T)dT \qquad (6.16)$$

It is illuminating to compare the results of heat capacity calculations using the full VDOS, the Kieffer model, and the Debye approximation for a system for which the VDOS has been determined by neutron scattering, namely SiO_2 (quartz) (Ross 1993). Alpha-quartz, SiO_2, crystallizes in space group $P3_12$. There are nine atoms in the primitive unit cell and hence 27 branches to the

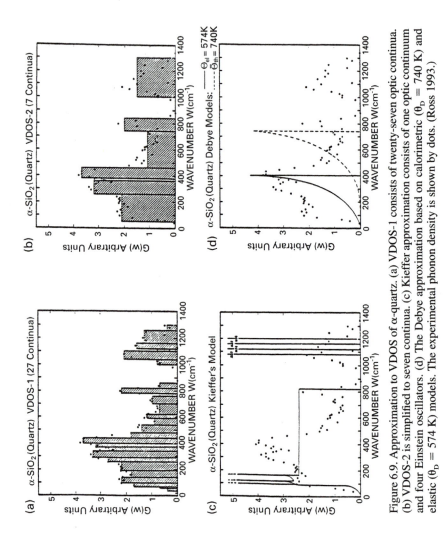

Figure 6.9. Approximation to VDOS of α-quartz. (a) VDOS-1 consists of twenty-seven optic continua. (b) VDOS-2 is simplified to seven continua. (c) Kieffer approximation consists of one optic continuum and four Einstein oscillators. (d) The Debye approximation based on calorimetric (θ_D = 740 K) and elastic (θ_D = 574 K) models. The experimental phonon density is shown by dots. (Ross 1993.)

phonon dispersion curves. Acoustic modes constitute 11% (3/27) of the total number of vibrational modes.

Models for α-quartz are shown in Figure 6.9. In the first model (Fig. 6.9a), VDOS-1, a close approximation to $g(\nu)$ is obtained by constructing a series of twenty-seven continua that mimic the experimental VDOS. The high frequencies (> 1000 cm^{-1}) account for 21.6% of the vibrational modes. The second model, VDOS-2, consists of seven continua. This model approximates $g(\nu)$ less rigorously than VDOS-1 but more rigorously than a Kieffer vibrational model, which greatly simplifies $g(\nu)$ by approximating the optic mode distribution with a single continuum extending from 90 cm^{-1} to 810 cm^{-1}. The high-frequency modes in the Kieffer model (Fig. 6.9c) are represented with four Einstein oscillators at 1080, 1117, 1162, and 1200 cm^{-1}. Twenty-two percent of the total number of vibrational modes are assigned to the Einstein oscillators. The lower and upper limits of the optic continuum and placement of Einstein oscillators are consistent with far-infrared, mid-infrared, and Raman spectroscopic data. Acoustic modes that constitute 11% of the modes are characterized by directionally averaged acoustic velocities ($u_1 = 3.76$, $u_2 = 4.46$, $u_3 = 6.05$ kms^{-1}) and are sinusoidally dispersed to 103, 122, and 165 cm^{-1} at the Brillouin zone boundary.

Two Debye models are shown in Figure 6.9d. The elastic Debye model (solid line) is based on the averaged shear wave and compressional velocities of α-quartz, 4.48 kms^{-1}, which gives an elastic Debye temperature of 574 K and a cutoff Debye frequency of 399 cm^{-1}. The thermal Debye model (dashed line) is based on an averaged thermal Debye temperature of 1062 K calculated from measured heat capacities between 350 K and 700 K (Robie, Heminway, and Fisher 1978). The cutoff Debye frequency for this model is 737 cm^{-1}. Both models deviate markedly from the measured $g(\nu)$. The high-frequency modes are not accounted for in either Debye model.

Heat capacities and entropies calculated from the models are compared with experimental values for α-quartz in Figures 6.10 and 6.11, respectively. Between 120 K and 700 K, heat capacities calculated from the VDOS-1, VDOS-2, and Kieffer models are within 2% of the experimental values. Deviations from the experimental values above 800 K may be due to increased anharmonic effects approaching the α-β quartz transition at 844 K. These three models also predict entropies within 2% of the experimental values from 298 K to 800 K. The elastic Debye model overestimates heat capacities by more than 10% from 100 K to 800 K. It also overestimates entropies by more than 25%. The thermal Debye model predicts heat capacities within 4% of the experimental data between 298 K and 800 K, but it underestimates entropies by more than 10%.

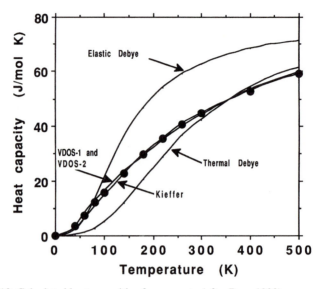

Figure 6.10. Calculated heat capacities for α-quartz (after Ross 1993).

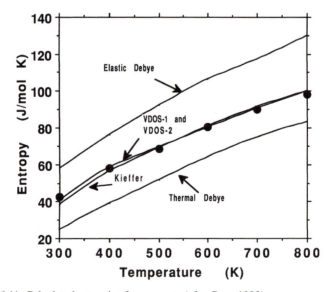

Figure 6.11. Calculated entropies for α-quartz (after Ross 1993).

Above 300 K, heat capacities calculated from the Kieffer, VDOS-2, VDOS-1, calorimetric Debye, and elastic Debye models increase in that order. Entropies above 300 K calculated from the calorimetric Debye model, however, are much lower than those calculated from the other models. Kieffer's model predicts slightly higher entropies than the VDOS models above 300 K.

The preceding calculations demonstrate how various approximations to $g(\nu)$ for α-quartz affect heat capacities and entropies. The elastic Debye model greatly overestimates heat capacities and entropies because all of the vibrational modes are concentrated below 400 cm^{-1}. The calorimetric Debye model, which provides reasonable estimates of heat capacities between 300 K and 800 K, underestimates entropies because the number of modes at low frequencies is underestimated.

Heat capacities and entropies change very little whether calculated from a model with twenty-seven continua that closely approximates $g(\nu)$ of α-quartz, an intermediate model with seven continua, or Kieffer's model with one continuum. Kieffer's model is correct in properly enumerating the acoustic modes and representing the optic mode distribution. All three models show excellent agreement with the experimental data. Thus these models all show vast improvement over the Debye approximation.

Kieffer's vibrational modeling approach is useful for estimating the heat capacities and entropies of complex mineral phases from vibrational and elastic data. There is, in general, no such thing as a unique Kieffer model for a given phase, but a family of slightly different models, all consistent with the available data, can be constructed and generally give C_p and $S°$ values within a few percent of each other. Models grossly at variance with the spectroscopic, elastic, and structural data generally yield significantly different thermodynamic parameters.

These families of vibrational models are useful in three ways. First, entropies and heat capacities can be estimated accurately for materials for which they have not been measured. These include both known materials (such as high-pressure phases available in small amounts) and hypothetical materials whose vibrational and elastic properties can be estimated from considerations of systematics. Second, the relations among structure, bonding, and thermodynamics can be probed by seeing the effect of different coordination geometry and linkage on lattice vibrations, heat capacity, and entropy. One can predict entropy systematics and ask what factors of structure and bonding lead to, for example, anomalously high or low entropies. These two applications will be discussed further in conjunction with the high-pressure magnesium silicates in Section 6.4. Third, the modeling can be used to refine an approximate VDOS most in accord with known thermochemical data. This refinement should be done with caution; it is usually unproductive to use the macroscopic thermodynamic parameters, which are grossly averaged samplings of the microscopic parameters, to try to win back the microscopic description in detail. The success of the Kieffer model lies in its relative insensitivity to fine details of the VDOS. But this strength turns to a weakness if one wishes to explore these fine details using the model.

6.3 Trends in compound formation

6.3.1 Formation of binary compounds from elements

Table 6.4 shows data for the enthalpies and entropies of formation of some oxides and silicates from the elements. These illustrate several general points. When compound formation involves the consumption of a gas (oxygen, halogens), a large negative entropy change is seen (about -120 ± 40 J/mol·K per mole of O_2, F_2, Cl_2, or other diatomic gas). When the reaction involves all condensed phases, $\Delta S°$ is small and can be either positive or negative. $\Delta H°$ in general parallels the lattice energy of the crystal, being most negative for the most ionic compound having the smallest lattice parameter. The ease of reduction of the compound to metal is generally determined by this value of $\Delta H°$ because $\Delta S°$ of formation does not vary much. Plots of $\Delta G°$ of formation versus temperature are generally straight lines whose slopes are determined by $\Delta S°$ and that show changes in slope when either reactants or products undergo a phase change. This is shown in Figure 6.12, which can also be used to find the equilibrium partial pressure for reduction.

In Table 6.4, the compound "FeO" is listed as $Fe_{0.95}O$; in fact, it has a homogeneity range that varies with temperature, total pressure, and oxygen partial pressure. The phenomenon of nonstoichiometry is a general one; the free-energy relations that determine the homogeneity range are shown in Figure 6.13a. For a very stable binary compound (such as NaCl) the free-energy curves show a deep and sharp minimum, and the resulting homogeneity range – defined by the common tangents to the free-energy curves for phases α, β, and γ – is small. For a less stable compound, the free-energy curves show a broader and shallower minimum, resulting in wider homogeneity ranges. This argument rationalizes the observation that stable ionic compounds (e.g., Mg_2SiO_4) generally show small deviations from stoichiometry, whereas compounds of relatively low stability (e.g., sulfides, intermetallic alloys) show extensive solid solution ranges. The exact forms of the free-energy curves are governed by the detailed energetics of substitution and/or defect formation.

Table 6.4. *Enthalpies and entropies of formation of selected compounds from elements and from oxides*

	Formation from elements				Formation from oxides			
	298 K		1000 K°		298 K°		1000 K°	
Compound	$\Delta H°$ kJ/mol	$\Delta S°$ J/mol·K	$\Delta H°$ kJ/mol	$\Delta S°$ J/mol·K	$\Delta H°$ kJ/mol	$\Delta S°$ J/mol·K	$\Delta H°$ kJ/mol	$\Delta S°$ J/mol·K
MgO (periclase)	−601.5[a]	−108.4[a]	−608.5[b]	−115.51[b]				
CaO (lime)	−635.1[a]	−106.5[a]	−634.3[b]	−103.6[b]				
Al_2O_3 (corundum)	−1675.7[a]	−313.8[a]	−1693.4[b]	−332.0[b]				
SiO_2 (quartz)	−910.7[a]	−182.6[a]	−905.1[b]	−174.9[b]				
SiO_2 (cristobalite)	−907.8[a]	−180.6[a]	−903.2[b]	−173.1[b]				
"FeO" (wüstite)	−266.3[a]	−70.9[a]	−263.3[b]	−63.9[b]				
Mg_2SiO_4 (forsterite)	−2174.4[a]	−400.7[a]	−2182.1[c]	−410.6[c]	−60.7[a]	1.4[a]	−58.2[c]	2.9[c]
$MgSiO_3$ (enstatite)	−1545.9[a]	−293.0[a]	−1552.9[c]	−296.5[c]	−33.7[a]	−2.1[a]	−38.2[c]	−4.9[c]
Fe_2SiO_4 (fayalite)	−1479.4[c]	−335.5[c]	−1472.3[c]	−321.4[c]	−24.6[a]	−12.7[a]	−28.7[c]	−19.9[c]
$CaSiO_3$ (wollastonite)	−1631.5[a]	−286.5[c]	−1630.4[c]	−278.2[c]	−85.7[a]	2.6[a]	−91.1[c]	0.1[c]
$CaSiO_3$ (pseudowollastonite)	−1627.4[a]	−283.0[a]	−1624.7[c]	−274.0[c]	−81.6[a]	6.1[a]	−85.3[c]	4.3[c]
$CaMgSi_2O_6$ (diopside)	−3200.5[a]	−585.2[a]	−3209.6[c]	−579.3[c]	−142.6[a]	−5.1[a]	−155.6[c]	−9.4[c]
$NaAlSi_3O_8$ (high albite)	−3924.2[c]	−730.6[c]	−3925.8[c]	−735.3[c]	−146.9[a]	39.0[a]	−156.0[c]	24.9[c]
$KAlSi_3O_8$ (sanidine)	−3959.6[c]	−737.5[c]	−3962.1[c]	−744.1[c]	−208.0[a]	36.1[a]	222.9[c]	12.3[c]
$CaAl_2Si_2O_8$ (anorthite)	−4228.7[c]	−756.7[a]	−4239.4[c]	−764.6[c]	−96.5[a]	28.7[a]	−100.9[c]	21.4[c]
$Mg_3Al_2Si_3O_{12}$ (pyrope)	−6286.5[c]	−1176.3[a]	−6317.0[c]	−1211.3[c]	−74.2[c]	−19.9[a]	−79.2[c]	−4.0[c]
$Mg_2Al_4Si_5O_{18}$ (cordierite)	−9158.7[a]	−1702.0[a]	−9200.3[c]	−1750.7[c]	−50.8[c]	55.2[a]	−67.0[c]	23.9[c]
$Ca_3Al_2Si_3O_{12}$ (grossular)	−6632.9[c]	−1214.4[c]	−6649.7[c]	−1214.4[c]	−319.8[c]	−33.3[a]	337.9[c]	−43.0[c]

[a] Berman (1988)
[b] JANAF (1986)
[c] Robie et al. (1978)

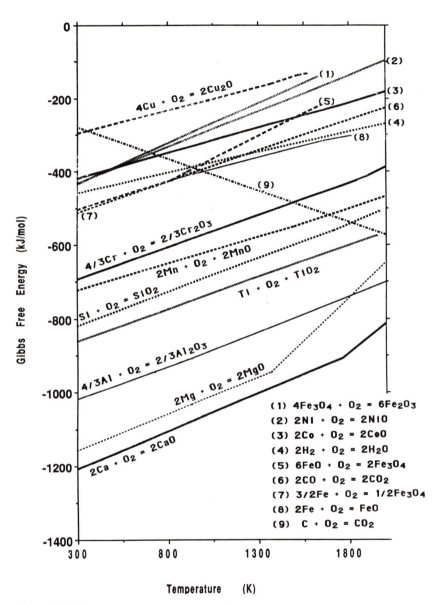

Figure 6.12. Free energy versus temperature for some binary oxides.

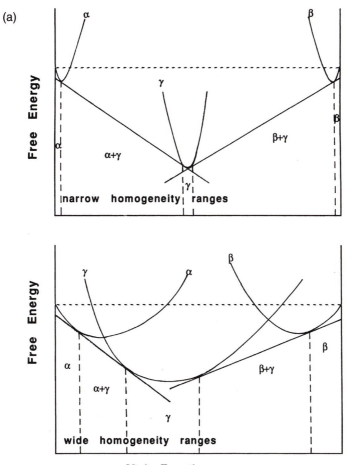

Figure 6.13. (*pp. 297–298*) (a) Free energy versus composition for compound showing narrow and wide homogeneity range. (b) Temperature composition section for the system Fe–O at 1 atm. Various oxygen fugacities are indicated. Note the wide homogeneity range of the wüstite phase (after Muan and Osborn 1965).

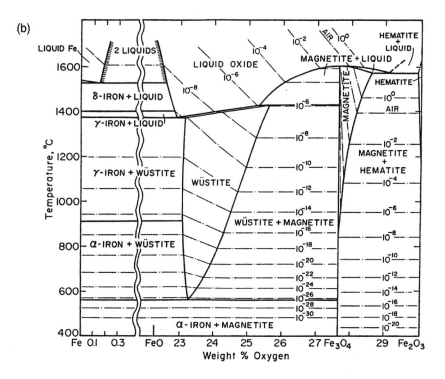

Figure 6.13b delineates the homogeneity range of "FeO" at 1 atm total pressure as a function of temperature and oxygen fugacity.

6.3.2 Formation of ternary compounds from binary compounds

Consider the enthalpy of the reaction

$$AO + B_aO_b = AB_aO_{1+b} \qquad (6.17)$$

with A = Ba, Sr, Ca, Mg, Mn, Fe, Co, Ni, and B = Si, Ge, W, C, S (see Fig. 6.14). The following patterns are evident. All enthalpies of formation are in the range 0 kJ to −300 kJ, with the aluminates, silicates, and germanates in the less negative portion of that range. Thus the stabilization of the ternary compounds is generally considerably less in magnitude than the formation enthalpies of the binary oxides. The entropies of formation, referred to solid binary oxides, are also quite small (from −15 to +15 JK^{-1} mol^{-1}). For a given B, the enthalpies of formation become more exothermic in the series Ni, Co,

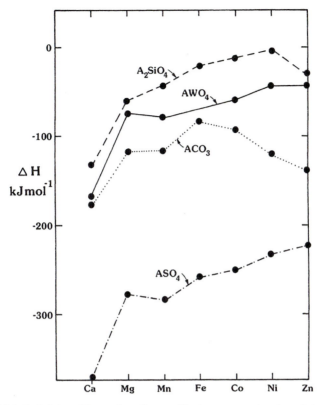

Figure 6.14. Enthalpies of formation of some silicates, germanates, tungstates, carbonates, and sulfates from their binary oxides (from Navrotsky 1974).

Fe, Mn, Mg, Ca, Sr, Ba, that is, with increasing basicity of the oxide AO. For a given A, the enthalpies of formation become more exothermic in the series Al, Si, Ge, W, C, S, that is, with increasing acidity of the oxide Al_2O_3, SiO_2, GeO_2, WO_3, CO_2, SO_3. Thus, the most stable compounds form when the most complete transfer of oxide ion from base to acid occurs, and the ternary structures then contain well-defined covalently bonded anions (SiO_4, GeO_4, WO_4, CO_3, SO_4). For compounds between more similar binary oxides (e.g., Al_2O_3 + SiO_2, CuO + Fe_2O_3, Fe_2O_3 + TiO_2), the enthalpies of formation are much smaller in magnitude and in the three cases mentioned are actually endothermic. Thus the stabilities of $Al_6Si_2O_{13}$ (mullite), $CuFe_2O_4$ (spinel), and Fe_2TiO_5 (pseudobrookite) derive from their entropies (configurational and/or vibrational), and at low temperature these compounds generally are metastable or decompose. Such *entropy stabilized* materials are quite common at high temperature in both mineralogical and ceramic systems.

6.4   High-pressure phase transitions and the Earth's interior

6.4.1   Geophysical background

We know about the interior of our planet from several types of evidence – direct sampling of rock brought to the surface by geologic process from depths no greater than 200 km, remote sensing especially by seismology (the study of natural or anthropogenic sound waves passing through the Earth), inferences from the chemistry of meteorites and from other geochemical arguments, and laboratory and computational simulations of the conditions at depth. In a sense, the basic question of deep earth geophysics is a peculiar sort of inverse problem in materials science; rather than determining the properties of a given material, one seeks to find materials, under constraints of natural elemental abundances, that have properties consistent with observations.

What are the hard facts about planet Earth as a material system? Its radius and mass are well constrained, as are pressure and density as a function of depth (see Fig. 6.15). Its vertical temperature distribution is less well known, but it is clear that the interior is hot, with temperatures in the mantle reaching perhaps 3000 K and in the core perhaps 6000–7000 K. The vertical stratification into a metallic solid inner core, metallic liquid outer core, nonmetallic (and presumably silicate) mantle, and silicate crust is well documented. The transitions between these regions are fairly sharp (on the order of several km) and the mantle is further divided into upper mantle (occupying depths of 30–400 km), transition zone (400–670 km), and lower mantle (670–2900 km). The latter is, in terms of both volume and mass, the largest part of the planet. The compositions of the various mantle regions are constrained within broad bounds, but present models can not pin down the composition of any region to better than a few percent variation in parameters such as $Fe/(Fe+Mg)$, $Si/(Fe+Mg)$, or Al_2O_3 content.

Recent experiments using arrays of seismometers to produce tomographic images of the interior have discovered lateral as well as vertical heterogeneities. Processes of plate tectonics, perhaps powered ultimately by convection in the metallic core, exchange matter between the crust and mantle. Downgoing slabs of cold near-surface material, initially containing significant amounts of water and carbon dioxide, perturb the temperature distribution and may remain distinct for millions of years to depths greater than 700 km. Upwellings reflecting mantle hot spots and spreading centers create new real estate at the

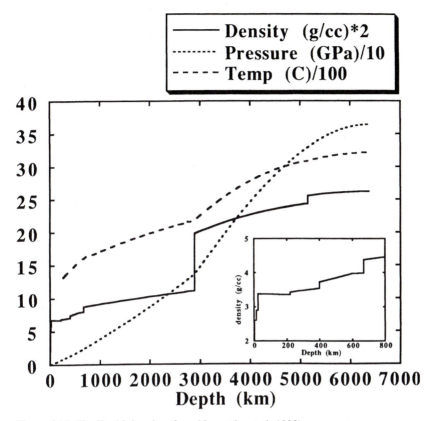

Figure 6.15. The Earth's interior (from Navrotsky et al. 1992).

surface in active volcanic regions such as Hawaii and Iceland. The Earth is thus a convecting and constantly chemically reacting dynamic system.

The change from crust to upper mantle (the Mohorovicic discontinuity, or Moho) is associated with a change in bulk composition to more silica-poor material and with the disappearance with increasing pressure of architecturally open structures such as quartz and feldspar. This transition and others as a function of depth are summarized in Table 6.5. As early as the 1930s, phase transitions were proposed to be associated with and responsible for the sharp increases in density that separate upper mantle, transition zone, and lower mantle. The transformation of Mg_2SiO_4 olivine (α) to a denser spinel (γ) structure was predicted on the basis of crystal chemical arguments and from its observation in *analog* materials such as Mg_2GeO_4 and Ni_2SiO_4. This use of chemical analogs, which show the desired structural changes at more readily

Table 6.5 *Seismic discontinuities, phase transitions, and chemical changes in the Earth*

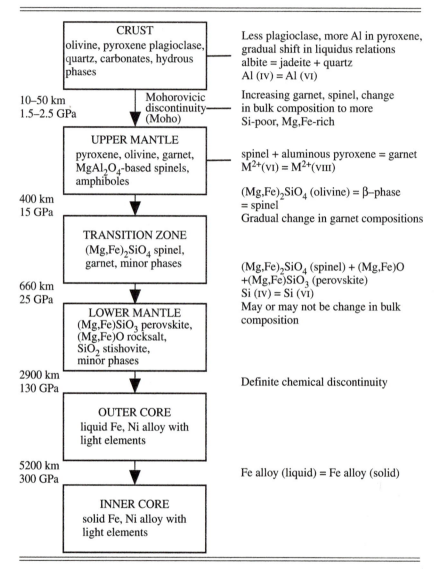

CRUST olivine, pyroxene plagioclase, quartz, carbonates, hydrous phases		Less plagioclase, more Al in pyroxene, gradual shift in liquidus relations albite = jadeite + quartz Al (IV) = Al (VI)
10–50 km 1.5–2.5 GPa	Mohorovicic discontinuity (Moho)	Increasing garnet, spinel, change in bulk composition to more Si-poor, Mg,Fe-rich
UPPER MANTLE pyroxene, olivine, garnet, $MgAl_2O_4$-based spinels, amphiboles		spinel + aluminous pyroxene = garnet $M^{2+}(VI) = M^{2+}(VIII)$
400 km 15 GPa		$(Mg,Fe)_2SiO_4$ (olivine) = β–phase = spinel Gradual change in garnet compositions
TRANSITION ZONE $(Mg,Fe)_2SiO_4$ spinel, garnet, minor phases		$(Mg,Fe)_2SiO_4$ (spinel) + (Mg,Fe)O +(Mg,Fe)SiO_3 (perovskite)
660 km 25 GPa		Si (IV) = Si (VI) May or may not be change in bulk composition
LOWER MANTLE $(Mg,Fe)SiO_3$ perovskite, (Mg,Fe)O rocksalt, SiO_2 stishovite, minor phases		
2900 km 130 GPa		Definite chemical discontinuity
OUTER CORE liquid Fe, Ni alloy with light elements		
5200 km 300 GPa		Fe alloy (liquid) = Fe alloy (solid)
INNER CORE solid Fe, Ni alloy with light elements		

accessible pressures and temperatures, widens the range of materials interesting to geophysics and blurs the boundary between earth science and materials science. The olivine-spinel transition in Mg_2SiO_4 and in natural olivine compositions containing 10–20% of the magnesium replaced by iron was confirmed when appropriate high-pressure equipment became available, but an intermediate phase (spinelloid, β, or wadsleyite) was also found. It is generally accepted that the olivine–beta phase–spinel transitions dominate the 400 km discontinuity, though garnet phases and hydrous magnesium silicates may also be important, the latter particularly in cooler regions associated with downgoing slabs.

At higher pressure, silicates undergo more radical structural rearrangements to phases with octahedrally coordinated silicon. The perovskite structure predominates and the reaction

$$(Mg,Fe)_2SiO_4 \text{ (spinel)} = (Mg,Fe)SiO_3 \text{ (perovskite)} \qquad (6.18)$$
$$+ (Mg,Fe)O \text{ (rocksalt)}$$

probably represents the dominant phase change at the 670 km discontinuity. Silicate perovskite may be the most abundant mineral in the lower mantle (80–100% by volume) and therefore the most abundant single phase in the planet. It is amusing that the most common mineral in the whole Earth may belong to a class of compounds (perovskites) whose structure and properties are far more important to materials scientists than to geologists ($CaTiO_3$ perovskite, the mineral after which the whole group is named, is a minor accessory mineral in the crust). It is equally ironic that $MgSiO_3$ perovskite, made laboriously in, at most, milligram quantities in the laboratory and subject to easy decomposition near ambient conditions, may exist in megaton quantities at a distance below our feet about the same as the distance between Boston and Washington (or London and Edinburgh).

The properties of $MgSiO_3$ perovskite may determine whether the 670 km discontinuity represents a change in bulk composition as well as a phase change. If so, movement of material across this boundary must be restricted to maintain the stratification and whole-mantle convection would be limited. Silicate perovskites may also be hosts for large ions, including the rare earths and potassium. This could seriously affect trace element geochemistry and the distribution of heat-generating natural radioactive elements in the Earth. Structural distortions, defect chemistry, and impurities (dopants) affect elastic constants, thermal and electrical conductivity, rheology (mantle silicates are generally in the ductile rather than brittle regime), diffusion rates, and other physical properties. Such properties determine the manner in which the planet evolves.

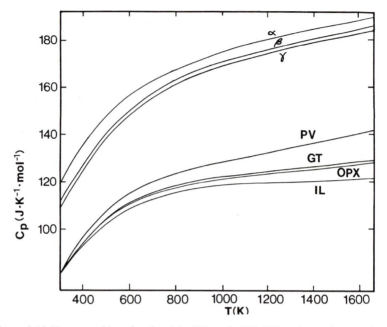

Figure 6.16. Heat capacities of α, β, γ Mg_2SiO_4 and of $MgSiO_3$ polymorphs, consistent with both direct measurements and calculations using Kieffer's model (after Navrotsky 1988a).

6.4.2 Systematics of relations among olivine (α), wadsleyite (β), and spinel (γ) structures

Heat capacities of α, β, and γ Mg_2SiO_4 are shown in Figure 6.16. They increase in the order $\gamma < \beta < \alpha$ at all temperatures, implying that entropy and volume are closely correlated and there are no crossovers in C_P and S°. The ΔH°, ΔS°, ΔV° of a given transition (e.g., olivine $\rightarrow$ spinel) have very different values for different compositions (see Table 6.6 and Fig. 6.17). The molar volume decreases in the order olivine, beta phase, spinel, and their heat capacities, bulk moduli, and thermal expansivities differ slightly (see Table 6.7). The contribution to the free energy of the ΔH°, $T\Delta S^\circ$, and $P\Delta V$ terms are of comparable magnitude. Thus the location of a phase boundary in P-T space can not be approximated from ΔH° alone. For example (see Fig. 6.18), although ΔH° for the olivine–spinel transition increases in the series Fe_2SiO_4, Ni_2SiO_4, Co_2SiO_4, Mg_2SiO_4, the pressure at which the spinel is first formed (near 1273 K) increases in the series Ni_2SiO_4 (near 3 GPa), Fe_2SiO_4 (near 6 GPa), Co_2SiO_4 (near 7 GPa), Mg_2SiO_4 (near 15 GPa).

Table 6.6. *Transformation energetics among* α, β, γ *polymorphs*

	$\Delta G°(T)$ (kJ/mol)	$\Delta H°$ (kJ/mol)	$\Delta S°$ (J/mol·K)	$\Delta V°$ (cm³/mol)
α = γ				
Mg_2SiO_4	54.1 (1000)	39.1	−15.0	−4.14
Fe_2SiO_4	17.8 (1000)	3.8	−14.0	−4.24
Co_2SiO_4	24.4 (1000)	11.3	−13.1	−3.92
Ni_2SiO_4	11.8 (1000)	6.0	−6.8	−3.42
Mg_2GeO_4	0 (1086)	−12.7	−11.7	−3.52
Fe_2GeO_4	−29.7 (1300)			−4.07
Co_2GeO_4	−14.6 (1473)			−3.89
Ni_2GeO_4	−34.3 (1473)			−3.12
α = β				
Mg_2SiO_4	36.1 (1000)	27.1	−9.0	−3.16
Fe_2SiO_4	20.5 (1000)	9.6	−10.9	−3.20
Co_2SiO_4	18.0 (1000)	9.0	−9.0	−2.90
Mn_2GeO_4	15.9 (1273)			−3.61

Source: Data from Navrotsky (1987a); Fei, Saxena, and Navrotsky (1990).

Table 6.7. *Thermal expansivity* (α), *bulk modulus* (K), *and its pressure derivative* (K' = (∂K/∂P)) *for some oxides and silicates*

Phases	V^0_{298} cm³/mol	$a(T) = a_0 + a_1T + a_2T^2$			K_0a GPa	K'	$(\partial K_T/\partial T)_p$ GPa/K
		a_0 (10⁴)	a_1 (10⁸)	a_2			
MgO							
Periclase	11.250	0.3681	0.9283	−0.7445	160.3	4.13	−0.0272
SiO₂							
Coesite	20.640	0.0543	0.8315	−0.0605	96.0	8.40	−0.0200
Stishovite	14.010	0.1023	1.3500	0.0000	314.0	6.00	−0.0470
Mg₂SiO₄							
Olivine	43.670	0.3052	0.8504	−0.5824	128.0	5.37	−0.0224
β-phase	40.540	0.2711	0.6885	−0.5767	172.0	4.30	−0.0323
Spinel	39.650	0.2367	0.5298	−0.5702	183.0	4.30	−0.0348
MgSiO₃							
Pyroxene	31.330	0.2947	0.2694	−0.5588	107.0	4.20	−0.0200
Garnet (cubic)	28.500	0.2966	0.2381	−0.5865	154.0	4.00	−0.0220
Ilmenite	26.350	0.2439	0.0000	0.0000	210.0	4.00	−0.0100
Perovskite	24.500	0.2627	1.5198	−0.0429	247.0	4.00	−0.0550

Source: Data from Fei et al. (1990).

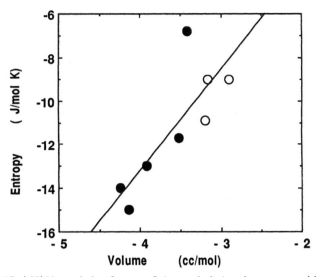

Figure 6.17. $\Delta S/\Delta V$ correlation for $\alpha \rightarrow \beta$ (open circles) and $\alpha \rightarrow \gamma$ transitions (solid circles).

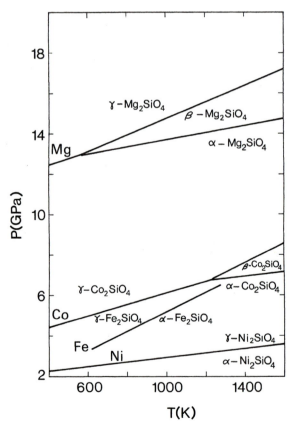

Figure 6.18. Phase relations in M_2SiO_4 where M = Mg, Fe, Co, Ni (from Navrotsky 1987a).

Table 6.8. *Bond lengths in olivine, modified spinel,*
and spinel polymorphs

	Average metal–oxygen distance (Å)	Average silicon–oxygen distance (Å)
Mg_2SiO_4		
α	2.112	1.636
β	2.081	1.651
γ	2.090	1.655
Co_2SiO_4		
α	2.121	1.639
β	2.081	1.642
γ	2.100	1.654
Fe_2SiO_4		
α	2.169	1.636
γ	2.136	1.654
Ni_2SiO_4		
α	2.090	1.640
γ	2.061	1.657

Source: Data from Sasaki et al. (1982).

The β-phase occurs between olivine and spinel; its stability field is limited to high temperatures by the α, β, γ triple point (see Fig. 6.18). $\Delta V°$, $\Delta H°$, and $\Delta S°$ for the α–β transition in Mg_2SiO_4 are 78%, 81%, and 63%, respectively, of the corresponding values for the α–γ transition. Similar relations hold for Co_2SiO_4. The β-phase is more similar in thermodynamic properties to the spinel than to the olivine. The α–β–γ transitions do not involve changes in the coordination numbers of any cations. However, the structures of β-phase and spinel are based on a slightly distorted cubic close packing of oxygen atoms, whereas the olivine structure is based on distorted hexagonal close packing. A small increase of mean Si–O bond length and a decrease of average M–O distance are observed in the α → β → γ transitions and these changes are larger in the α → β transition (see Table 6.8). Therefore, in terms of structure as well as in terms of thermochemistry data, the β-phase is more similar to spinel than to olivine for both Mg_2SiO_4 and Co_2SiO_4.

The β-phase has been observed in Mg_2SiO_4, Co_2SiO_4, Mn_2GeO_4, and Li_2WO_4, but not in the other silicates and germanates. A phase of similar structure is found at intermediate compositions in the systems Ni_2SiO_4–$NiAl_2O_4$

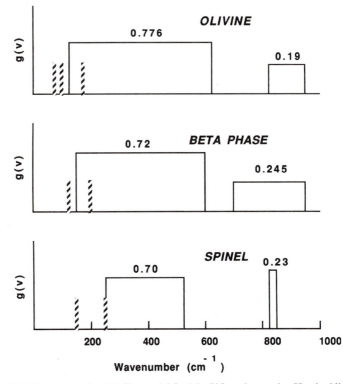

Figure 6.19. Representative Kieffer model for Mg_2SiO_4 polymorphs. Hatched lines represent acoustic modes, solid boxes represent optic continua (marked with fraction of modes therein). Each model is representative of a group of similar models (after Akaogi et al. 1984; Akaogi, Ito, and Navrotsky 1989).

and Mg_2GeO_4–$MgGa_2O_4$ but not for those end-members. The β-phase is one member of a family of structures called spinelloids, which appear to be higher in enthalpy and volume than the spinels and which probably owe their stability at high temperature to vibrational and/or configurational entropy (Akaogi and Navrotsky 1984; Leinenweber and Navrotsky 1989). Why the β-phase is stable for Co_2SiO_4 and Mg_2SiO_4 but not for Fe_2SiO_4 and Ni_2SiO_4 is not clear.

Application of Kieffer's vibrational models to the Mg_2SiO_4 polymorphs (see Fig. 6.19) leads to the following conclusions (Akaogi et al. 1984). A number of vibrational models can be constructed that are consistent with the infrared and Raman spectra of each phase. These models predict heat capacities and entropies of transition that are in good agreement with those calculated from combining high-pressure equilibria and calorimetry. The negative $\Delta S°$ values for $\alpha \to \beta$ and $\beta \to \gamma$ are based on rather subtle changes in the vibrational density of states and largely result from changes in the region of the density of states treated as an optic continuum. In particular, the low-frequency cutoff for this continuum, consistent with spectra, thermodynamics, and models for dispersion of low-frequency modes, is 130–140 cm^{-1} for olivine, 165–180 cm^{-1} for modified spinel, and 180–250 cm^{-1} for spinel. It is tempting to associate this shift to higher frequency with increasing regularity of the MgO_6 octahedron in the series α, β, γ. This regularity is in turn related to decreasing deviation from ideal close packing of the oxygens and changes in the pattern of occupancies of octahedral and tetrahedral interstices, which also result in increasing density in the sequence α, β, γ. Although the olivine–spinel transition does not involve any increase in the number of anions around a cation (conventional coordination number), it does involve an increase of the number of next-nearest cations around a given cation, $SiMg_9$ to $SiMg_{12}$. Cation–cation nonbonded interactions may make an important energetic contribution to the energetics (O'Keeffe and Hyde 1981). It appears reasonable that such an overall increase in packing efficiency results in additional vibrational constraints that shift some of the lowest frequency modes.

6.4.3 Systematics of ABO_3 structures at high pressures

At pressures considerably above those for the olivine–spinel transition, silicates and germanates either decompose to mixtures of MO (rocksalt) and SiO_2 or GeO_2 (rutile) or form new phases (garnet, ilmenite, perovskite) in which some or all of the Si or Ge is octahedrally coordinated (see Fig. 6.20 and Table 6.9). The large volume decrease associated with this increase in coordination number is the driving force for these transitions to "postspinel" phases. The wealth of phases observed is evidence for very closely balanced enthalpy, entropy, and volume factors. Table 6.9 lists available thermochemical data and Figure 6.21 compares the enthalpies relative to the phase stable at ambient conditions of a number of silicates and germanates. In the germanates, the wollastonite and garnet phases are very similar in enthalpy; the garnet is a high-pressure phase not because of significantly unfavorable energetics but because of its lower entropy and volume. $MnSiO_3$ and $MgSiO_3$ garnets, in contrast, are energetically much higher than the corresponding pyroxenes. $MgSiO_3$

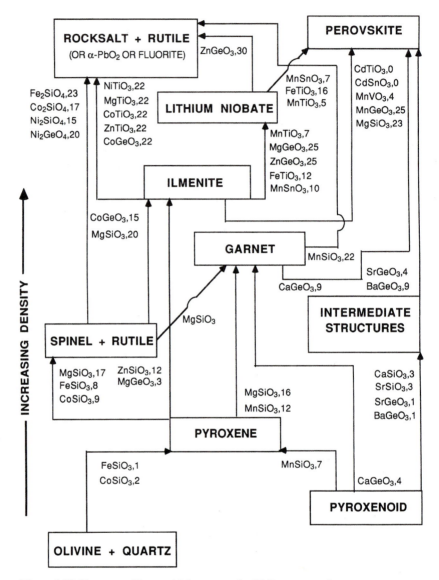

Figure 6.20. Phase transitions at high pressure in ABO_3 compounds.

Table 6.9. *Thermochemical parameters for high pressure phase transitions in ABO_3 compounds*

	$\Delta H°$ (kJ mol^{-1})	$\Delta S°$ (J/mol·K)	$\Delta V°$ (cm^3/mol)
pyroxene or pyroxenoid → garnet			
CaGeO$_3$[a]	-4.90 ± 4.2	-5.9 ± 1.5	-5.97
CdGeO$_3$[a]	$+0.6 \pm 2.7$	-8.4 ± 2.0	-5.30
MnSiO$_3$[a]	$+34.6 \pm 2.5$	-6.7 ± 2.1	-4.00
MgSiO$_3$[b]	35.1 ± 3.0	-2.0 ± 2.0	-2.83
pyroxene → ilmenite			
MgGeO$_3$[a]	7.5 ± 0.6	-6.3 ± 2.0	-5.11
MgSiO$_3$[c]	59.0 ± 4.3	-15.5 ± 2.0	-4.94
ZnSiO$_3$[g]	52.4 ± 2.7	-14.3 ± 5.9	-6.36
garnet → perovskite			
CaGeO$_3$[a]	$+43.3 \pm 5.0$	$+10.9 \pm 3.8$	-5.35
CdGeO$_3$[a]	$+43.1 \pm 5.0$	-1.7 ± 3.0	-4.88
MgSiO$_3$[b]	75.0 ± 3.0	-7.5 ± 3.0	
ilmenite → perovskite			
CdTiO$_3$[a]	15.0 ± 0.8	$+14.2 \pm 3.0$	-2.94
CdGeO$_3$[a]	34.3 ± 4.0	$+2.6 \pm 2.0$	-3.00
MgSiO$_3$[c]	51.1 ± 6.6	$+6.0 \pm 4.0$	-1.89
NaNbO$_3$[d]	-5.5 ± 1.3		-2.86
ilmenite → lithium niobate			
LiNbO$_3$[d]	-9.8 ± 4.1		-1.64
MnTiO$_3$[e]	5.3 ± 1.0	2.0 ± 1.0	-0.44
FeTiO$_3$[f]	18 ± 5		-0.34
lithium niobate → perovskite			
MnTiO$_3$[e]	1.0 ± 1.0	-8.0 ± 1.0	-1.3

[a] Navrotsky (1987a)
[b] Fei et al. (1990)
[c] Ito et al.(1990)
[d] Mehta et al. (1993)
[e] Ko et al. (1989)
[f] Leinenweber et al. (pers. comm.)

ilmenite is much higher in enthalpy relative to pyroxene than is MgGeO$_3$ ilmenite. The perovskites CaGeO$_3$, CdGeO$_3$, and MgSiO$_3$ are all energetically very unfavorable phases. The energies of transition become larger as the molar volumes become smaller.

The ilmenite MnTiO$_3$, MnSnO$_3$, and FeTiO$_3$ undergo phase transitions above 15 GPa. Similar transitions also probably happen in MgGeO$_3$ and ZnSiO$_3$ high-pressure ilmenites. The products, initially thought to be dis-

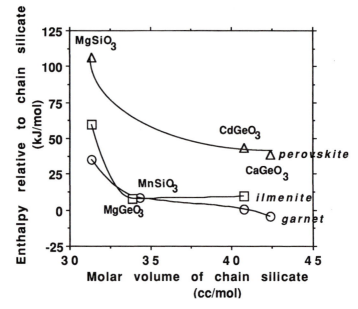

Figure 6.21. Enthalpies of ABO_3 polymorphs relative to chain silicate phase stable at 1 atm versus molar volume of chain silicate.

ordered ilmenite or corundum, have recently been shown to be of the lithium niobate structure type (Ko et al. 1989). This is an ordered superstructure based on the corundum type, with ordering of both cations within each plane (see Fig. 2.24) rather than in alternate planes as in the ilmenite structure. For a given composition, the lithium niobate structure is slightly denser than the ilmenite form. Thermochemical parameters for known transitions are given in Table 6.9. The lithium niobate form appears able to transform to and from the perovskite by a cooperative mechanism not involving long-range diffusion and reconstruction. This provides an easy pathway for the transitions in the diamond cell, even at room temperature but, at the same time, makes quenching the perovksite form impossible.

Distorted perovskites are classic examples of ferroelastic materials. As temperature increases, the tilting angles generally decrease and the orthorhombic crystal transforms first to tetragonal and then to cubic symmetry. Increasing pressure probably favors greater distortion, however. Because a macroscopic shear strain is associated with these distortions, one shear modulus often becomes soft in the vicinity of such transformation. The change in angle with temperature can also give rise to an excess temperature dependence of shear modulus. $MgSiO_3$ perovskite exhibits several of the characteristics of a ferro-elastic material (Navrotsky et al. 1992). Crystals are often twinned. Untwinned single crystals are twinned on heating to 373 K, under deviatoric stress, and even with exposure to low-power laser light. Polycrystals exhibit glasslike Raman spectra on heating to 373 K.

Since the high-temperature phase in a ferroelastic transformation is not generally quenchable, it is difficult to determine the crystal structure of the high pressure and temperature material. However, a recent TEM study on recovered $MgSiO_3$ perovskite suggests that such a ferroelastic transformation may occur and define a phase boundary that is very close to the pressure and temperature conditions at the top of the lower mantle. Thus the structural state of $MgSiO_3$ perovskite under lower mantle conditions remains unclear.

6.4.4 Phase relations in $MgO–FeO–SiO_2$ at high pressure

An internally consistent thermodynamic data base for $MgO–SiO_2$ at high pressure has been developed using available phase equilibrium data, thermochemical data (heat capacity, enthalpy of formation and transition), and thermophysical data (molar volume, compressibility, thermal expansion) (Fei, Saxena, and Navrotsky 1990). Phase relations calculated using these data are shown in Figure 6.22. The topology of the experimental relations is consistent with the thermochemical data, and the calculated diagram shows a wealth of phases. Mg_2SiO_4, an olivine (α) at low pressure, transforms first to spinelloid (β) and then to spinel (γ), which in turn decomposes to a mixture of $MgSiO_3$ perovskite and MgO. At $MgSiO_3$ composition, the pyroxene first decomposes to a mixture of β-phase plus stishovite, which then recombine to form first $MgSiO_3$ ilmenite and finally $MgSiO_3$ perovskite. At high temperature, a field of $MgSiO_3$ tetragonal garnet appears at intermediate pressures. A striking feature of the phase diagram is the negative $P–T$ slope of the perovskite-forming reactions. Since the perovskite-bearing assemblage is denser, it must have the higher entropy. This is almost certainly related to the simultaneous presence of silicon in octahedral coordination (with longer and weaker Si–O bonds than in an SiO_4 tetrahedron) and of magnesium in a roughly eight-coordinated central site in the orthorhombic perovskite. Negative $P–T$ slopes for high-pressure changes involving coordination number increase may be a general phenomenon (Navrotsky 1980).

Phase relations in $MgO–FeO–SiO_2$ are shown in Figures 6.23 and 6.24. At low pressure, the diagram is dominated by three solid solutions (magnesiowustite, olivine, pyroxene) plus quartz, at intermediate pressure by two solid solutions (magnesiowustite, spinel) plus stishovite, and at high pressure by a mag-

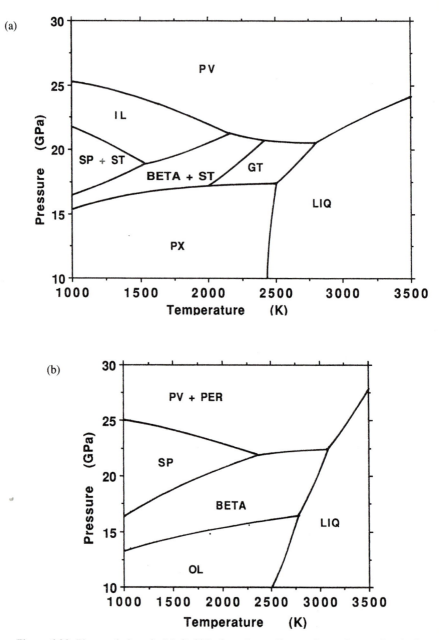

Figure 6.22. Phase relations in MgO–SiO$_2$ from internally consistent thermochemical data set: (a) MgSiO$_3$ composition, (b) Mg$_2$SiO$_4$ composition. (PX = pyroxene, BETA = wadsleyite, LIQ = liquid, SP = spinel, ST = stishovite, IL = ilmenite, PV = perovskite, GT = garnet.) (After Fei, Saxena, and Navrotsky 1990.)

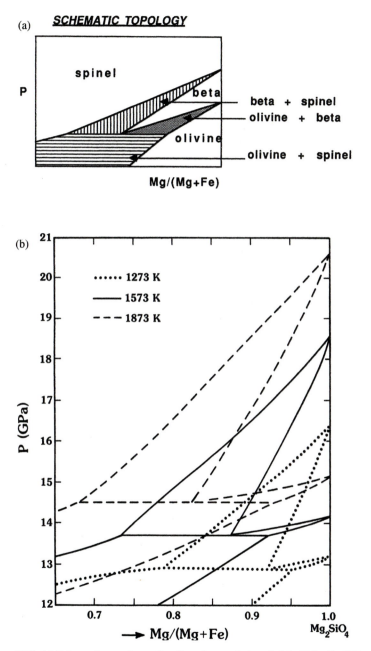

Figure 6.23. (a) Schematic topology of α, β, γ phase relations in Mg_2SiO_4–Fe_2SiO_4. (b) Variation with temperature of two- and three-phase regions in *P*-X diagram of α, β, γ phases of Mg_2SiO_4–Fe_2SiO_4 system (after Akaogi et al. 1989).

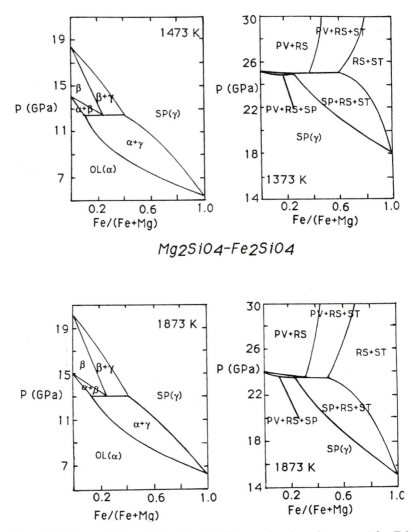

Figure 6.24. Phase relations in Mg_2SiO_4–Fe_2SiO_4 as a function of pressure (after Fei, Mao, and Mysen 1991). (α (OL) = olivine, β = wadsleyite, γ (SP) = spinel, PV = perovskite, RS = rocksalt, ST = stishovite.)

nesiowustite solid solution, a perovskite phase limited in extent of Fe substitution, and stishovite. A striking feature is the less complete substitution of Fe for Mg with increasing pressure.

Because β-Fe_2SiO_4 has no P–T stability field, the phase field of the spinelloid pinches out with increasing Fe/(Fe+Mg), giving a characteristic "rabbit ears" topology (see Fig. 6.23 and left-hand side of Fig. 6.24). The widths of the two phase fields are generally consistent with the thermochemical data, which include both the enthalpies and entropies of transition and the mixing properties of the olivine, spinelloid, and spinel solid solutions (see Chap. 7). With increasing temperature, the phase relations shift to higher pressures and the two-phase fields to more iron-rich compositions. This shift and known seismic properties of the 400 km discontinuity have been used to estimate the temperature in this region of the Earth as being about 1800 K (Akaogi et al. 1989).

At higher pressure, the spinel phases decompose to perovskite-bearing

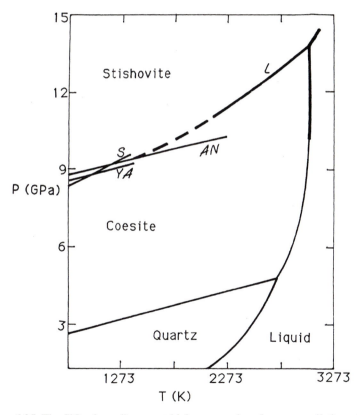

Figure 6.25. The SiO_2 phase diagram at high pressure based on a compilation of various data.

6   Mineral thermodynamics

assemblages (see Fig. 6.24). Iron is strongly fractionated out of the perovskite, and a (Mg,Fe)O magnesiowustite phase forms that is quite iron-rich (Fei, Mao, and Mysen 1991). The instability of $FeSiO_3$ perovskite is not readily explained, since Fe^{2+} occupies similar coordination in many garnets. Similarly, the apparent instability of $FeSiO_3$ in the ilmenite structure is not well understood. Nor are the roles of minor elements, especially Ca and Al, in stabilizing various high-pressure phases well characterized at present.

Phase relations at high pressure in SiO_2 are shown in Figure 6.25. Stishovite, the octahedrally coordinated rutile phase, dominates at high pressure.

6.4.5 Lattice vibrational models for post-spinel phase transitions

The Kieffer approach is particularly useful for these high-pressure phases, which certainly will not be available in quantities suitable for low-temperature C_P measurements in the foreseeable future. The input data needed to constrain the model are space group, molar volume, elastic constants, bulk modulus, thermal expansion, and vibrational (infrared and Raman spectra). All these data are obtainable for ultrahigh-pressure phases and indeed are of interest to geophysics for other reasons.

Figure 6.26a shows schematic vibrational spectra for the pyroxene, garnet, ilmenite, and perovskite forms of $MgSiO_3$. Figure 6.26b shows representative Kieffer-type VDOS models consistent with the spectra. Figures 6.26c and 6.26d show heat capacities and entropies consistent with these models (also see Fig. 6.16). The heat capacity and entropy curves show crossovers at low temperature, and at high temperature the order of decreasing entropy is pyroxene, garnet, perovskite, ilmenite, whereas that of increasing density is pyroxene, garnet, ilmenite, perovskite. The complex heat capacity relations reflect the balance of several factors. With increasing density (smaller unit cell), the entropy will decrease (all other factors being similar). This reflects "normal" entropy–volume systematics as seen in the α–β–γ transitions. Superimposed on this trend are two other effects. As Si becomes six-coordinated (half of it is octahedral in garnet, all of it in ilmenite and perovskite), the Si–O stretching vibrations at 900–1100 cm^{-1} of the strongly bonded SiO_4 tetrahedra are transformed to lower frequency (600–800 cm^{-1}) vibrations of the SiO_6 octahedra, which are not readily identifiable and form part of the optic continuum. This transformation increases C_P and $S°$ at a given temperature and counteracts the effect of increasing density. The low-frequency cutoff of the optic continuum is highest for ilmenite, intermediate for pyroxene and garnet, and lowest for perovskite. Though it is tempting to describe this as increasing "rattling" of

Mg in a larger and more distorted coordination polyhedron, the real dynamical situation is more complex and involves lattice modes rather than localized vibrations. Nonetheless, this effect tends to raise the heat capacity and entropy of perovskite and lower that of ilmenite. Finally, the somewhat larger thermal expansivity of perovskite than of other phases increases the term for converting C_V to C_P and raises the entropy of perovskite.

The calculated results suggest the following. Pyroxene is the phase of highest entropy, ilmenite of lowest, and garnet and perovskite of intermediate. Above 1000 K, perovskite has a slightly higher entropy than garnet. The IL→PV transition has a definitely positive $\Delta S°$ and a negative dP/dT. Above 1000 K, $\Delta S°$ of phase transitions is virtually independent of temperature.

It is important to note that the vibrational modeling used here has certain limitations. First, the vibrational density of states is only a very simplified approximation. Second, dispersion across the Brillouin zone is hard to predict, adding uncertainty to the location of the low-frequency cutoff. Third, if some group of frequencies is very temperature dependent (i.e., soft modes), this can not be readily incorporated. Thus distortional transitions in perovskite can not be modeled. Nevertheless, models like those discussed appear to be very useful in making predictions of heat capacities and entropies.

From the preceding calculations and from experimental observations, one can draw the following general conclusions. For phase transitions involving no changes in cation coordination numbers, ΔS generally scales with ΔV and phase transitions have positive P–T slopes. The lower entropies of denser phases generally result from small shifts in the vibrational density of states, with the denser phase showing fewer low-frequency modes. When a change in coordination occurs, much more pronounced changes occur in the vibrational spectrum. Since an increase in coordination number lengthens and weakens the bonds in the first coordination sphere (while increasing their number), this change generally implies a shift in the vibrational density of states toward lower frequencies. This shift is especially pronounced when tetrahedral Si–O bonds are converted to octahedral Si–O linkages. The result is that entropy–volume systematics break down, and the denser phase has a substantially higher entropy than otherwise expected and, in extreme cases, has an entropy higher than that of the lower pressure polymorph (e.g., for the ilmenite–perovskite transition). Anomalies in the relation of other physical properties (elastic constants, thermal expansion, compressibility) to density may also occur in phase transitions involving coordination number changes. Such effects, though especially pronounced in silicates, can be expected to occur for many solid-state systems undergoing phase transitions with changes in nearest neighbor coordination.

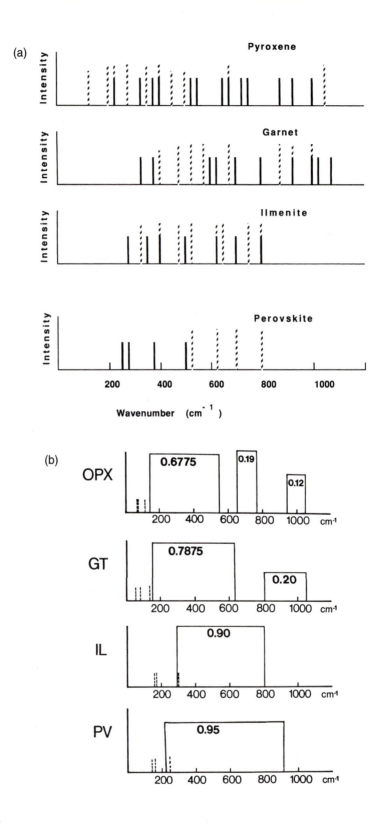

(a)

Pyroxene

Intensity

Garnet

Intensity

Ilmenite

Intensity

Perovskite

Intensity

200   400   600   800   1000

Wavenumber   (cm^{-1})

(b)

OPX

0.6775   0.19

0.12

200   400   600   800   1000   cm^{-1}

GT

0.7875

0.20

200   400   600   800   1000   cm^{-1}

IL

0.90

200   400   600   800   1000   cm^{-1}

PV

0.95

200   400   600   800   1000   cm^{-1}

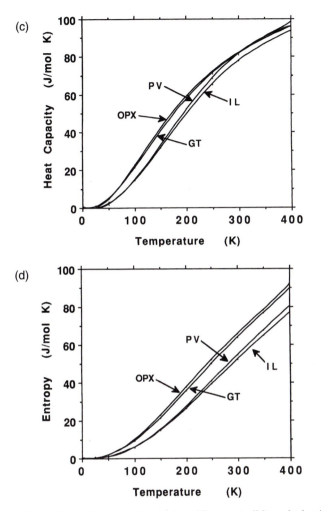

Figure 6.26. (a) Positions of infrared (hatched) and Raman (solid) peaks in vibrational spectra of $MgSiO_3$ polymorphs. Heights of lines have no significance. (Data from Hofmeister and Ito 1992; Jeanloz 1981; Kieffer 1979; McMillan et al. 1989; Williams, Jeanloz, and McMillan 1987). (b) Kieffer vibrational models used by Fei et al. (1990) for $MgSiO_3$ polymorphs, generally consistent with spectra in (a). (c) Heat capacities calculated from Kieffer model. (d) Entropies calculated from Kieffer model.

6.5 Phase transitions in AX compounds

The major structures to consider are rocksalt, wurtzite or sphalerite, and nickel arsenide. Table 6.10 gives thermochemical parameters. The transition from NaCl (octahedral coordination) to CsCl (cubic coordination) is well known in halides. This transformation also occurs in BaO and CaO and in BaS, although

Table 6.10. *Thermodynamic parameters for phase changes in AX compounds*

Compound	$\Delta H°$(kJ/mol)	$\Delta S°$(J/mol·K)	$\Delta V°_{298}$(cm³/mol)
Wurtzite or sphalerite = rocksalt			
InP[a]	56.0[d]		−5.5
CdTe[a]	16.5[d]		−6.3
CdSe[a]	17.4[d]		−5.6
CdS[a]	13.5[d]		−5.9
ZnTe[a]	53.0[d]		
ZnS(sph)[a]	55.0[d]		
ZnS(w)[a]	67		
ZnSe[a]	54.0[d]		−4.7
MgS[a]	~0[d]		−4.6
MgSe[a]	~0[d]		−5.7
AgI[a]	1.9[d]		−7.3
CuBr[a]	18.3[d]		
ZnO(w)[b]	24.5	+0.5	−2.7
MgO[e]	40[f]		
CoO(w)[e]	−18[d]		
CoO(sph)[e]	−37		
NiO	−36		
Rocksalt = cesium chloride			
NaCl[c]	27.3	+6.3	0.864
KF[c]	5.3	−3.9	3.747
KCl[c]	8.5	+0.1	4.315
KBr[c]	7.9	−0.3	4.517
KI[c]	8.3	+0.8	4.489
RbCl[c]	3.7	+1.8	6.045
RbBr[c]	2.3	−1.4	6.029
Rb5[c]	9.0	−1.5	7.478
CsCl[c]	−3.2	−4.4	8.115
CsBr[c]	−4.9	−4.3	(9.41)[e]
AgF[c]	4.9	+1.5	1.647
BaS[c]	30.8	−0.9	4.826
BaO[c]	51.5	−1.1	3.351
CaO[c]	121.4	+0.1	1.844
Rocksalt = nickel arsenide			
MnTe[c]	0[d]		−0.77
CaTe[c]	27.0[d]		−1.089
MnSe[c]	3.6[d]		−0.826
FeSe[c]	−8.1[d]		
CoSe[c]	−3.3[d]		
NiSe[c]	−5.9[d]		
MgS[c]	41.1[d]		−0.594
MnS[c]	8.3[d]		−0.603
FeS[c]	−1.2[d]		

[a] Phillips (1971)
[b] Davies and Navrotsky (1981)
[c] Navrotsky (1980)
[d] $\Delta G°$, not ΔH
[e] DiCarlo and Navrotsky (1992)
[f] Navrotsky and Muan (1971)
[g] Kenny and Navrotsky (1972)

BaO transforms to a slightly distorted form of the CsCl structure. The radius ratio seems to determine the volume change (see Fig. 6.27a). Thus ΔV for the transition is not constant, but ranges from near -15% (CsCl, KF) to about -3% (NaCl). MgO in the rocksalt and cesium chloride structure would have virtually the same volume, making that transition highly unlikely.

For transition metal oxides, the CsCl structure would be destabilized both by the predicted small volume change and by the loss of crystal field stabilization energy. This would make a transition to the CsCl structure unlikely for MgO, FeO, CoO, and NiO.

The entropy change of the NaCl $\rightarrow$ CsCl transition is strongly correlated with the percentage volume change, and thus with radius ratio. Indeed, their correlation shows that for halides with small cations $\Delta S°$ for the NaCl $\rightarrow$ transition is positive, but for halides with cations and anions of comparable size it is negative (see Fig. 6.27a).

The enthalpies of the NaCl $\rightarrow$ CsCl transition are shown in Figure 6.27b as a function of the product of the radius ratio and molar volume of the rocksalt phase. This product, rather than the radius ratio alone, is used because the enthalpy (and free energy and pressure) for the transition generally increases in magnitude from bromides to chlorides to oxides. This increase is presumably related to the larger magnitude of the lattice energy in materials having a smaller interatomic separation, which leads to a larger difference in energy between polymorphs. BaO, BaS, and CaO fall on the same linear trend if their enthalpies of transition are divided by four, the product of the ionic charges. The data suggest an enthalpy of transition of greater than 200 kJ mol^{-1} for MgO, which coupled with the small (or maybe even positive) ΔV further supports the contention that for MgO the CsCl phase is not likely to exist.

The transition rocksalt $\rightarrow$ NiAs occurs at atmospheric pressure in MnSe and MnTe. The NiAs structure occurs in transition metal chalcogenides; the rocksalt structure is limited to the alkaline-earth chalcogenides and to oxides and halides.

Can an analogous transition occur in oxides at high pressure? Compared to the rocksalt structure, the NiAs structure represents a smaller degree of ionicity and a larger contribution of metal–metal bonding. For a series of MX oxides or chalcogenides, this decrease in ionicity, uncompensated by any strong covalency as seen in tetrahedral compounds, can be expected to lead to a diminished stability of the compound relative to the elements. Therefore one may expect the NiAs structure to become more competitive relative to the rocksalt structure as the enthalpy for formation becomes less negative, since both reflect a relative decrease in ionicity and an increase in the contributions of electron delocalization. This is supported by the band structure calculations discussed in Chapter 5.

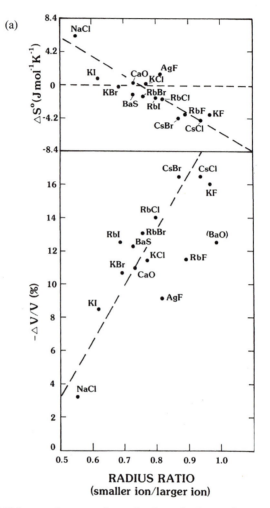

Figure 6.27. (a) Volume and entropy change for the rocksalt to cesium chloride transformation plotted against radius ratio. (b) Calculated enthalpy of the rocksalt to cesium chloride structure divided by the product of the ionic charges plotted against the product of the radius ratio and the molar volume of the rocksalt phase. (c) Standard free energy of the rocksalt to nickel arsenide transition plotted against the standard enthalpy of formation of rocksalt phase at 298 K. (From Navrotsky and Davies 1981.)

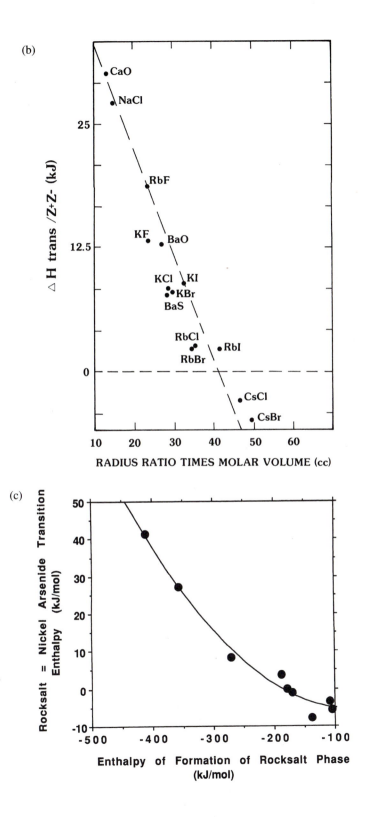

(b)

$\triangle$ H trans /Z$^+$Z$^-$ (kJ)

CaO

NaCl

25

RbF

KF   BaO

12.5

KCl   KI

KBr

BaS

RbCl

RbI

RbBr

0

CsCl

CsBr

10   20   30   40   50   60

RADIUS RATIO TIMES MOLAR VOLUME (cc)

(c)

Rocksalt = Nickel Arsenide Transition
Enthalpy (kJ/mol)

50

40

30

20

10

0

-10

-500   -400   -300   -200   -100

Enthalpy of Formation of Rocksalt Phase
(kJ/mol)

Table 6.11. *Thermodynamic parameters for other phase transitions*

Transition	ΔH° (kJ/mol)	ΔS° (J/K·mol)	ΔV° (cm³/mol)
SiO$_2$ (α-quartz = β-quartz)	0.47[a]	0.35	0.101
SiO$_2$ (β-quartz = cristobalite)	2.94[e]	1.93	0.318
GeO$_2$ (rutile = quartz)	5.6[c]	4.0	11.51
CaSiO$_3$ (wollastonite = pseudowollastonite)	5.0[b]	3.6	0.12
Al$_2$SiO$_5$ (andalusite = sillimanite)	3.88[e]	4.50	−0.164
Al$_2$SiO$_5$ (sillimanite = kyanite)	−8.13[e]	−13.5	−0.571
MgSiO$_3$ (ortho = clino)	−0.37[e]	0.16	−0.002
MgSiO$_3$ (ortho = proto)	1.59[e]	1.27	0.109
FeSiO$_3$ (ortho = clino)	−0.17[d]	−0.03	−0.06
MnSiO$_3$ (rhodonite = pyroxmangite)	0.25[d]	−1.03	−0.39
MnSiO$_3$ (pyroxmangite = pyroxene)	0.88[d]	−2.66	−0.3
NaAlSi$_3$O$_8$ (low albite = high albite)	13.5[b]	14.0	0.40
KAlSi$_3$O$_9$ (microcline = sanidine)	11.1[e]	15.0	0.027

[a] Treated as though all first order, though a strong higher order component.
[b] Robie et al. (1978)
[c] Navrotsky (1971)
[d] Navrotsky (1987a)
[e] Berman (1988)

Figure 6.27c shows the available data for the standard free energy, ΔG° (transitional), for the rocksalt $\rightarrow$ NiAs transformation plotted against the standard enthalpy of formation, ΔH_f°, of the rocksalt phase. A smooth trend can be used to estimate the free energies of transition for the oxides. The results suggest that the NiAs structure might be plausible for FeO under lower mantle conditions, but not for MgO.

The energy difference between octahedrally (rocksalt) and tetrahedrally (wurtzite or sphalerite) coordinated phases has been discussed in terms of ligand field effects for transition metal oxides (Navrotsky and Muan 1971) and in terms of the Phillips ionicity scale for semiconductors (Phillips 1970, 1971). CoO has been synthesized with both zincblende and sphalerite structure; both are metastable with respect to the rocksalt form (DiCarlo and Navrotsky 1993). Their existence is consistent with the tetrahedral CFSE of Co^{2+}.

6.6 Other phase transitions

Table 6.11 shows enthalpy, entropy, and volume changes associated with several other classes of phase transitions. Those between chain silicate polymorphs (pyroxene, pyroxenoid) have very small ΔH, ΔS, and ΔV, as do distortional phase transitions among different perovskite types. More substantial changes are seen among the ZrO_2, TiO_2, SiO_2, and GeO_2 polymorphs and for silicate polymorphism involving two totally different structures.

6.7 References

6.7.1 General references and bibliography

Anderson, D. L. (1989). *Theory of the Earth*. Oxford: Blackwell.

Babushkin, V. I., G. M. Matveyev, O. P. Mchedlov-Petrossyan. (1985). *Thermodynamics of silicates*. Ed. O. P. Mchedlov-Petrossyan; trans. B. N. Frenkel and V. A. Terentyev. Heidelberg, Germany: Springer-Verlag.

Bodsworth, C., and A. S. Appleton. (1965). *Problems in applied thermodynamics*. London: Longmans, Green.

Carmichael, I. S. E., and H. P. Eugster, eds. (1987). *Thermodynamics modeling of geologic materials: Minerals, fluids, and melts*. Reviews in Mineralogy, vol. 17. Washington, DC: Mineralogical Society of America.

Darken, L. S., and R. W. Gurry. (1953). *Physical chemistry of metals*. New York: McGraw-Hill.

Denbigh, K. (1966). *The principles of chemical equilibrium*. 2nd ed. Cambridge: Cambridge University Press.

Gaskell, D.R. (1973). *Introduction to metallurgical thermodynamics*. New York: McGraw-Hill.

Greenwood, H. J., ed. (1977). *Application of thermodynamics to petrology and ore deposits*. Short Course Handbook, vol. 2. Toronto: Mineralogical Association of Canada.

Kieffer, S. W., and A. Navrotsky, eds. (1985). *Microscopic to macroscopic, atomic environments to mineral thermodynamics*. Reviews in Mineralogy, vol. 14. Washington, DC: Mineralogical Society of America.

Kubaschewski, O., E. L. Evans, and C. B. Alcock. (1967). *Metallurgical thermochemistry*. 4th ed. New York: Pergamon Press.

Lewis, G. N., and M. Randall. (1961). *Thermodynamics*. 2nd ed. Rev. K. S. Pitzer and L. Brewer. New York: McGraw-Hill.

Nordstrom, D. K., and J. L. Munoz. (1985). *Geochemical thermodynamics*. Palo Alto: Benjamin/Cummings.

Pauling, L. (1960). *The nature of the chemical bond*. 3rd ed. Ithaca, NY: Cornell University Press.

Poirier, J. P. (1991). *Introduction to the physics of the Earth's interior*. New York: Cambridge University Press.

Putnis, A. (1992). *Introduction to mineral sciences*. Cambridge: Cambridge University Press.

Salje, E. K. H. (1988). *Physical properties and thermodynamic behaviour of minerals*. Dordrecht, Holland: Reidel.

Schmalzried, H., and A. Navrotsky. (1975). *Festkorperthermodynamik*. Weinheim, Germany: Verlag Chemie.

Schmalzreid, H., and A. Navrotsky. (1978). *Festkorperthermodynamik Chemie des festen Zustandes*. Berlin: Akademie-Verlag.

Wagner, C. (1952). *Thermodynamics of alloys*. Reading, MA: Addison-Wesley.

6.7.2 Tabulations of thermodynamic data

Berman, R. G. (1988). Internally consistent thermodynamic data for minerals in the system Na$_2$O–K$_2$O–CaO–MgO–FeO–Fe$_2$O$_3$–Al$_2$O$_3$–SiO$_2$–TiO$_2$–H$_2$O–CO$_2$. *J. Petrol. 29*, 445–522.

Coughlin, J. P. (1954). *Heats and free energies of formation of inorganic oxides.* U.S. Bureau of Mines Bulletin 542. Washington, DC: GPO.

Elliott, J. F., and M. Gleiser. (1960). *Thermochemistry for steelmaking.* vol. 1. Reading, MA: Addison-Wesley.

Ghiorso, M. S., I. S. E. Carmichael. (1987). *Modeling magmatic systems: Petrologic applications.* Reviews in Mineralogy, vol. 17, 467–99.

Helgeson, H. C., J. Delany, H. W. Nesbitt, and D. K. Bird. (1978). Summary and critique of the thermodynamic properties of rock-forming minerals. *Amer. J. Science 278A*, 1–229.

Holland, T. J. B., and R. Powell. (1985). An internally consistent thermodynamic dataset with uncertainties and correlations: 2. Data and results. *J. Metamorphic Geol. 3*, 343–70.

Holland, T. J. B., and R. Powell. (1990). An enlarged and updated internally consistent thermodynamic dataset with uncertainties and correlations: The system K$_2$O–Na$_2$O–CaO–MgO–MnO–FeO–Fe$_2$O$_3$–Al$_2$O$_3$–TiO$_2$–SiO$_2$–C–H$_2$–O$_2$. *J. Metamorphic Geol. 8*, 89–124.

Hultgren, R., P. Desai, D. Hawkins, M. Gleiser, K. Kelley, and D. Wagman. (1973). *Selected values of the thermodynamic properties of the elements.* Metals Park, OH: American Society for Metals.

Hultgren, R., et al. (1973). *Selected values of the thermodynamic properties of binary alloys.* Metals Park, OH: American Society for Metals.

JANAF thermochemical tables. 2nd ed. (1971). Ed. D. R. Stull and H. Prophet. U.S. National Bureau of Standards Rep. NSRDS-NBS 37. Washington, DC: GPO.

JANAF thermochemical tables. 3rd ed. (1986). New York: American Institute of Physics.

Kelley, K. K. (1960). *High-temperature heat-content, heat-capacity, and entropy data for the elements and inorganic compounds.* U.S. Bureau of Mines Bulletin 584. Washington, DC: GPO.

Kelley, K. K., and E. G. King. (1961). *Entropies of the elements and inorganic compounds.* U.S. Bureau of Mines Bulletin 592. Washington, DC: GPO.

Kubaschewski, O. (1970). *The thermodynamic properties of double oxides.* National Physics Laboratory DSC Report 7. Washington, DC: GPO.

Kubaschewski, O., and C. B. Alcock. (1967). *Metallurgical thermochemistry.* Oxford: Pergamon Press.

Robie, R. A., B. S. Heminway, and J. R. Fisher. (1978). Thermodynamic properties of minerals and related substances at 298.15 K and 1 bar (10^5 pascals) and at high temperatures. *U.S. Geol. Survey Bull. 1452.*

Rosenquist, T. (1970). *Thermochemical data for metallurgists.* Trondheim, Norway: Tapir-Verlag.

Rossini, F. D., et al. (1952). Selected values of chemical thermodynamic properties. U.S. National Bureau of Standards Circular 500. Washington, DC: GPO.

Samsonov, G. V. (1973). *The oxide handbook* (Englische Übersetzung). New York: IFL-Plenum.

Wagman, D. D., et al. (1968). *Selected values of chemical thermodynamic properties – Tables for the first thirty-four elements.* National Bureau of Standards Technical Note 270-3. Washington, DC: GPO.

Wagman, D. D., et al. (1969). *Selected values of chemical thermodynamic properties*

– *Tables for elements 35 through 53*. National Bureau of Standards Technical Note 270-4. Washington, DC: GPO.

6.7.3 Tabulations of phase-diagram information

Elliot, J. F., and M. Gleiser. (1960). *Thermochemistry for steelmaking*. Vols. 1 and 2. Reading, MA: Addison-Wesley.

Hansen, M. (1958). *Constitution of binary alloys*. New York: McGraw-Hill. First Supplement (1965), Second Supplement (1969).

Levin, E. M., C. R. Robbins, and H. F. McMurdie. (1964). *Phase diagrams for ceramists*. Columbus, OH: American Ceramic Society. Supplement (1969), vol. 3 (1975), vol. 4 (1980).

Morey, G. W. (1964). Data of geochemistry, phase equilibrium relations of the common rock-forming oxides except water. 6th ed. U.S. Geological Survey Paper 440-L. Washington, DC: GPO.

Muan, A., and E. F. Osborn. (1965). *Phase equilibria among oxides in steelmaking*. Reading, MA: Addison-Wesley.

Samsonov, G. V. (1972). *Handbook of phase diagrams of silicate system*. Vol. 1, *Binary systems*; Vol. 2, *Metal oxygen compounds*. Springfield, VA: National Technical Information Service.

6.7.4 Specific references

Akaogi, M., and A. Navrotsky. (1984). Calorimetric study of the stability of spinelloids in the system $NiAl_2O_4$–Ni_2SiO_4. *Phys. Chem. Minerals 10*, 166–72.

Akaogi, M., E. Ito, and A. Navrotsky. (1989). Olivine-modified spinel-spinel transitions in the system Mg_2SiO_4–Fe_2SiO_4: Calorimetric measurements, thermochemical calculations, and geophysical application. *J. Geophys. Res. 94*, 15671–86.

Akaogi, M., N. L. Ross, P. McMillan, and A. Navrotsky. (1984). The Mg_2SiO_4 polymorphs (olivine, modified spinel and spinel) – Thermodynamic properties from oxide melt solution calorimetry, phase relations, and models of lattice vibrations. *Amer. Mineral. 69*, 499–512.

Davies, P. K., and A. Navrotsky. (1981). Thermodynamics of solid solution formation in NiO–MgO and NiO–ZnO. *J. Solid State Chem. 38*, 264–76.

DiCarlo, J., and A. Navrotsky. (in press). Energetics of cobalt monoxide with the zinc-blende structure. *J. Amer. Ceram. Soc.*

Elcombe, M. M. (1967). Some aspects of the lattice dynamics of quartz. *Proc. Phys. Soc. 91*, 947–58.

Fei, Y., H.-K. Mao, and B. O. Mysen. (1991). Experimental determination of element partitioning and calculation of phase relations in the MgO–FeO–SiO_2 system at high pressure and high temperature. *J. Geophys. Res. 96*, 2157–69.

Fei, Y., S. K. Saxena, and A. Navrotsky. (1990). Internally consistent thermodynamic data and equilibrium phase relations in the system MgO–SiO_2 at high pressure and high temperatures. *J. Geophys. Res. 95*, 6913–28.

Hofmeister, A. M., and E. Ito. (1992). Thermodynamic properties of $MgSiO_3$ ilmenite from vibrational spectra. *Phys. Chem. Minerals 18*, 423–32.

Ito, E., M. Akaogi, L. Topor, and A. Navrotsky. (1990). Negative *P-T* slopes for reactions forming $MgSiO_3$ perovskite confirmed by calorimetry. *Science 249*, 1275–8.

Jeanloz, R. (1981). Majorite: Vibrational and compressional properties of a high-pressure phase. *J. Geophys. Res. 86*, 6171–9.

Kenny, D. S., and A. Navrotsky. (1972). Approximate activity-composition relations in the system MgO–ZnO at 1205±5 °C. *J. Inorg. Nucl. Chem. 34*, 2115–19.

Kieffer, S. W. (1979). Thermodynamics and lattice vibrations of minerals: 2. Vibrational characteristics of silicates. *Rev. Geophys. Space Phys. 17*, 20–34.

Kieffer, S. W. (1985). *Heat capacity and entropy: Systematic relation to lattice vibrations.* Reviews in Mineralogy, vol. 14, 65–126. Washington, DC: Mineralogical Society of America.

Ko, J., N. E. Brown, A. Navrotsky, C. T. Prewitt, and T. Gasparik. (1989). Phase equilibrium and calorimetric study of the transition of $MnTiO_3$ from the ilmenite to the lithium niobate structure and implications for the stability field of perovskite. *J. Phys. Chem. Minerals 16*, 727–33.

Leinenweber, K., P. F. McMillan, and A. Navrotsky. (1989). Transition enthalpies and entropies of high-pressure zinc metasilicates and zinc metagermanates. *Phys. Chem. Minerals 16*, 799–808.

Leinenweber, K., and A. Navrotsky. (1989). Thermochemistry of phases in the system $MgGa_2O_4$–Mg_2GeO_4. *Phys. Chem. Minerals 16*, 497–502.

McMillan, P., M. Akaogi, E. Ohtani, Q. Williams, R. Nieman, and R. Sato. (1989). Cation disorder in garnets along the $Mg_3Al_2Si_3O_{12}$–$Mg_4Si_4O_{12}$ join: An infrared Raman and NMR study. *Phys. Chem. Minerals 16*, 428–35.

Mehta, A., A. Navrotsky, N. Kumada, and N. Kinomura. (1993). Structural transitions in $LiNbO_3$ and $NaNbO_3$. *J. Solid State Chem. 102*, 213–25.

Navrotsky, A. (1971). Enthalpies of transformation among the tetragonal, hexagonal, and glassy modifications of GeO_2. *J. Inorg. Nucl. Chem. 33*, 1119–24.

Navrotsky, A. (1974). *Thermodynamics of binary and ternary transition metal oxides in the solid state. MTP international reviews of science,* ed. D. W. A. Sharp. Inorganic chemistry, 2nd ser., vol. 5, 29–70. Baltimore, MD: Butterworths-University Park Press.

Navrotsky, A. (1980). Lower mantle phase transitions may generally have negative pressure–temperature slopes. *Geophys. Res. Lett. 7*, 709–11.

Navrotsky, A. (1987a). High-pressure transitions in silicates. *Prog. Solid St. Chem. 17*, 53–86.

Navrotsky, A. (1987b). Thermodynamic aspects of solid-state chemistry. In *Solid state chemistry,* ed. P. Day and A. Cheetham, 362–93. Oxford: Oxford University Press.

Navrotsky, A. (1988a). Experimental studies of mineral energetics. In *Physical properties and thermodynamic behaviour of minerals,* ed. E. K. H. Salje, 403–32. Dordrecht, Holland: Reidel.

Navrotsky, A. (1988b). Thermodynamic aspects of inorganic solid-state chemistry. In *Solid state chemistry techniques,* ed. A. K. Cheetham and Peter Day, 362–93. Oxford: Oxford Science Publications.

Navrotsky, A., and P. K. Davies. (1981). Cesium chloride versus nickel arsenide as possible structures for (Me,Fe)O in the lower mantle. *J. Geophys. Res. 86*, 3689–94.

Navrotsky, A., and A. Muan. (1971). Activity-composition relations in the systems CoO–ZnO and NiO–ZnO at 1050 °C. *J. Inorg. Nucl. Chem. 33*, 35–47.

Navrotsky, A., D. J. Weidner, R. C. Liebermann, and C. T. Prewitt. (1992). Materials science of the earth's deep interior. *MRS Bull. 17*, 30–7.

O'Keeffe, M., and B. G. Hyde. (1981). Why olivine transforms to spinel at high pressure. *Nature 293*, 727–8.

Phillips, J. C. (1970). The chemical bond and solid-state physics. *Physics Today 23*, 23–30.

Phillips, J. C. (1971). Covalent-ionic and Covalent-metallic transitions of tetrahedrally coordinated $A^N B^{8-N}$ crystals under pressure. *Phys. Rev. Lett. 27*, 1197–200.

Ross, N. L. (1993). Lattice vibration and mineral stability. In *Stability of minerals*, ed. G. D. Price and N. L. Ross, 132–71. London: Chapman and Hall.

Sasaki, S., C. T. Prewitt, Y. Sato, and E. Ito. (1982). Single crystal X-ray study of γ-Mg_2SiO_4, *J. Geophys. Res. 87*, 7829–32.

Williams, Q., R. Jeanloz, and P. McMillan. (1987). Vibrational spectrum of $MgSiO_3$ perovskite: Zero-pressure raman and mid-infrared spectra to 27 GPa. *J. Geophys. Res. 92*, B8, 8116–28.

7

Solid solutions and order–disorder

7.1 Introduction

Disorder in a solid may arise from several different sources: (1) The periodicity may be broken by displacing atoms from their equilibrium positions in a nonperiodic fashion. These defects can be point defects, single atoms in wrong positions, missing atoms, or impurities; line defects, chains of missing, extra, or misplaced atoms, or dislocations; planar defects, twin planes, or sheets of missing atoms; or three-dimensional or structural disorder or lack of periodicity leading eventually to amorphization. (2) A solid solution or alloy may form by replacing one atom, more or less at random, by another with reasonably similar properties, for example, Mg, Fe substitution in olivines. (3) Substitutional disorder may occur when similar atoms ordered onto crystallographic sites at low temperature are partially or completely randomized at high temperature. This situation creates positional disorder but maintains overall stoichiometry. (4) Nonstoichiometry, namely, deviation from ideal atomic ratios, for example, $Fe_{0.97}O$ instead of FeO, creates another kind of disorder. (5) The orientation of polyhedra (octahedral tilting in perovskites, CO_3 orientation in carbonates) may lose long-range order. (6) Magnetic moments or electron spins may become disordered.

These various types of disorder are frequently related. In all cases, the equilibrium degree of disorder increases with increasing temperature because of the balance of energy and entropy.

In many substitutional solid solutions formed at high temperature, the disorder may decrease as temperature is lowered by one or both of two mechanisms: (1) The atoms may order into a superstructure. (2) Two phases of different compositions may exsolve; one or both of these may be ordered. Thus order–disorder, the extent of solid solubility, and the phase diagram are closely related and will be considered together in this chapter. For more detailed dis-

332

cussions of thermodynamic formalism and properties, see the general references and bibliography section at the end of the chapter.

7.2 Solid solution formation

7.2.1 Thermodynamic formalism

Minerals frequently show a wide range of substitutions in their structures. If two end-members have the same structure (e.g., Mg_2SiO_4 and Fe_2SiO_4 olivines), they can form either complete solid solution or show a solvus, depending on pressure, temperature, and thermodynamic parameters describing their mixing properties.

The free energy of mixing can be written at given pressure P, and temperature, T, as

$$\Delta G_{mix} = \Delta H^0_{mix} - T\Delta S^0_{mix} + \int_{1\ atm}^{P} \Delta V_{mix} dP \tag{7.1}$$

where the superscript zeroes on ΔH and ΔS refer to standard states (1 atm, T), and the ΔV term (itself a function of P and T) describes the pressure dependence.

An ideal solution is defined as one in which $\Delta H_{mix} = 0$ and ΔS_{mix} is the ideal configurational entropy of mixing for one mole of species (ions) being mixed:

$$\Delta S_{mix} = -R \sum_i X_i \ln X_i \tag{7.2}$$

where X_i is the mole fraction of each ion. For a binary system, this gives

$$\Delta S_{mix} = -R[X \ln X + (1 - X) \ln (1 - X)] \tag{7.3}$$

with X the one independent mole fraction.

These expressions, derived through statistical mechanics, tacitly assume that (1) there is one mole total of ions being mixed and (2) the distribution of ions on the sites is random, with no tendency toward clustering (like ions preferentially occupying adjacent sites) or ordering (unlike ions preferentially occupying adjacent sites). Thus for Mg_2SiO_4–Fe_2SiO_4 olivines, the approximation of ideality would rest on two other assumptions: (1) The components would be $MgSi_{0.5}O_2$ and $FeSi_{0.5}O_2$, so that one mole of $Mg_xFe_{1-x}Si_{0.5}O_2$ would contain one mole of ions being mixed; and (2) there is no ordering of Fe and Mg between M1 and M2 sites, and the arrangement of iron and magnesium on each sublattice is random.

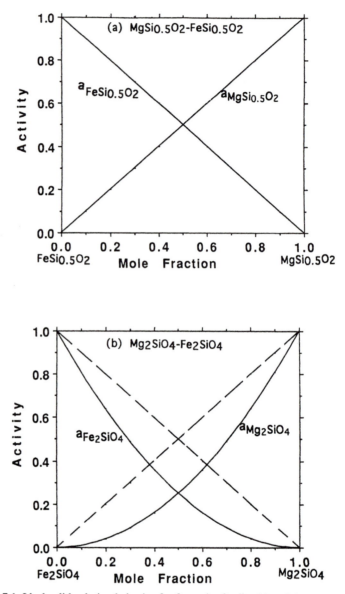

Figure 7.1. Ideal solid solution behavior for forsterite-faydite (a) activity–composition relations for components chosen as $MgSi_{0.5}O_2$ and $FeSi_{0.5}O_2$ and (b) activity–composition relations for components chosen as Mg_2SiO_4 and Fe_2SiO_4.

The thermodynamic activity, a_i, of each component in an ideal solution is then equal to its mole fraction, X_i, and its activity coefficient unity:

$$a_i = X_i, \quad \gamma_i = 1 \tag{7.4}$$

For the assumed ideal olivine solid solution, this would mean

$$a_{MgSi_{0.5}O_2} = X, \quad a_{FeSi_{0.5}O_2} = 1 - X \tag{7.5}$$

(see Fig. 7.1a), but, were one to write the same ideal relations per mole of $(Mg_xFe_{1-x})_2SiO_4$ (see Fig. 7.1b),

$$a_{Mg_2SiO_4} = X^2, \quad a_{Fe_2SiO_4} = (1 - X)^2 \tag{7.6}$$

Yet both relations refer to the same microscopic assumptions and are consistent because the entropy of mixing is an extensive parameter (depends on the mass of the system), and

$$\Delta S_{mix}(Mg_2SiO_4\text{–}Fe_2SiO_4) = -2R[X \ln X + (1 - X) \ln (1 - X)] \tag{7.7}$$
$$= 2\Delta S_{mix} (MgSi_{0.5}O_2\text{–}FeSi_{0.5}O_2)$$

For more complicated charge-coupled substitution involving several sublattices, expressions for statistically ideal solid solutions can be deduced, but they need not take the simple form $a_i = X_i^n$. A detailed example, the plagioclase feldspars, is discussed in Section 7.3.4.

In many systems, ideality is not an adequate assumption. The next level of approximation is the *regular solution*. In the *strictly regular* case, the assumption of ideal entropy of mixing is maintained, but the enthalpy of mixing is given symmetrical form

$$\Delta H_{mix} = \lambda X (1 - X) \tag{7.8}$$

where λ is a constant interaction parameter, which in geologic literature is given the symbol W_H (Thompson 1969). For a binary system, the free energy of mixing has the form

$$\Delta G_{mix} = \lambda X (1 - X) + RT[X \ln X + (1 - X) \ln (1 - X)] \tag{7.9}$$

and the activity coefficients are (if $X = X_B$ in a binary system A–B)

$$RT \ln \gamma_B = \lambda(1 - X)^2, \quad RT \ln \gamma_A = \lambda X^2 \tag{7.10}$$

Activity composition relations for various temperatures for $\lambda = 16.6$ kJ/mol ($\lambda/RT = 2$ at 1000 K) are shown in Figure 7.2 as are the appropriate free energies of mixing. For $\lambda/RT > 2$, the system can lower its free energy by separation into two phases. With decreasing temperature (see Fig. 7.3) this

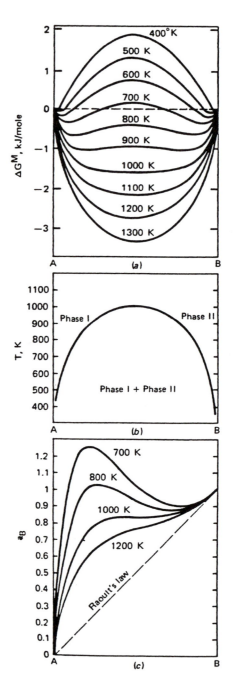

instability sets in at a critical temperature $T_c = \lambda/2R$ and, in the regular solution approximation, the miscibility gap (or solvus or binodal) is symmetric about $X = 0.5$ (see Fig. 7.2).

The free energy at a temperature below T_c is shown schematically in Figure 7.3a. Everywhere within the solvus, the system can lower its free energy by separating into a mixture of the two limiting compositions, α and β. All points on the free-energy curve lie above the tangent line joining α and β. However, diffusion over fairly long distances is required to obtain phases of macroscopically different compositions, so phase separation can be a slow process. If the top of the solvus is sufficiently low in temperature (below ~800 K in silicates), then exsolution may be kinetically hindered, even on a geologic time scale.

There is another region on the curve in Figure 7.3b, between the points marked S and S', where any fluctuation in composition lowers the free energy because $(\partial^2 G/\partial X^2) < 0$. Here composition fluctuations can grow spontaneously and lead to rapid phase separation and characteristic exsolution textures. This curve in T–X space (see Fig. 7.4) is called the *spinodal*.

The preceding discussion neglects the effect of the interfacial energy. Creation of new surface between exsolving phases costs energy, and the *coherent spinodal*, which includes this effect, lies below and within a narrower region of T–X space than the simple spinodal (see Fig. 7.4).

A digression on the mechanisms of exsolution and phase transformation is appropriate here. On crossing a solvus with decreasing temperature, exsolution may occur by normal nucleation and growth. However, for compositions within the coherent spinodal, any fluctuation will lower the free energy. Compositional fluctuations of a narrow range of amplitude and wavelength (spatial extent) may dominate and the result is spinodal decomposition, sometimes with a very characteristic texture.

In solid-solid reactions involving two different crystal structures, the interfacial energy will depend on crystallographic orientation. If one particular orientation is strongly preferred because the two structures share, for example, a common lattice spacing or oxygen array, the new phase will grow only or predominantly in that orientation. This condition is called *epitaxy*.

If a phase transformation requires extensive bond breaking and

Figure 7.2. (*Facing page*) Regular solution behavior in a binary system, A–B, for $\lambda = 16.6$ kJ/mol ($\lambda/RT = 2$ at 1000 K). (a) Free energy of mixing, (b) solvus or miscibility gap, (c) activity–composition relations. (From Gaskell 1981.)

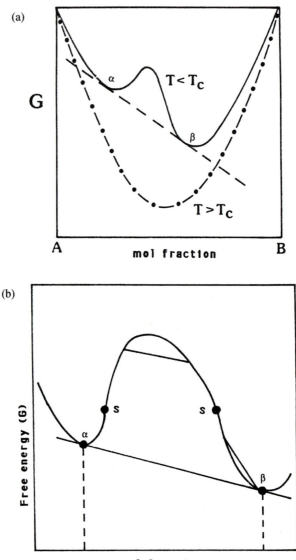

Figure 7.3. (a) Schematic free-energy curve for a system (symmetry lower than regular solution) at a temperature above and below the region of immiscibility. Points α and β define the two-phase region (binodal). (b) Expansion of unmixing region in (a) showing that fluctuation raises the free energy between α and S or β and S' but lowers the free energy within the spinodal between S and S'. S and S' define the spinodal.

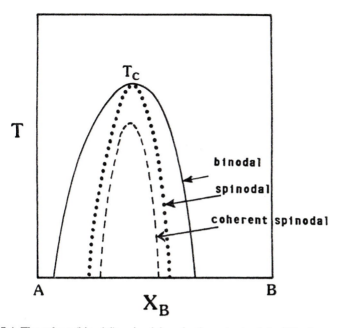

Figure 7.4. The solvus (binodal), spinodal, and coherent spinodal within the same system, schematic.

rearrangement of local coordination, it must proceed by nucleation and growth of the new phase. Such a transition is called *reconstructive*. However, sometimes one structure can be transformed into another by a coherent shift of atoms by very small distances without breaking any strong bonds. Such a transition is called *displacive* and is generally not quenchable. Examples are the α–β transition in quartz, perovskite distortional transitions, and the austenite-to-martensite transition in iron alloys. The latter has given its name, *martensitic*, to a subclass of displacive transformations, and the term martensitic is sometimes loosely used as a synonym for displacive. There has been extensive discussion of the conditions under which the olivine–β-phase–spinel transitions in silicates and germanates can occur by displacive versus reconstructive mechanisms and the role of these transitions in deep-focus earthquakes (Burnley, Green, and Prior 1991).

A phase transformation or chemical reaction in the solid-state reaction will occur at a rate that is controlled by (1) particle size, (2) temperature and pressure, (3) the driving force for the reaction, ΔG, and (4) diffusion coefficients and defect chemistry. Thus, in general, to get rapid reaction one should be well into the *P–T* stability field of a phase, at high temperature, and use fine-grained materials. A rigorous kinetic formalism does not exist but the Avrami equation is a useful semiempirical expression:

$$\% \text{ transformation} = 1 - \exp[-a(t)]^n \tag{7.11}$$

where t is time, n is an integer related to the dimensionality of crystal growth, and a is a rate constant related to nucleation, morphology, and mechanism.

A variation on the regular solution theme that is not strictly regular is described by a formalism that keeps the symmetry about $X = 0.5$ but lets the interaction parameter vary with temperature, $\lambda = a + bT$, or $W_G = W_H - TW_S$. This in effect introduces an additional entropy of mixing term that can be regarded as an empirical correction to take into account some degree of deviation from randomness, which always decreases the entropy, whether by ordering or clustering. W_S also serves as a correction resulting from differences in the vibrational entropy of the solid solution from that of a weighted average of end-member lattice entropies (this excess entropy can be positive or negative but is more frequently positive, reflecting looser bonding in a solid solution containing ions of different sizes) or as a combination of both factors. Though solvi calculated by this formalism are still symmetric, their shape depends on the value of W_S.

Most real systems show asymmetric solvi, implying that the regular solution formalism is inadequate. The next level of complexity in formalism is the *subregular* (or Margules parameter) description. In it, λ is a linear function of composition

$$\lambda = a + bX \tag{7.12}$$

or

$$W = W_B X_A + W_A X_B \tag{7.13}$$

Equation 7.13 gives clearer meaning to the two W parameters: W_1 is the limiting excess free energy or enthalpy of dissolving a small amount of component 1 in dilute solution in component 2; W_2 is the enthalpy of dissolving component 2 at infinite dilution into component 1. If the W terms are allowed to depend on temperature, then

$$W_A = W_{GA} = W_{HA} - TW_{SA} \tag{7.14a}$$

$$W_B = W_{GB} = W_{HB} - TW_{SB} \tag{7.14b}$$

These four parameters (W_{HA}, W_{HB}, W_{SA}, W_{SB}) are sufficient to fit almost any observed solvus or observed activity–composition relations. Lamentably, these are often used simply as empirical fitting parameters; a good fit is frequently obtained by W_H and W_S terms that are each unphysically large and that almost cancel each other in W_G. Keeping the meaning of enthalpy and entropy in mind, one should certainly be skeptical of the physical reality of W_H parameters greater than about 100 kJ/mol in magnitude and of W_S parameters greater than about 50 J/mol·K in magnitude. One should also bear in mind my earlier statement that with enough parameters you can fit an elephant.

Yet such empirical fits are useful for storing (and using) thermodynamic data in phase diagram calculations and are often encountered in geothermometry and geobarometry. They can describe a given system quite well, and as long as one is cautious about physical interpretation and doubly cautious about extrapolation beyond the *P-T-X* limits of the data used to obtain parameters, the Margules formalism can be useful in practical applications.

7.2.2 Systematics in mixing properties of solid solutions

In solid solutions without order–disorder phenomena, enthalpies of mixing are generally zero or positive. At low temperature, a positive enthalpy of mixing will outweigh the $T\Delta S$ term and result in a miscibility gap. Three main factors affect enthalpy of mixing and therefore the range and stability of solid solution. The first and usually most important is the size difference of the ions or atoms being mixed (Davies and Navrotsky 1983). This disparity results in a strain energy that is larger the greater the difference in size. For a given size difference, it is often easier to put a smaller atom into a larger site than vice versa. The second factor affecting enthalpy of mixing is differences in bonding character. In cases where the ionicity (however defined) of the end-members differs significantly, solubility can be limited, even when the ions are similar in size. Examples are NaCl–AgCl and NaBr–AgBr, which show large positive heats of mixing. Similar effects are expected in sulfides, though differences in crystal structure are frequent complications. If the differences in bonding between FeO and MgO become accentuated at high pressure, similar complexity might be seen. Specific electronic effects resulting from incompletely filled d shells may influence the thermodynamics of solid solutions containing transition metal ions. The third factor is charge. The higher the charge when species of the same charge are mixed, the more positive the heat of mixing for comparable size difference. When species of different charges mix in one sublattice (e.g., Fe^{2+}, Fe^{3+}, and Ti^{4+} in the iron-titanium oxides), short-range order may dominate the thermodynamics.

Of these factors, the role of size mismatch has been most thoroughly explored. Models based on calculating a lattice strain due to bond length mismatch have been moderately successful in simple systems such as the alkali halides. For many minerals the ions being mixed occupy only a small fraction of the total volume of the structure. If one compares solid solubility in CaO–MgO (very limited), $CaCO_3$–$MgCO_3$ (extensive) and $Ca_3Al_2Si_3O_{12}$–$Mg_3Al_2Si_3O_{12}$ (complete), one is drawn to the conclusion that greater size mismatch can be tolerated by a structure with larger molar volume in which the ions being mixed are embedded in a matrix that can itself change geometry slightly to absorb the strain. In a thermodynamic sense, the volume of a phase is a more general macroscopic parameter than any individual bond length.

With these considerations in mind, a correlation can be obtained between the thermodynamics of mixing (excess free energy in a regular or subregular model) and a term describing the volume mismatch for a variety of oxide, chalcogenide, halide, and silicate systems (Davies and Navrotsky 1983). Values of the excess free energy (W_G or W_{GA} and W_{GB}) are obtained from measured heats of mixing, measured activity–composition relations, or solid solubilities in systems showing a miscibility gap. A volume mismatch term is defined as follows: For a system showing regular solution behavior, W_G be correlated with $\Delta V = (V_A - V_B)/[0.5(V_A + V_B)]$, which is an average volume mismatch with $V_A > V_B$, where V_A and V_B are molar volumes. For subregular behavior W_{GA} is correlated with $(V_B - V_A)/V_B$ and W_{GB} with $(V_B - V_A)/V_A$. Thus the limiting interaction parameter for the dissolution of a component in the opposite end-member is correlated with the volume mismatch for substitution into that end-member. The resulting correlations are shown in Figure 7.5.

The use of a volume mismatch term rather than a bond length mismatch enables one to group together many diverse systems in which divalent ions are being mixed (oxides, chalcogenides, spinels, garnets, olivines, other silicates) into one correlation, which gives, for $\lambda = W_G$, W_{GA}, or W_{GB} as just defined,

$$\lambda = 100.8\Delta V - 0.4 \text{ kJ mol}^{-1} \qquad (7.15)$$

Within this correlation are points for cation mixing on four-, six-, and eight-coordinated sites and for anion mixing. Thus, once the effect of coordination number is included in the molar volume of the phases involved, it does not appear explicitly in the correlation. These correlations also suggest that the data for cation mixing and for anion mixing follow the same trend. Figure 7.6 shows the correlations for alkali halides, rocksalt oxides and chalcogenides, and tungstates and molybdates. The correlation segregates these systems into three distinct groups. The alkali halide systems as a group show much smaller positive deviations from ideality than the oxide and chalcogenide systems. The different slopes of these two correlations confirm the expectation that more highly charged ions mix less easily than ions of lower charge. A few points

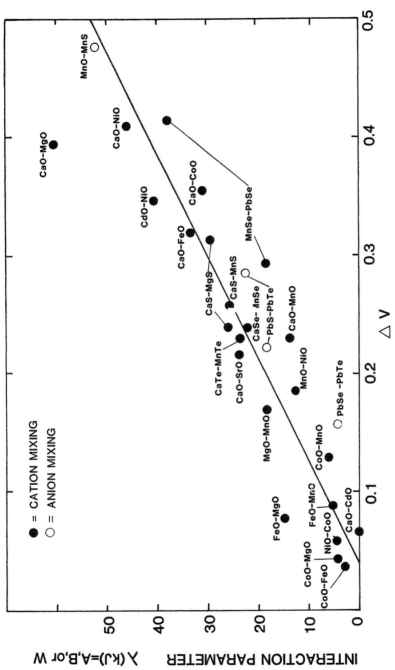

Figure 7.5. Correlation of interaction parameters and volume mismatch terms for divalent cations and anions (from Davies and Navrotsky 1983).

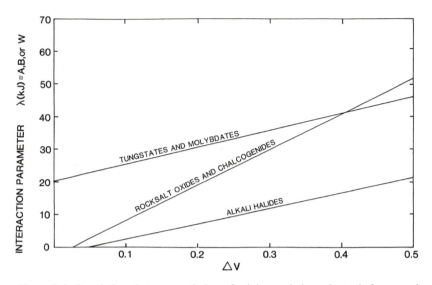

Figure 7.6. Correlations between enthalpy of mixing and size mismatch for several groups of compounds (Davies and Navrotsky 1983).

(e.g., Cr_2O_3–Al_2O_3, TiO_2–SnO_2) for trivalent and tetravalent ions also support this trend. The molybdate and tungstate systems appear anomalous for unknown reasons.

The main value of such empirical correlations is that they can be used to estimate deviations from ideality in phases where such measurements are difficult. For example, Table 7.1 lists predicted W_H parameters (based on the regular solution approximation) for a number of (Mg,Fe) mineral solid solutions including some high-pressure phases for which mixing data are not readily available. The calculations suggest the following: "FeO"–MgO shows the largest positive deviation from ideal mixing. In reality this is compounded by nonstoichiometry, which is much larger in this system than in any iron-bearing silicate solid solution. Olivine, pyroxene, β-phase, and silicate spinels are all predicted to have small but significant positive heats of mixing. Garnet (both aluminous and Al-free) and ilmenite are predicted to have smaller heats of mixing mainly because the molar volume difference between Mg end-members and Fe end-members is smaller. These predictions agree reasonably well with available experimental data. However, at high pressure, the ilmenite and perovskite solid solutions do not extend far toward the Fe end-member.

Table 7.1. *Predicted and experimental interaction parameter for Fe, Mg mixing in several structures*

System	Structure	Volume (cm³/mol)		W (kJ/mol)	
		Mg	Fe	calc.[a]	expt.
MgO–FeO	rocksalt	11.25	12.25	8.2	15.9
$1/2(Mg_2SiO_4–Fe_2SiO_4)$	olivine	21.84	23.14	5.4	6.3
$1/2(Mg_2SiO_4–Fe_2SiO_4)$	β-phase	20.26	21.61	6.1	6.3
$1/2(Mg_2SiO_4–Fe_2SiO_4)$	spinel	19.83	21.02	5.5	6.3
$MgSiO_3–FeSiO_3$	orthopyroxene	31.33	32.96	4.7	4.0
$MgSiO_3–FeSiO_3$	garnet	28.53	29.38	2.6	
$MgSiO_3–FeSiO_3$	ilmenite	26.35	26.85	1.5	
$MgSiO_3–FeSiO_3$	perovskite	24.46	25.49	3.8	
$1/3(Mg_3Al_2Si_3O_{12}–Fe_3Al_2Si_3O_{12}$	garnet	37.73	38.41	1.4	

[a] Calculated using Eq. 7.15.

7.2.3 Relation of mixing properties to phase diagram

It is informative to pursue qualitatively the relation of the magnitude of excess mixing terms (λ or W_G) to the corresponding binary phase diagram. Figure 7.7 shows the evolution of the phase diagram as λ in the solid becomes more positive. Starting with an ideal solution binary loop, one first sees a solvus develop at temperatures far below the solidus. As this immiscibility dome grows, the shapes of the solidus and liquidus are perturbed. The system gradually evolves into a eutectic with wide terminal solid solubility and finally into a simple eutectic.

The opposite case, one of negative λ, implies a tendency toward ordering in the solid. The system may evolve to a series of ordered phases as seen in alloys (see Fig. 7.8). If this tendency toward maximizing A–B interactions in the solid also is seen in the liquid, one can trace evolution of compound formation, as seen in Figures 7.8 and 7.9. Here, as an ordered compound (of a different structure from the two end-members) becomes more stable, its melting point rises. The diagram transforms from a simple eutectic to a peritectic to compound formation with a domed maximum to compound formation with a cusp. These relations can be quantified and phase diagrams calculated from thermodynamic data or thermodynamic data estimated from phase diagrams. A difficulty in the latter is the nonuniqueness of thermodynamic models for silicate melts (see Chap. 8).

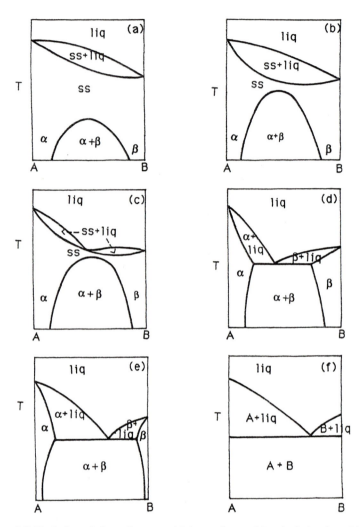

Figure 7.7. Evolution of phase diagram with increasing positive deviation from ideality in the solid in a solid solution forming system.

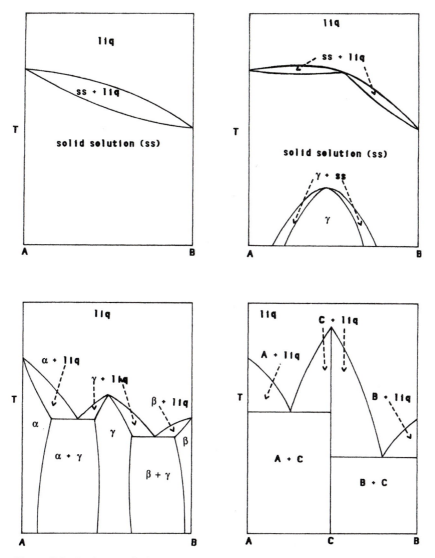

Figure 7.8. Evolution of phase diagram with increasing tendency for forming an ordered solid phase from the solid solution.

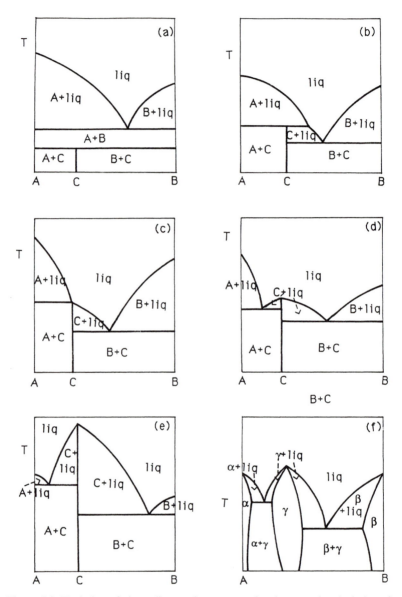

Figure 7.9. Evolution of phase diagram in a system showing negative deviations from ideality in the liquid phase indicative of compound formation.

7.2.4 Phases with different structures

Partial solid solution can exist among end-members of different structure. Consider two substances, A with structure α and B with structure β. Writing formula units corresponding to one mole of ions being mixed, one has

$$\mu(A, \alpha) = \mu^0(A, \alpha) = RT \ln a(A, \alpha) \qquad (7.16a)$$

$$\mu(A, \beta) = \mu^0(A, \alpha) + \Delta\mu(A, \alpha{\to}\beta) + RT \ln a(A, \beta) \qquad (7.16b)$$

$$\mu(B, \alpha) = \mu^0(B, \beta) + \Delta\mu(B, \beta{\to}\alpha) + RT \ln a(B, \alpha) \qquad (7.16c)$$

$$\mu(B, \beta) = \mu^0(B, \beta) + RT \ln a(B, \beta) \qquad (7.16d)$$

where $\mu^0(A, \alpha)$ is the free energy per mole of pure A in structure α, $\Delta\mu^0$ (A, $\alpha{\to}\beta$) is the free-energy change for the transformation of A from structure α to structure β, $a(A, \alpha)$ refers to the activity of A in structure α (unity for pure A) and $a(A, \beta)$ refers to the activity of A in structure β (also unity for pure A, with structure β taken as standard state). Analogous definitions hold for the terms in component B. Pure A in structure β can be a real phase at some P and T (e.g., $MgSi_{0.5}O_2$ in the spinel structure) or a physically unrealizable *crypto-modification* or *fictive phase* unstable at all P and T (e.g., $FeSi_{0.5}O_2$ in the modified spinel structure). The limiting solubilities are given by equating chemical potentials:

$$\mu(A, \alpha) = \mu(A, \beta), \qquad \mu(B, \alpha) = \mu(B, \beta) \qquad (7.17)$$

In general, these equations must be solved iteratively to obtain the coexisting compositions of the two phases at a given pressure and temperature. Figure 7.10 shows schematically the topologies of phase diagrams when the end-members have different structures. The two solubility curves in Fig. 7.10c are not limbs of a solvus but are the locus of common tangents to two different free-energy curves (see Fig. 7.10d). Such a solubility curve usually terminates in *T–X* space by melting. When each component can undergo a phase transition, topologies like those in Figure 7.10a,b are possible. A more complex version of such a diagram, with three possible phases (α, β, γ) in *P–X* space, has been shown in Figure 6.24 for the Mg_2SiO_4–Fe_2SiO_4 system.

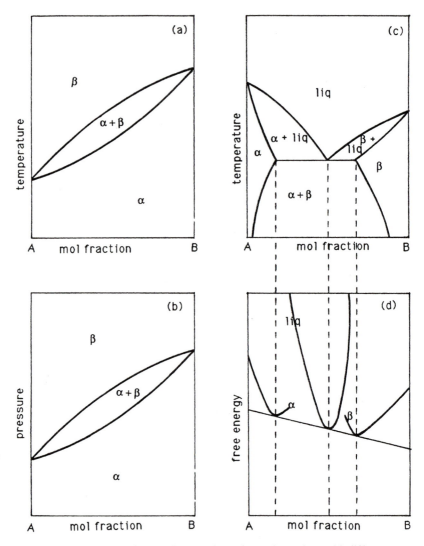

Figure 7.10. Topologies of phase diagrams involving end-members with different structures. (a) Both end-members transform as a function of temperature. (b) Both end-
members transform as a function of pressure. (c) Neither end-member transforms as a
function of temperature before melting. (d) Free-energy curves corresponding to temperature in (c).

7.3 Order–disorder

7.3.1 General comments

Ordering phenomena complicate the thermodynamic behavior of minerals and their solid solutions in several ways: (1) If the order–disorder process is slow, several series of solid solutions, each with a different degree of order, can be prepared or observed in nature. Then the thermodynamics of mixing of other species (e.g., Na, K in feldspars) will depend on the degree of order in the aluminosilicate framework. (2) The extent of ordering depends on temperature in a complex fashion. Thus the equilibrium properties of a solid solution have a complex dependence on temperature that is not modeled easily using conventional formulations such as Margules parameters. Extrapolation of measured properties to higher or lower temperatures is far more risky than in simple systems. Thermodynamic models that specifically include order–disorder must be developed. (3) The observed degree of ordering is often kinetically controlled. At low temperature, changes in the degree of order are hindered, thus both ordering and disordering occur slowly and many metastable states are preserved. At intermediate temperature, equilibrium is attained and that state can be quenched to ambient conditions. At high temperature, equilibrium is attained, but it may be impossible to quench to the full extent of disorder back to ambient conditions. The effect of pressure on the rates of order–disorder reactions has only begun to be explored. Thus the interpretation of experimental observations must be made with both thermodynamics and kinetics in mind. Specifically, the extent to which order–disorder contributes to experimentally measured heat capacities poses a complex problem. (4) A strong tendency to ordering is manifested in significant negative heats of mixing and sometimes leads to compound formation (e.g., dolomite, omphacite pyroxenes, etc.). Both the heat and entropy of mixing in such systems depend strongly on the degree of order and, therefore, on temperature. Destabilizing energetics arising from mixing of ions in the disordered state may be present at the same time as stabilization arising from ordering, leading to complex interplay between ordering and exsolution. (5) Order–disorder reactions often involve a change in the symmetry of the structure. Such changes may be ordinary first-order phase transitions, or they may be higher-order phase transitions leading to unconventional phase diagram topologies. The following sections will describe some examples and present thermodynamic analyses for several systems.

7.3.2 Cation interchange equilibria, especially in spinels

Many minerals (e.g., spinel, olivine, pyroxene, feldspar, melilite, pseudobrookite) have two or more crystallographically nonequivalent sites. If these sites
represent positions of very different energy, a completely ordered structure
results. If the energetics of cation interchange are not prohibitive, a partially
or completely randomized structure will form, and the degree of disorder will
vary with thermal history. Since the sites remain crystallographically distinct,
even in the fully disordered state, the disordering process is nonconvergent.
Octahedral–tetrahedral cation disorder in spinels is an example of this situation. The spinel structure has been described in Section 2. 4. The degree of
disorder, x, is the mole fraction of tetrahedral sites occupied by B cation, that
is, the cation distribution is $(A_{1-x}B_x) [A_xB_{2-x}]O_4$.

A simple thermodynamic treatment of cation distributions in simple spinels
is formulated as follows (Navrotsky and Kleppa 1967): The enthalpy of disordering is assumed to be the extent of disordering, x, multiplied by an interchange energy, ΔH_{int}. The latter may be thought of as the enthalpy associated
with the reaction

$$A_{tet} + B_{oct} = A_{oct} + B_{tet} \tag{7.18a}$$

or

$$(A) + [B] = [A] + (B) \tag{7.18b}$$

and is assumed to be independent of the degree of disorder. Then

$$\Delta H_{dis} = x \, \Delta H_{int} = H(x = x) - H(x = 0) \tag{7.19}$$

The entropy of disordering is assumed to be just the configurational entropy

$$\Delta S_{dis} = S_{conf} = -R[x \ln x + (1 - x) \ln (1 - x) + x \ln (x/2) \tag{7.20}$$
$$+ (2 - x) \ln (1 - x/2)]$$

S_{conf} is shown in Figure 7.11, which clearly shows the random distribution
($x = 2/3$) as being that of highest entropy. The free energy of disordering is
then

$$\Delta G_{dis} = x \, \Delta H_{int} - RT[x \ln x + (1 - x) \ln (1 - x) \tag{7.21}$$
$$+ x \ln (x/2) + (2 - x) \ln (1 - x/2)]$$

and, at any temperature

$$-RT \ln \left[\frac{x^2}{(1 - x) (2 - x)} \right] = \Delta H_{int} \tag{7.22}$$

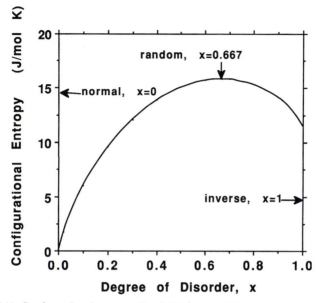

Figure 7.11. Configurational entropy (Eq. 7.20) for spinel disordering.

From cation distributions observed in spinels, a series of values of interchange enthalpies can be calculated and a set of octahedral site preference energies for cations in the spinel sructure obtained.

This simple model can be refined in two ways. Lattice energy arguments suggest that the interchange energy should depend on the degree of disorder.

$$\Delta H_{int} = \alpha + \beta x \qquad (7.23)$$

and there may be nonconfigurational contributions to the entropy of disordering. Then (O'Neill and Navrotsky 1983, 1984),

$$\Delta G_{dis} = x (\alpha + \beta x) - T (\Delta S_{conf} + \Delta S_{nonconf}) \qquad (7.24)$$

In general, experimental data are too sparse to determine both the nonconfigurational entropy term and the dependence of enthalpy on degree of disordering, and O'Neill and Navrotsky chose to ignore the former as being smaller in magnitude.

At equilibrium $(dG_D/dx) = 0$, and (neglecting the nonconfigurational entropy)

$$-RT \ln \left[\frac{x^2}{(1 - x) (2 - x)} \right] = \alpha + 2\beta x \qquad (7.25)$$

The coefficients α and β are generally of approximately comparable magnitude and opposite sign. The effects of crystal field stabilization energies are implicitly included in this model, since they can be accommodated in α. The coefficients α and β will depend mainly on the differences in radii and charge of the two ions, but these coefficients must be determined empirically from observed cation distributions.

O'Neill and Navrotsky (1984) applied this model to spinel solid solutions. After examining data for a number of spinels, they concluded that the value of the coefficient β depended mainly on the charge type, with $\beta = \sim -20$ kJ/mol for 2–3 spinels and $\beta = \sim -60$ kJ/mol for 2–4 spinels. Assuming these constant values of β, a series of αs, which maintain a meaning of cation site preference energies, could be calculated and are shown in Figure 7.12. For a spinel solid solution at any given temperature, the cation distribution is calculated by solving a set of equilibria involving several pairs of cations. The lattice parameters in the solid solution are calculated from the site occupancies. Activity–composition relations are then calculated using the cation distributions at each temperature, the change in configurational entropy for solid solution formation at that temperature, the negative change in enthalpy resulting because the site occupancy is not a simple average of the site occupancies of the end-members, and, finally, an empirical regular solution enthalpy of mixing term reflecting destabilization from size mismatch.

Several general conclusions can be drawn: (1) The enthalpy and entropy of mixing depend strongly on temperature because the degree of order is temperature dependent. (2) The lattice parameter can vary rather nonlinearly with composition, especially at low temperature for end-members with different cation distributions. A sigmoid variation of lattice parameter with composition is commonly seen in solid solutions between a normal and an inverse spinel (e.g., $FeCr_2O_4$–Fe_3O_4) (see Fig. 7.13a). (3) The calculated entropies of mixing can be quite asymmetric, especially for the mixing of a normal and an inverse spinel. This results in significantly asymmetric free energies of mixing, not because one invokes Margules parameters and asymmetry in the enthalpies and vibrational entropies of mixing, but because the configurational entropy term itself departs from symmetrical behavior. At low temperature this asymmetry is pronounced and a solvus between a largely normal and largely inverse spinel develops (see Fig. 7.13b).

Cation distribution data for $MgAl_2O_4$ are shown in Figure 7.14. Measurements on quenched samples were obtained both by NMR (Millard, Peterson, and Hunter 1992; Wood, Kirkpatrick, and Montez 1986) and by analysis of Cr ESR signals (Schmocker and Waldner 1976). The inversion parameter, x, for quenched samples show a sigmoid variation, characteristic of the inability to

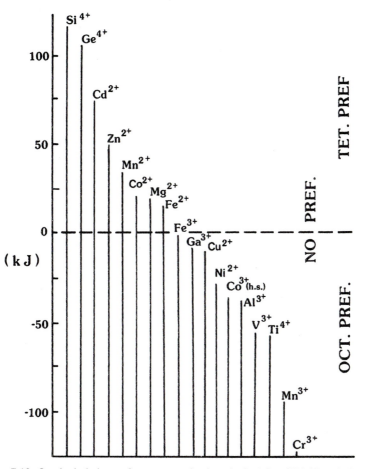

Figure 7.12. Octahedral site preference energies in spinels (after O'Neill and Navrotsky 1984).

obtain equilibrium below about 800 K and the inability to quench the high-temperature disorder above about 1100 K. This behavior is quite common. In situ neutron diffraction data [Peterson, Lager, and Hitterman 1991] show greater disorder and no evidence of a plateau, though they extend only to 1273 K. The in situ data are consistent with a constant intercharge enthalpy of 24 ± kJ/mol as the simple disordering model or with $\alpha = 31 \pm 1$ kJ and $\beta = -10 \pm 3$ kJ as the two-parameter model (Peterson et al. 1991). These values of disordering enthalpy are consistent with an observed enthalpy of annealing of about 1 kJ/mol for samples quenched from 1200 K and dropped into a calorim-

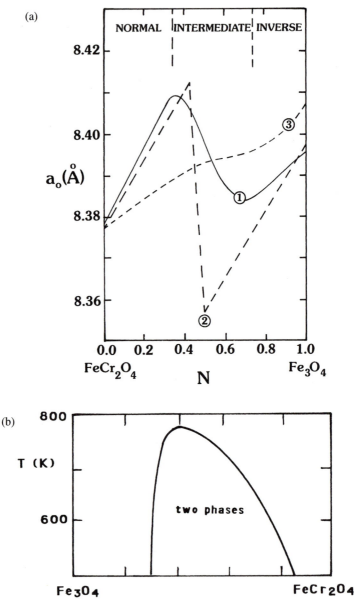

Figure 7.13. Thermodynamics of cation distribution in Fe_3O_4–$FeCr_2O_4$ (from O'Neill and Navrotsky 1984). (a) Lattice parameters: (1) experimental data, (2) calculated using O'Neill and Navrotsky model for 298 K, (3) calculated for 1000 K. (b) Calculated solvus.

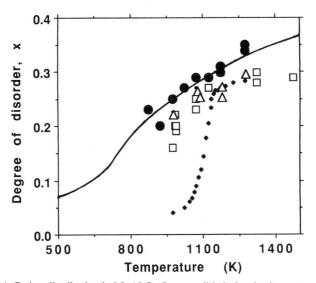

Figure 7.14. Cation distribution in $MgAl_2O_4$: Large solid circles, in situ neutron diffraction data of Peterson, Lager, and Hitterman (1991); open triangles, NMR data on quenched samples (Millard, Peterson, and Hunter 1992); open squares, NMR data of Wood, Kirkpatrick, and Montez (1986), corrected for pulse repeat times as suggested by data of Millard et al. (1992); small solid circles, ESR data on a heat-treated natural spinel (data from Schmocker and Waldner 1976); curve, calculated from simple disordering model, $\Delta H_{int} = 24$ kJ/mol.

eter at 973 K (Navrotsky 1986). $MgAl_2O_4$ is complicated by two other factors. First, above 1500 K there is considerable solubility of excess Al_2O_3, leading to nonstoichiometry in synthetic samples (Navrotsky 1986). Second, there have been numerous but inconclusive indications of a phase transition (involving a possible spare group change) somewhere between 1000 K and 1200 K (see Peterson et al. 1991). Thus the type-mineral of the spinel group is surprisingly complex.

7.3.3 Convergent disordering in carbonates

The major crystal–chemical difference between the calcite and dolomite structures is that calcite contains only one distinct cation layer and dolomite has two. The chemical difference between these layers results in two geometrically inequivalent cation sites. Furthermore, a continuous range of structures between the ordered dolomite and disordered calcite structures is possible because the space group of dolomite ($R\bar{3}$) is a subgroup of that calcite ($R\bar{3}c$). Indeed, some natural dolomites (i.e., $CaMg(CO_3)_2$) are reported to have disordered structural states, and partially disordered samples may be obtained by annealing ordered material at high temperature. However, the high temperatures involved, CO_2 evolution, and possible problems in quenching the disordered phase make detailed study of the order–disorder transition in $CaMg(CO_3)_2$ difficult.

The system $CdCO_3$–$MgCO_3$ is analogous in that similar ordered and disordered phases exist. Because chemical homogenization of distinct Cd and Mg layers proceeds within an experimentally accessible range of pressure and temperature, samples can be prepared that, on quenching, possess greatly different states of order. Phase diagrams for both the $CaCO_3$–$MgCO_3$ and the $CdCO_3$–$MgCO_3$ systems are shown in Figure 7.15.

The Cd,Mg disordering reaction is convergent, that is, alternate cation layers are crystallographically distinct only when the cations are ordered. The simplest model that can be applied to the order–disorder reaction is of the Bragg-Williams type in which the long-range order parameter, s, is related to one energy parameter, W, by the equation (at the stoichiometric dolomite composition)

$$s = \tanh\,(-Ws/2R) \qquad (7.26)$$

W is negative and its numerical value reflects the energy of ordering per mole $CdMg(CO_3)_2$ (Capobianco et al. 1987).

The enthalpy of mixing (formation of $CdMg(CO_3)_2$ from $CdCO_3$ and $MgCO_3$) is then

$$\Delta H_{mix} = W[0.5 - 0.25\,(1 - s^2)] \qquad (7.27)$$

The critical temperature at which s becomes zero is given by $T_c = -W/2R$. It is about 1100 K for $CdMg(CO_3)_2$ and about 1500 K for $CaMg(CO_3)_2$.

A striking feature of both Cd and Ca systems is that the ordered dolomite phase has a negative enthalpy of mixing from the end-member carbonates, whereas the disordered phase at dolomite composition has a positive enthalpy of mixing. This can not be accounted for by a simple Bragg-Williams model because its one parameter, W, must be negative for ordering to take place.

To include positive heats of mixing in the disordered phase, one can apply more sophisticated models that take next-nearest neighbor interactions into account (Capobianco et al. 1987). In such models there are two distinct interactions: an attractive (interlayer) interaction promoting unlike pair formation (parameterized as W_{inter}) and a repulsive (intralayer) interaction promoting segregation (parameterized as W_{intra}).

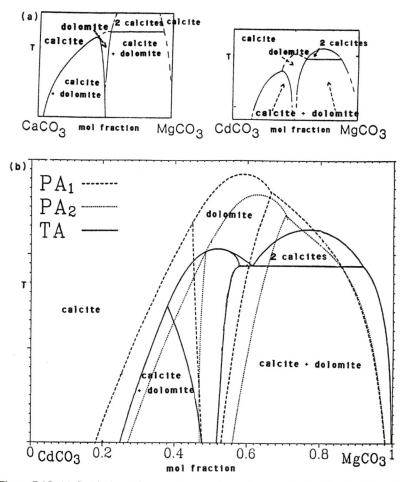

Figure 7.15. (a) Semischematic experimental phase diagram for $CaCO_3$–$MgCO_3$ and $CdCO_3$–$MgCO_3$ (Burton 1987). (b) Calculated $CdCO_3$–$MgCO_3$ phase relations using three different cluster variation method approximations (Capobianco et al. 1987).

The Gibbs free energy of mixing (formation of $(Cd_{1-x}Mg_x)_2(CO_3)_2$ from $CdCO_3$ and $MgCO_3$) per formula unit containing two carbonates is given by

$$\Delta G_{mix} = W_{inter} [X - X^2(1 - s^2)] + W_{intra} [X - X^2(1 + s^2)] \quad (7.28)$$
$$+ RT\sum_i \sum_j x_{ij} \ln x_{ij}$$

where x_{ij} is the cation site occupancy of the jth atom on the ith sublattice. Thus for $i = \alpha$ (layer preferred by Mg), β (layer preferred by Cd) and $j = $ Mg, Cd,

$x_{\alpha,Mg} = X + Xs$, $x_{\alpha,Cd} = 1 - X - Xs$, $x_{\beta,Mg} = X - Xs$, and $x_{\beta,Cd} = 1 - X + sX$. The equilibrium value of s is found by minimizing ΔG_{mix} with respect to variations in s, and the critical temperature for disordering is obtained by setting the second derivative of ΔG_{mix} to zero for $s = 0$. For $W_{inter} < 0$ and $W_{intra} > 0$, intersite ordering decreases ΔH_{mix} while intrasite mixing increases it, and at low temperature the ordered phase is stable. Above T_c, $s = 0$, so that ΔH_{mix}, given by $(W_{inter} + W_{intra}) (X - X^2)$ is positive for $W_{intra} > \mid W_{inter} \mid$. W_{inter} and/or W_{intra} may depend on composition, leading to asymmetry in solution properties.

A model based on the tetrahedral approximation (TA) in the cluster variation method (CVM) also can be applied (Capobianco et al. 1987). The CVM approach includes the effects of short-range order (which the preceding approximations neglect) by considering the possible arrangements of clusters of atoms (tetrahedra of cations in this case) (Burton 1987). The calculated equilibrium proportions of the chemically distinct tetrahedra are constrained by the composition and the energetic interactions between the atoms of the tetrahedron. These calculations provide information on both long-range and short-range order. Because of short-range order, the configurational entropy in the CVM calculations is generally smaller than in the previous models.

Calculations using CVM models correctly predict the topology and qualitatively fit the observed phase boundaries and account qualitatively for positive heats of mixing in the disordered phases. However, quantitative fit to the structural and thermochemical data in $CdCO_3$–$MgCO_3$ is not possible. This failure may be related to several factors. First, the carbonate group may undergo shifts of position and small deviations from a planar configuration as a result of Cd, Mg ordering. Such involvement of CO_3 groups might give the transition more cooperative character and compress it, as observed, into a smaller temperature interval. Second, the compositional asymmetry of inter- and intralayer interactions may be more complex. Third, differences in lattice vibrations between ordered and disordered phases may result in excess heat capacities, excess entropies, and temperature dependent enthalpies.

Nevertheless, it is important to note the essential features of a model that can qualitatively explain the variation of long-range order with temperature, the enthalpy of mixing, and the topology of the phase diagram. Such a model must incorporate three main factors: a negative (stabilizing) interaction for ordering of cations between layers, a positive (destabilizing) interaction for the mixing of cations within a layer, and compositional asymmetry for at least one parameter. These features lead to a strong temperature dependence of the order parameters and of the enthalpy and entropy of mixing. Empirical models based on a Margules formulation simply can not build in the appropriate temperature dependence and can not account for the phase diagram topology.

7.3.4 Al–Si order–disorder in feldspar

The ability of the feldspar structure to undergo fairly rapid alkali exchange at low temperature while preserving a state of Al–Si order enables one to study the influence of the latter on the energetics of Na, K mixing. For a series with constant Al,Si order, the enthalpy of Na, K mixing becomes less positive with increasing framework disorder, that is, in the series maximum microline–orthoclase–high sanidine. The slightly larger volumes of the disordered feldspars, coupled with a larger variety of local environments within the framework, may make it energetically easier for Na and K to find local geometries that minimize the strain. The plagioclase feldspars, with the charge coupled substitution $Na_A + Si_T = Ca_A + Al_T$, are more complex.

Several approaches to the thermodynamics of $NaAlSi_3O_8$–$CaAl_2Si_2O_8$ solid solutions are possible. Two sorts of experimental thermodynamic data are available: (1) activity–composition in plagioclase of "high" structural state derived from hydrothermal and ion-exchange equilibria and (2) solution calorimetric determinations of enthalpies of mixing in solid solutions of several states of Al,Si order. The interpretation of these data depends on the thermodynamic model one assumes. Saxena and Ribbe (1972) and Blencoe, Merkel, and Seil (1982) assumed that high plagioclases represent a continuous solid solution represented by one free-energy surface and derived empirical polynomials to describe the excess free energy, relative to ideal ($a_{Ab} = X_{Ab}$, $a_{An} = X_{An}$) behavior. These polynomials fit the observed positive deviations from Raoult's law at 700–1000 °C and extrapolate to ideal one-site mixing ($a = X$) at the high-temperature limit (see Fig. 7.16).

However, the stoichiometry and crystallography preclude simple one-site mixing in plagioclase, since Na and Ca mix on A-sites and Si and Al mix on T-sites. The nature of the configurational entropy depends on the degree of Al,Si order and the extent of local charge balance observed in the Na + Si = Ca + Al substitution. The activity data can be described reasonably well by combining the observed positive heats of mixing with entropies of mixing calculated assuming aluminum avoidance in the high plagioclases. This leads to positive deviations from Raoult's law at 873–1273 K, virtually Raoultian behavior at 1273–1473 K, and negative deviations from Raoult's law at higher temperature. This crossover occurs because positive heats of mixing are balanced by entropies of mixing larger than $-R[X \ln X + (1 - X) \ln (1 - X)]$.

An interesting sidelight to plagioclase thermodynamics is that at temperatures of interest to igneous petrology (1173–1573 K), the numerical deviations of activity–composition relations from Raoultian behavior are rather small. This leads to the use of ideal solution models ($a_{NaAlSi_3O_8} = X_{NaAlSi_3O_8}$) in petro-

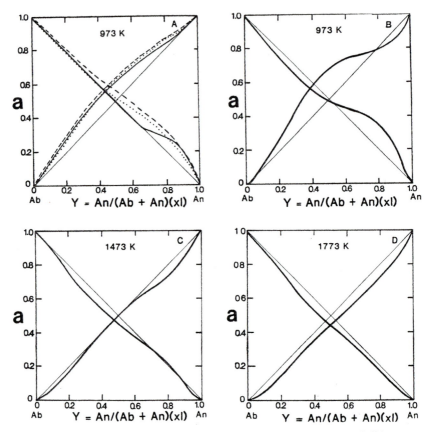

Figure 7.16. Activity composition relations in plagioclase feldspars. (a) Solid curves represent two-Henrian region model of Orville (1972); dotted curves represent subregular solution model of Saxena and Ribbe (1972); dashed curves represent subregular solution model of Seil and Blencoe (1978). These are all based on ion-exchange equilibria near 973 K. (b, c, d) Calculated activity composition relations based on enthalpies of solution of Newton, Charlu, and Kleppa (1980) and an Al-avoidance entropy model at various temperatures. (From Henry, Navrotsky, and Zimmerman 1982.)

logic applications. The preceding discussion suggests that such behavior is fortuitous and results from the approximate cancellation of complex and temperature-dependent heats and entropies within a restricted range. However, their extrapolation to lower or high temperature or to high pressure may lead to significant errors.

The original study of ion exchange equilibria (Orville 1972) chose to separate activity plots into three regions: (1) $0 \leq X_{An} < {\sim} 0.5$, $\gamma_{An} = 1.27$, $\gamma_{Ab} = 1$

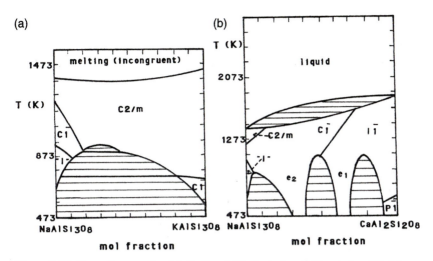

Figure 7.17. Phase diagrams for (a) alkali and (b) plagioclase feldspars (modified from Carpenter 1988). Areas with horizontal hatching represent two-phase regions.

(Henry's law for An, Raoult's law for Ab); (2) $0.5 < X_{An} < \sim 0.9$, both γ_{Ab} and γ_{An} change with composition; and (3) $\sim 0.9 < X_{An} < 1$, $\gamma_{An} = 1$, $\gamma_{Ab} = 1.89$. This separation into two Henry's law regions separated by an intermediate region implies that in albite-rich compositions the anorthite component is substituting into a rather constant albite-like environment, at anorthite-rich compositions an anorthite-like environment persists, and a fairly rapid change in structure and thermodynamic properties occurs at intermediate compositions that are close to those for the $C\bar{1}$–$I\bar{1}$ transition. This transition, which is not first order, separates albite-rich feldspars with a largely disordered Al,Si distribution from anorthite-rich feldspars with a largely ordered Al,Si distribution.

Carpenter and Ferry (1984) pointed out that the $C\bar{1}$–$I\bar{1}$ transition, which is probably second order, complicates the phase diagram (see Fig. 7.17) and means that activity–composition relations can not be derived from a simple free-energy curve. Rather, the region of high albite-like ordering (with Al–Si disorder) gives way to a region of anorthite-like ordering with increasing Al/(Al + Si). This situation would produce activity–composition relations something like those of Orville (1972).

The enthalpies of solution in molten $2PbO \cdot B_2O_3$ near 973 K of ordered and disordered plagioclases are shown in Figure 7.18 (Carpenter, McConnell, and Navrotsky 1985). A continuous curve through the data for high plagioclases implies a positive heat of mixing. The data can also be fit by two lines, one for

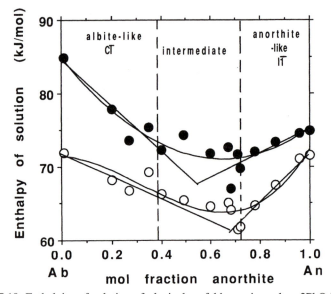

Figure 7.18. Enthalpies of solution of plagioclase feldspars in molten $2PbO·B_2O_3$ at 973 K (simplified from Carpenter, McConnell, and Navrotsky 1985). Solid circles represent data for ordered low structural state feldspars; open circles represent data for disordered high structural state feldspars. Curves represent subregular fits to data treated as continuous solid solutions. Lines represent trends in C$\bar{1}$ and I$\bar{1}$ phase fields if these are considered separate phases. Vertical dashed lines separate terminal regions of albite-like and anorthite-like structures separated by an intermediate transitional region.

I$\bar{1}$ and one for C$\bar{1}$. These two different descriptions fit not only the enthalpy data but also the activity–composition relations (as discussed) reasonably well in the temperature range where they have been measured. But they extrapolate to very different behavior at higher and lower temperature and yield different phase diagram topologies. Thus the interpretation of the same macroscopic thermodynamic observations can indeed be quite different depending on the picture one favors for the microscopic structural behavior. The situation is further complicated by intermediate ordering states, microstructures, and exsolution.

7.3.5 *Landau theory*

The Landau theory has become quite popular in describing phase transitions and other phenomena related to order–disorder (Carpenter 1988; Salje 1987).

It is basically a macroscopic thermodynamic formulation, which utilizes a general series expansion of the free energy

$$G - G_o = AQ^2 + BQ^4 + CQ^6 \ldots \tag{7.29}$$

where G_o is a reference free energy, Q is an order parameter (see following), and A, B, C are constants. The physical idea is that as one approaches a phase transition (first or second order), one can define some macroscopic parameter (order parameter) that is a measure of how far the system is from the disordered state. As the system disorders, its free energy is perturbed (lowered) by this disordering and above the transition, $Q = 0$. Commonly, Q can be associated with a lattice strain (difference is two lattice parameters that become identical in the high-temperature phase of higher symmetry) but other choices are possible. The task, then, is to choose using symmetry arguments which coefficients in the Landau expansion are nonzero, and to obtain expressions for heat capacity, entropy, and other functions. The attractiveness of this approach lies in its generality and in that it links macroscopic observables (lattice parameters, other physical properties) with thermodynamic parameters. The disadvantage is that, being macroscopic, these order parameters do not give immediate meaning to what atomistic interactions are driving the transition. Further discussions are given in the general references and bibliography section and in the specific applications to feldspars.

Salje's group has applied Landau formalism to many minerals, including feldspars (Salje et al. 1985; Salje 1985, 1987). The approach may be applied to albite as follows: Two simultaneously occurring processes that reduce the symmetry of Na-feldspar are known – very sluggish Al,Si ordering and rapid displacive transformation. As a result, a two-order parameter theory using order parameters Q_{od} and Q (representing the preceding processes, respectively) is appropriate. Based on symmetry considerations, Q and Q_{od} are not independent of each other due to their common strain components. If one order parameter is known, the other can be calculated if the crystal is in thermal equilibrium. If, however, the crystal is in a metastable state, Q_{od} is independent of temperature and the observed thermodynamic behavior is controlled entirely by the temperature dependence of Q.

The symmetry-breaking transition (C2/m $\rightarrow$ C$\bar{1}$) under thermal equilibrium is a function of both Q and Q_{od} and is accompanied by a smooth crossover between high and low albite. Finally, simple formulas are written as functions of the order parameters, Q and Q_{od}, and of temperature, T, from which thermodynamic functions, lattice parameters, and other observable quantities may be calculated. By deriving numerical parameters for their equations from experimental data, expressions are obtained for the free energy, entropy, and enthalpy of sodium feldspar in thermal equilibrium and in metastable states as a function of the two order parameters, Q and Q_{od}, and the temperature T. Then the lattice strains of all the stable and metastable states are calculated and a phase diagram constructed.

An attractive feature of this order theory is that its parameters can be determined from independent thermodynamic and physical measurements, rather than merely varied in computer simulations and obtained by a "best fit" to a phase diagram. Salje (1987) has formulated a Landau theory approach to the $I\bar{1} \rightarrow P\bar{1}$ phase transition in anorthite and in Ca-rich plagioclases. The theory uses two coupled order parameters, one related to rapid lattice distortion at the phase transition itself (which occurs at 510 K in anorthite), the other related to the degree of "frozen in" Al,Si disorder arising from the high temperature $C\bar{1}$–$P\bar{1}$ equilibrium near the melting point. Thus the low-temperature transition is sensitive to plagioclase composition and thermal history, and behavior at low-temperature offers a tool for studying the fairly inaccessible high-temperature disordering. Such an approach can be expected to lead toward a full description of plagioclase thermodynamics in the future. In the meantime, the most petrologically useful simplified description of plagioclase solid-solution behavior is still in doubt.

Each of the approaches used has some advantages and some drawbacks. Empirical (Margules-type) formulations are convenient and easily calibrated but they are unreliable for extrapolation. Models based on Al–Si mixing entropies approximate the microscopic behavior but are too crude to capture local ordering, exsolution, and intergrowths. Landau-type models describe the macroscopic behavior leading to both ordering and exsolution but are too cumbersome to give simple expressions for activities to put into petrologic computer codes. For the feldspars, we still see only parts of the elephant and not the whole beast at once.

7.4   Defects in solids

A perfect crystal would be virtually unreactive; diffusion of atoms through a solid and solid-state reactions proceed through its imperfections. These defects can be grouped into two classes, intrinsic and extrinsic. *Intrinsic defects* represent thermodynamic equilibrium in the pure solid. No matter how endothermic the energy of a given defect, it will be present at some concentration because its entropy of formation will counter the enthalpy until a minimum in free energy is reached. This concentration will increase with temperature. The defects present in greatest concentrations (majority defects) are those that are thermodynamically most favorable. Bear in mind that at any temperature above absolute zero, a solid with its equilibrium concentration of defects has a lower free energy than one with no defects at all; therefore, an absolutely perfect crystal is not realizable. *Extrinsic defects* are introduced by impurities, whether in nature or in the laboratory. Because their concentrations are arbitrary (or historical, reflecting past but not present conditions), extrinsic defects can occur at much higher concentrations than intrinsic defects and, especially at low temperature, extrinsic defects control physical properties.

One can further classify defects according to their dimensionality: point defects, line defects, planar defects, and three-dimensional defects. The latter three are often grouped together as *extended defects*. Because their entropy of formation is much smaller than that of point defects, their equilibrium concentrations are generally much smaller and their effects on bulk thermodynamic properties much less. However, they are tremendously important in diffusion, sintering, dissolution and precipitation, exsolution, viscosity, creep, fracture, and other behavior.

Chemical reactivity, diffusion, electrical and thermal conductivity – transport and kinetic properties in general – are governed by defect equilibria. In iron-bearing minerals, the predominant defect reaction is usually related to the oxidation of ferrous iron and often involves creation of cation vacancies for charge balance. However, perovskite structures appear to tolerate oxygen vacancies. Furthermore, defect chemistry at high pressures and temperature is very poorly constrained at present.

As an illustration of these concepts for $(Mg,Fe)SiO_3$ perovskite, the incorporation of iron may take place by several mechanisms. Ferrous iron may be included by

$$Mg_A^{2+} \rightarrow Fe_A^{2+} \tag{7.30}$$

where A is the large eight-fold central site and B is the octahedral site. This is the straightforward Mg–Fe substitution along the $MgSiO_3$–$FeSiO_3$ perovskite join. Ferric iron may be accommodated by a number of mechanisms. One is

$$Mg_A^{2+} + Si_B^{4+} = Fe_A^{3+} + Fe_B^{3+} \tag{7.31}$$

This represents solid solution along the hypothetical join between $MgSiO_3$ perovskite and Fe_2O_3 in the perovskite structure as well. This substitution would change the perovskite composition to $(Mg+Fe)/Si > 1$ and expel SiO_2 (stishovite) from an initially stoichiometric $(Mg,Fe)SiO_3$ composition. No vacancies would be created.

A second possible mode of incomposition of Fe^{3+} could be

$$2Si_B^{4+} + O_O^{2-} = 2Fe_B^{3+} + \square_O \qquad (7.32)$$

where $\square$ represents a vacancy. This would also make $(Fe+Mg)/Si > 1$, but would put iron only on the octahedral sites and balance its charge by oxygen vacancies.

At present, there is increasing evidence that iron magnesium silicate perovskites (and spinels as well) contain significant ferric iron. The richness of defect chemistry is just beginning to be explored.

7.5 References

7.5.1 General references and bibliography

Carmichael, I. S. E., and H. P. Eugster, eds. (1987). *Thermodynamic modeling of geological materials: Minerals, fluids and melts.* Reviews in Mineralogy, vol. 17. Washington, DC: Mineralogical Society of America.

Chatterjee, N. D. (1991). *Applied mineralogical thermodynamics.* Heidelberg, Germany: Springer-Verlag.

Eugster, H. P., and I. S. E. Carmichael, eds. (1987). *Models of crystalline solutions in thermodynamic modelling of geologic systems: Minerals, fluids, and melts.* Reviews in Mineralogy, vol. 17, 35–70. Washington, DC: Mineralogical Society of America.

Ganguly, J., and S. K. Saxena. (1987). *Mixtures and mineral reactions.* Heidelberg, Germany: Springer-Verlag.

Gaskell, D. R. (1981). *Introduction to metallurgical thermodynamics.* 2nd ed. New York: McGraw-Hill.

Gokcen, N. A. (1975). *Thermodynamics.* Parker and Son.

Greenwood, H. J., ed. (1977). *Application of thermodynamics to petrology and ore deposits.* Short Course Handbook, vol. 2. Toronto: Mineralogical Association of Canada.

Kubaschewski, O., and C. B. Alcock. (1979). *Metallurgical thermochemistry.* 5th ed., vol. 24. New York: Pergamon Press.

Nordstrom, D. K., and J. L. Munoz. (1985). *Geochemical thermodynamics.* Palo Alto: Benjamin/Cummings.

Putnis, A. *Introduction to mineral sciences.* (1992). Cambridge: Cambridge University Press.

Schmalzreid, H., and A. Navrotsky. (1978). *Festkorperthermodynamik Chemie des festen Zustandes.* Berlin: Akademie-Verlag.

Thompson, J. B., Jr. (1969). Chemical reactions in crystals. *Amer. Mineral. 54,* 341–75.

7.5.2 Specific references

Blencoe, J. G., G. A. Merkel, and M. K. Seil. (1982). Thermodynamics of crystal-fluid equilibria, with applications to the system $NaAlSi_3O_8$–$CaAl_2Si_2O_8$–SiO_2–$NaCl$–$CaCl_2$–H_2O. In *Advances in physical geochemistry,* ed. S. K. Saxena, vol. 2, 191–222. New York: Springer-Verlag.

Burnley, P. C., H. W. Green, II, D. J. Prior. (1991). Faulting associated with the olivine to spinel transformation in Mg_2GeO_4 and its implications for deep-focus earthquakes. *J. Geophys. Res. 96*, 425–43.

Burton, B. P.. (1987). Theoretical analysis of cation ordering in binary rhombohedral carbonate systems. *Amer. Mineral. 72*, 329–36.

Capobianco, C., B. P. Burton, P. M. Davidson, and A. Navrotsky. (1987). Structural and calorimetric studies of order–disorder in $CdMg(CO_3)_2$. *J. Solid State Chem. 71*, 214–23.

Carpenter, M. A. (1985). *Order–disorder transformation in mineral solid solutions*. In "Microscopic to Macroscopic," S. W. Kieffer and A. Navrotsky, Eds., Reviews in Mineralogy, vol. 13, 187–224.

Carpenter, M. A. (1988). Thermochemistry of aluminum/silicon ordering in feldspar minerals. In *Physical properties and thermodynamic behaviour of minerals*, ed. E. K. H. Salje, 265–323. Dordrecht, Holland: D. Reidel.

Carpenter, M. A., and J. M. Ferry. (1984). Constraints on the thermodynamic mixing properties of plagioclase feldspars. *Contrib. Mineral. Petrol. 87*, 138–48.

Carpenter, M. A., J. D. C. McConnell, and A. Navrotsky. (1985). Enthalpies of Al–Si ordering in the plagioclase feldspar solid solution. *Geochim. Cosmochim. Acta 49*, 947–66.

Davies, P. K., and A. Navrotsky. (1983). Quantitative correlations of deviations from ideality in binary and pseudo-binary solid solutions. *J. Solid State Chem. 46*, 1–22.

Henry, D. J., A. Navrotsky, and H. D. Zimmermann. (1982). Thermodynamics of plagioclase-melt equilibria in the system albite–anorthite–diopside. *Geochim. Cosmochim. Acta 46*, 381–91.

Millard, R. L., R. C. Peterson, and B. K. Hunter. (1992). Temperature dependence of cation disorder in $MgAl_2O_4$ spinel using ^{27}Al and ^{17}O magic-angle spinning NMR. *Amer. Mineral. 77*, 44–52.

Navrotsky, A. (1986). Cation distribution energetics and heats of mixing in $MgFe_2O_4$–$MgAl_2O_4$, $ZnFe_2O_4$–$ZnAl_2O_4$, and $NiAl_2O_4$–$ZnAl_2O_4$ spinels: Study by high-temperature calorimetry. *Amer. Mineral. 71*, 1160–79.

Navrotsky, A., and O. J. Kleppa. (1967). The thermodynamics of cation distributions in simple spinels. *J. Inorg. Nucl. Chem. 29*, 2701–14.

Newton, R. C., T. V. Charlu, and O. J. Kleppa. (1980). Thermochemistry of the high structural state plagioclase. *Geochim. Cosmochim. Acta 44*, 933–41.

O'Neill, H. St. C., and A. Navrotsky. (1983). Simple spinels: Crystallographic parameters, cation radii, lattice energies, and cation distributions. *Amer. Mineral. 68*, 181–94.

O'Neill, H. St. C., and A. Navrotsky. (1984). Cation distributions and thermodynamic properties of binary spinel solid solutions. *Amer. Mineral. 69*, 733–55.

Orville, P. M. (1972). Plagioclase cation exchange equilibria with aqueous chloride solution: Results at 700 °C and 2000 bars in the presence of quartz. *Amer. J. Sci. 272*, 236–72.

Peterson, R. C., G. A. Lager, and R. L. Hitterman. (1991). A time-of-flight neutron powder diffraction study of $MgAl_2O_4$ at temperatures up to 1273 K. *Amer. Mineral. 76*, 1455–8.

Salje, E. (1985). Thermodynamics of sodium feldspars, I: Order parameter treatment and strain induced coupling effects. *Phys. Chem. Minerals 12*, 93–8.

Salje, E. (1987). Thermodynamics of plagioclases, I: Theory of $I\bar{1}$–$P\bar{1}$ phase transition in anorthite and Ca-rich plagioclase. *Phys. Chem. Minerals 14*, 181–8.

Salje, E., B. Kuscholke, B. Wruck, and H. Kroll. (1985). Thermodynamics of sodium feldspar, II: Experimental results and numerical calculations. *Phys. Chem. Minerals 12*, 132–40.

Saxena, S. K., and P. H. Ribbe. (1972). Activity-composition relations in feldspars. *Contrib. Mineral. Petrol. 37*, 131–8.

Schmocker, U., and F. Waldner. (1976). The inversion parameter with respect to the space group of $MgAl_2O_4$ spinels. *J. Phys. C: Solid State Phys. 9*, L235–7.

Seil, M. K., and J. G. Blencoe. (1978). Activity-composition relations of $NaAlSi_3O_8$– $CaAl_2Si_2O_8$ feldspars at 2kb, 600–800 °C. *Geol. Soc. Am. Abst. Progr. 11*, 513.

Wood, B. J., R. J. Kirkpatrick, and B. Montez. (1986). Order–disorder phenomena in $MgAl_2O_4$ spinel. *Amer. Mineral. 71*, 999–1006.

8

Melts, glasses, and amorphous materials

8.1 Introduction

Silicate melts, as magmas at depth, lavas on the surface, and transient phases on meteor impact, play important roles in geochemical and cosmochemical processes. A natural silicate melt contains SiO_2, Al_2O_3, MgO, CaO, iron oxides, and alkalis as major components, with minor to trace amounts of TiO_2, P_2O_5, transition metals, and rare earths, and minor but significant amounts of volatiles, notably H_2O, CO_2, SO_2, and HF. The complete description of the structure, properties, and reactions of such a complex multicomponent melt at temperatures of 800 K to 2200 K and pressures of 1 atmosphere to 30 GPa is an immense task; the understanding of these properties in terms of fundamental atomic-level chemical interactions is even more formidable. Therefore many different approaches have been taken, ranging from theoretical and experimental studies of simple model systems to largely empirical studies of simulated or actual rock compositions.

Many silicate melts form glasses, rigid but noncrystalline materials, upon reasonably rapid cooling. Glasses and glass-crystal mixtures (so called glass ceramics) are technologically important. The relation of this dynamically "frozen" state to the liquid and the glass transition separating these two regions raise fundamental questions of physics and chemistry. In addition, noncrystalline (amorphous) materials can be made by a variety of methods other than quenching from a liquid. These include chemical vapor deposition, sol-gel processing and precipitation from aqueous solution, radiation damage, and pressure amorphization. This chapter addresses some of the properties and behavior of noncrystalline materials.

8.2 Structure of silicate glasses and liquids

A silicate melt is, first of all, a molten salt. Thus cations are surrounded by anions and vice versa. The anions, primarily SiO_4^{4-} and AlO_4^{3-}, can share corners to form dimmers, chains, rings, and two- and three-dimensional networks. Thus Si^{4+}, Al^{3+} (and B^{3+}, Ge^{4+}, Ga^{3+} and sometimes Fe^{3+}, Ti^{4+}, and Mg^{2+}) are *network formers*. Other cations (alkalis, alkaline earths, large high-charged cations) do not form linked tetrahedra and are called *network modifiers*. Al^{3+}, Ga^{3+}, Fe^{3+}, Ti^{4+}, and Mg^{2+} can also act as network modifiers, depending on melt composition, pressure, and temperature. The transfer of an oxide ion from a basic oxide (network modifier) to an acidic oxide (network former) can be described as follows. If the initial melt composition is very poor in silica, the silicate tetrahedra that form are isolated, and the reaction is

$$2MO + SiO_2 = 2M^{2+} + SiO_4^{4-} \qquad (8.1)$$

As the silica concentration increases, the tetrahedra polymerize with each other by the equilibria

$$2SiO_4^{4-} = Si_2O_7^{6-} + O^{2-} \qquad (8.2a)$$
$$Si_2O_7^{6-} + SiO_4^{4-} = Si_3O_{10}^{8-} + O^{2-} \qquad (8.2b)$$
$$Si_nO_{3n+2}^{(2n+4)-} + SiO_4^{4-} = Si_{n+1}O_{3n+5}^{(2n+6)-} + O^{2-} \qquad (8.2c)$$

Thus chains of tetrahedra grow. They may also branch or close to form rings. This polymerization is formally analogous to the growth of organic polymers, and polymer theory has been applied with considerable success to molten slags, for which the ratio of network modifier to network former is high (mole fraction of SiO_2 is 0.1 to 0.4) (Masson 1965, 1968; Hess 1971). However, even in this case, two important differences between organic polymers and silicate melts must be kept in mind. The first is that organic polymer chains, once formed, remain together, whereas the silicate units in a melt constantly are forming and breaking bonds at high temperature, as confirmed by recent in situ NMR studies (Stebbins 1988, 1992). The second is that organic polymers are generally uncharged, but the silicate chains carry a large negative charge (see Eq. 8.2) that must be locally balanced by the network-modifying cations. Organic polymers can be described by strong interactions within the chains and very weak interactions between them, whereas in silicate melts the interactions between the polymer units are strong electrostatic repulsions while their interactions with the network-modifying cations are strong electrostatic attractions. The local coordination of cations in these slags appears to be similar to that in crystalline phases, though the average coordination number (C.N.) is a little smaller (see Table 8.1).

Geologically important silicate melts are far more silica-rich ($X_{SiO_2} > 0.5$, weight percent $SiO_2 > 45$) than metallurgical slags. The concentration of free

Table 8.1. *Cation–oxygen distance: Ab initio calculated values and experimental values in crystals, glasses, and melts*

Ion	C.N.	Calculated	Crystal 300 K	Glass 300 K	Melt > 1.473 K
			MO bond length		
Si^{4-}	4	1.63	1.62	1.62	1.62–1.63
Ge^{4-}	4	1.71	1.76	1.73	
Al^{3-}	4	1.72	1.75	1.72	
	6	1.79	1.89	1.76	
B^{3-}	3	1.36	1.38	1.37	
	4	1.46	1.48	1.48	
Li	4	1.80	1.95		
	3.7–4.1			2.07	2.08
	6	1.97	2.10		
Mg^{2-}	4	1.83	1.94		
	4.3–5.1			2.14	2.16
	4.1			2.08	
	6	1.91	2.08		
Na^-	4	2.02	2.35		
	4.1–6.1			2.36	2.36
				2.4	
	5			2.3	
	6	2.07	2.38		
Ca^{2-}	6		2.36		
	5.4–6.8			2.43	2.41
	7		2.43		
Fe^{2-}	4		1.99		
	3.9–4.2				2.02–2.05
	6			2.07	
	6		2.13		

oxide ions, O^{2-}, in magmas is vanishingly small, and rather than the foregoing polymerization equilibria, the melt is better described in terms of depolymerization of an initially completely linked three-dimensional tetrahedral framework of AlO_4 and SiO_4 tetrahedra (McMillan 1984).

$$Si–O–Si + MO = 2\ SiO^- + M^{2+} \qquad (8.3)$$

Thus a bridging oxygen is converted to a nonbridging oxygen, the network modifying cation "sticks around" for charge balance (perhaps forming a strong complex, perhaps a weak one), and the overall degree of polymerization decreases. A useful shorthand is the "Q" (quaternary) species notation. A tetrahedron bonded to four others is Q^4, that to three Q^3, and so on. Q^4 species represent a fully polymerized region of the structure, Q^3 species represent sheetlike regions, Q^2 chains, Q^1 Si_2O_7 groups, and Q^0 isolated tetrahedra. One can also distinguish between a silicate tetrahedron with four silicate neighbors (Q^4 with three SIO_4 and one AlO_4 neighbor is $Q^4(1Al)$, etc.) (Kirkpatrick 1988).

A glass represents a "frozen snapshot" of the melt at the glass transition temperature (see also Sec. 8.4). It contains the disorder of the liquid state but without the dynamics of constant rearrangement. Much of our knowledge of melt structure comes from spectroscopy on the corresponding glasses. The following picture emerges. Silica glass (vitreous SiO_2) consists of a fully polymerized (Q^4) network of linked tetrahedra. They are arranged with a distribution of ring sizes, although six-membered rings (six tetrahedra joined through corners) predominate. Both the average and the most common Si–O–Si angles are near the 143° predicted to be the most stable from molecular orbital theory, and the average bond length near 1.61 Å (with small variance) is also close to the expected minimum and varies little with pressure and temperature (Navrotsky et al. 1985). Thus the disorder is not in nearest and next-nearest neighbor configurations, but in third and higher neighbors, that is, on the 5–100 Å scale. Figures 8.1a and 8.1b show radial distribution functions (RDF) for crystalline and amorphous SiO_2. The RDF gives the probability of finding atoms at a given distance from a given atom. The point to note is that the nearest and next-nearest distances are well preserved. An almost identical RDF to that for SiO_2 glass has been obtained for molten silica near 2000 K, reflecting the highly polymerized nature of this liquid (Waseda 1981) (see Fig. 8.1b,c).

The substitution of aluminum, charge balanced by an alkali or alkaline earth ion, which can locate near the Al inside a ring of tetrahedra, can and does maintain a fully polymerized network. There is evidence from Raman spectroscopy (McMillan 1984), radial distribution functions (Taylor and Brown 1979), and NMR (Kirkpatrick 1988) that the network is more strongly perturbed by divalent than by monovalent cations, and that the ring size distribution depends on the Al/(Al + Si) ratio and on the nature of the charge-balancing cation. Thus it has been proposed that albite glass and liquid resemble a stuffed silica framework (though of course without long-range order), while anorthite glass and melt retain more four-membered rings reminiscent of the feldspar structure (Taylor and Brown 1979). However, the description of mid-range structure in glasses and melts is still rather crude. What is clear is that the factors that affect T–O–T angle distributions in crystalline solids (see Chap. 5) also play a role in the amorphous state. There is evidence from solid-state NMR that the number of Al–O–Al contacts is smaller than a random model would predict, but larger than the minimum number required by the

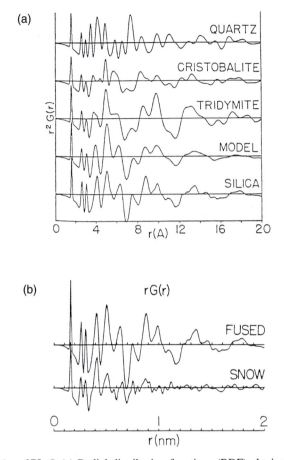

Figure 8.1. (*pp. 375–6*) (a) Radial distribution functions (RDF) obtained experimentally for SiO_2 glass and calculated from atomic coordinates for a 1412 atom disordered model, tridymite, cristobalite, and quartz (data from Konnert, D'Antonio, and Karle 1982). (b) Comparison of RDF for bulk SiO_2 glass and a chemically vapor-deposited (CVD) SiO_2 "snow" powder. The latter shows more disorder, especially beyond 5 Å than the former, though both are amorphous (from Konnert et al. 1987). (c) RDF of molten silica near 2000 K (from Waseda 1980).

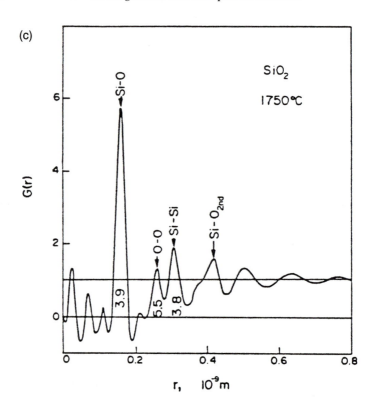

stoichiometry (Stebbins 1988, 1992). Thus an intermediate degree of aluminum avoidance probably applies and may change with temperature.

The description of melt and glass structure for compositions that contain
more network modifier than is required to balance aluminum is more problematic. These partially depolymerized melts must contain, stoichiometrically,
species less polymerized than Q^4. Furthermore, the reaction $Q^3 = Q^2 + Q^4$
may be favorable, and the regions of different degrees of polymerization may
cluster to produce incipient phase separation. There is no agreement on a general model for the structure of such systems or even for what questions are
meaningful. One of the difficulties is that it is not yet clear to what extent
various spectroscopic tools are sensitive to changes in structure beyond next-
nearest-neighbor. There has been lively discussion of these questions over the
past ten years (McMillan 1984; Seifert, Mysen, and Virgo 1982; Stebbins
1988, 1992).

Borosilicates are even more complex because boron can take both tetrahedral and trigonal coordination. The ratio of BO_3 to BO_4 groups depends on

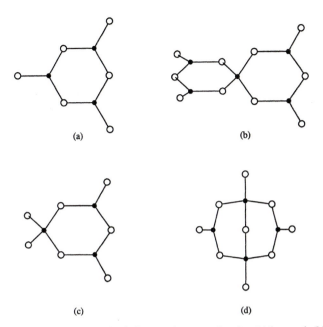

Figure 8.2. Possible structural units in borate glasses and melts: (a) boroxol, (b) pentaborate, (c) triborate, (d) diborate. Small solid circles represent boron, large open circles oxygen. These groups contain BO_3 triangles; BO_4 tetrahedra are also present. (From Griscom 1977.)

temperature, pressure, and bulk composition and may be sensitive to cooling rate. The BO_3 triangles and BO_4 tetrahedra can form a boroxol group (see Fig 8.2), which may be a major building block.

Immiscibility in borosilicate glasses is common. An early process for making almost pure silica glass at moderate temperature involved the phase separation of a borosilicate glass to borate-rich and silica-rich regions, followed by leaching the borate phase with acid, leaving behind a slightly translucent silica glass called Vycor.

8.3 Thermodynamics of silicate glasses and liquids

8.3.1 Source of thermodynamic data

The phase diagram itself is a major source of free-energy data. This is illustrated schematically in Figure 8.3 for a binary system A–B. If pure crystalline A melts to form pure liquid A, then the melting point (T_1) is the intersection of the free-energy curves for pure crystal (curve 1 in Fig. 8.3b) and pure liquid

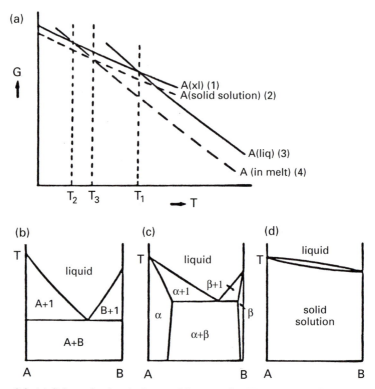

Figure 8.3. (a) Schematic chemical potential curves for (1) pure crystalline substance A, (2) crystalline A in solid solution at a given concentration, (3) pure molten A, and (4) A in a melt at some concentration. (b) Schematic phase diagram for binary system A–B, no solid solution, simple eutectic. (c) Modified eutectic with terminal α (A-rich) and β (B-rich) solid solutions. (d) Binary loop with complete solid and complete liquid solubility.

(curve 2). If B is added to the liquid and is insoluble in the crystal, then the chemical potential and activity of A in the liquid is lowered (curve 3) and the temperature where solid first appears is depressed (to T_2 for a given liquid composition). The result is a simple eutectic system (Fig. 8.3a). If the crystalline phase is a solid solution, then the chemical potential of A in the solid is also lowered (curve 4 in Fig. 8.3b) and for a given pair of solid and liquid compositions, equilibrium will occur at T_3. The resulting phase diagram will be a eutectic modified by terminal solid solubility or a continuous binary loop. Thus to calculate the phase diagram, one needs to know three sets of parameters: (1) free energies of fusion of pure A and B at various temperatures, that is, heats and entropies of fusion of A and B and their temperature dependence, (2) mixing properties in the crystals, and (3) mixing properties in the melts.

The liquidus surface provides a trajectory through temperature-pressure-composition space along which the activity of a component in the melt is fixed by its activity in a crystalline phase. The activity of components in the melt is governed by both enthalpy and entropy of mixing terms, which generally cannot be determined accurately from such a polythermal section. If enthalpies of mixing can be determined independently by calorimetric studies, then combining such enthalpies with liquidus data and heat of fusion data provides a potentially quite accurate estimate of entropies of mixing to test against the predictions of various models (Navrotsky et al. 1980, 1989; Navrotsky 1986).

For compounds that melt congruently, the pressure dependence of the melting point can be used to obtain the heat of fusion through the Clausius-Clapeyron equation

$$(dP/dT)_{\text{equil}} = \Delta S/\Delta V \qquad (8.4)$$

If the fusion curve deviates from linearity, ΔS and/or ΔV are not constant, which can be taken as evidence for a different (usually greater) compressibility of the melt than of the crystal or, in extreme cases, for changes in local coordination in the melt.

Several types of calorimetric studies (see Chap. 4) are applicable to glasses and melts. Heat capacities of glasses at low temperature (from near 0 K to somewhat above room temperature) can be measured by conventional adiabatic calorimetry, which requires several grams of sample. Amorphous materials often have anomalously high heat capacities below 10 K (Pohl, Love, and Stephens 1974), perhaps related to cooperative motion leading to closely spaced low-lying energy levels in the vibrational density of states (a so-called two-level system, see Sec. 6.1). Differential scanning calorimetry measures heat capacities on milligram quantities of samples in the range 200–1100 K. For geologically important aluminosilicate systems, T_g is tantalizingly close to the upper limit of the range covered by commercially available DSC equipment. At temperatures between 100 K and the glass transition, the heat capacities of glasses are usually quite similar to those of the corresponding crystalline phases.

Because glasses are disordered metastable materials, their absolute entropies contain a "frozen-in" configurational contribution in addition to the vibrational ($\int(C_p/T)dT$) term obtained by heat capacity measurements. To relate the thermodynamic properties of glass, liquid, and crystal, a thermochemical cycle involving their interconversion at high temperature is needed. Conventional drop calorimetry (furnace at high T, calorimeter at room T) can obtain heat contents and, by differentiation of $H_T - H_{298}$, heat capacities of liquids. If the melt crystallizes rapidly, heats of fusion can be measured directly. Glass-forming systems quench to amorphous materials in the calorimeter, so the heat of fusion is not measured directly by drop calorimetry.

One can measure the heat of fusion directly by melting the crystal by dropping it into a calorimeter above the liquidus or by scanning through the melting point. One can measure the heat of fusion indirectly by measuring heats of solution in hydrofluoric acid near room temperature or molten lead borate near 973 K, of both crystal and glass, and then applying appropriate heat capacity corrections.

8.3.2 Thermodynamic relations among crystal, glass, and melt

The relation among enthalpies of crystalline, molten, and glassy forms of diopside are shown in Figure 8.4 (Lange, DeYoreo, and Navrotsky 1991). At temperatures of 1200–1600 K the crystal has an almost constant heat capacity leading to roughly linear $H_T - H_{298}$. The melting point is 1665 K, the heat of fusion 138 kJ mol^{-1}. (A small degree of incongruent melting is ignored in this discussion.) The liquid has a higher heat capacity than the crystal, resulting in a steeper $H_T - H_{298}$ curve. If one cools the melt reasonably rapidly, it does not crystallize at the melting point but proceeds to supercool, with $H_T - H_{298}$ following an extrapolation of the curve for the liquid. At some temperature, the rates of configurational rearrangements in the liquid become too slow to keep up with the drop in temperature, and the liquid freezes in these configurations, loses the degrees of freedom associated with configurational change, and becomes a glass. This *glass transition* is a kinetic phenomenon, and the temperature at which it occurs and the properties of the glass produced depend on cooling rate. The glass transition is associated with pronounced changes in physical properties.

The glass transition is most pronounced when the melt is most fluid and/or most depolymerized. Thus the difference in heat capacity between glass and liquid increases in the sequence SiO_2, $KAlSi_3O_8$, $NaAlSi_3O_8$, $CaAl_2Si_2O_8$, $MgCaSi_2O_6$. Yet more depolymerized melts, for example, Fe_2SiO_4, do not readily form glasses.

Table 8.2 shows some trends in the enthalpies of vitrification, mainly of chain silicates. Pyroxenes such as $NaAlSi_2O_6$ and $CaAl_2SiO_6$ produce glasses that can be described as fully polymerized, with all the Al in the glass present in tetrahedral coordination and charge balanced by M^+ or M^{2+} in nonframework sites. Such pyroxenes have heats of vitrification between 55 and 67 kJ mol^{-1} whereas those forming glasses with insufficient Al for complete M–Al charge

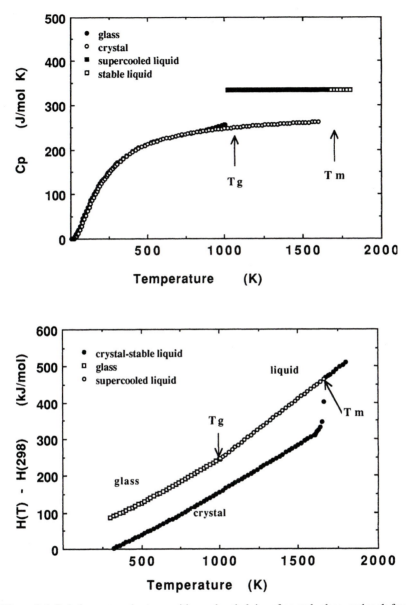

Figure 8.4. Relation among heat capacities and enthalpies of crystal, glass, and melt for diopside, $CaMgSi_2O_6$ (data from Lange, DeYoreo, and Navrotsky 1991).

Table 8.2. *Enthalpies of vitrification and fusion*

Compound	Vitrification ΔH (kJ/mol)	Fusion	
		Melting point T(K)	$\Delta H(T)$ (kJ/mol)
MgO	—		
CaO	—		
Al_2O_3	—	2323	107.5 ± 5.4
SiO_2 (quartz)	—	1700[b]	9.4 ± 1.0
SiO_2 (cristabalite)	—	1999	8.9 ± 1.0
"FeO" (wustite)	—	1652	31.3 ± 0.2
Mg_2SiO_4 (forsterite)	—	2163	114 ± 20[a]
$MgSiO_3$ (enstatite)	42 ± 1	1834[a]	77 ± 5
Fe_2SiO_4 (fayatite)	—	1490	89 ± 10
$CaSiO_3$ (wallastonite)	25.5 ± 0.4	1770[b]	62 ± 4
$CaSiO_3$ (pseudowollastonite)	—	1817	57 ± 3
$CaMgSi_2O_6$ (diopside)	85.8 ± 0.8	1665	138 ± 2
$NaAlSi_3O_8$ (high albite)	51.8 ± 0.8	1373	63 ± 20
$KAlSi_3O_8$ (sanidine)		1473[a]	56 ± 4
$CaAl_2Si_2O_8$ (anorthite)	77.8 ± 0.8	1830	134 ± 4
K_2SiO_3	9 ± 1	1249	20 ± 4
$Mg_3Al_2Si_3O_{12}$ (pyrope)	—	1500[a]	243 ± 8
$Mg_2Al_4Si_5O_{18}$ (cordierite)	209 ± 2	1740	346 ± 10

[a] Estimated metastable congruent melting.
[b] Melting of metastable phase.

coupling show heats of vitrification between 64 and 85 kJ mol^{-1}. Within the latter group, the heat of vitrification generally increases with decreasing Al-content and with decreasing basicity of the divalent oxide. Thus values of ΔH_{vit} increase in the series $Ca_2Si_2O_6$, $CaCoSi_2O_6$, $Mg_2Si_2O_6$, $CaMgSi_2O_6$, $CaNiSi_2O_6$. This latter trend is consistent with a relative destabilization of the glass as the ability of the divalent ion to bond to oxygen and perturb the tetrahedral Si–O–Si linkages increases. For pyroxenes containing transition metals, ligand field effects and specific crystal chemical factors may also play a role in determining ΔH_{vit} (Navrotsky 1986).

In framework aluminosilicates, which form nominally completely polymerized glasses and melts, the enthalpy of vitrification increases with increasing ionic field strength (z/r) of the framework cation (see Fig. 8.5) (Navrotsky et al. 1985).

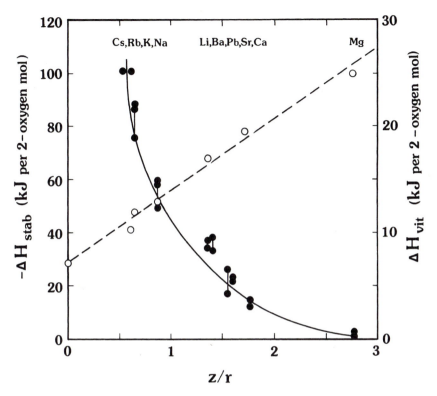

Figure 8.5. Dependence of heat of vitrification (right scale and open symbols) and heat of substitution $Si^{4+} = 1/n\ M^{n+} + Al^{3+}$ in glasses (left scale and solid symbols) on field strength of cation (from Navrotsky, Geisinger, McMillan, and Gibbs 1985).

8.3.3 *Heats of mixing in glasses and melts*

In binary metal oxide–silica systems, the thermodynamic mixing properties are dominated by the major acid–base reaction: the transfer of oxide ion from metal oxide to silica. This is a depolymerizing reaction, forming two nonbridging Si–O bonds from one Si–O–Si linkage. The more basic the oxide (smaller z/r), the more complete the oxide ion transfer and, for a given composition, the more depolymerized and less viscous the melt. These trends are reflected in enthalpies of mixing in binary molten silicates (see Fig. 8.6). Note the rather large enthalpies involved (up to -100 kJ mol^{-1}), the maximum stabilization near orthosilicate composition, and the increase in stabilization with increasing basicity of the metal oxide, that is, the order Pb, Ca, Na, K (Navrotsky 1986).

Heats of mixing in a number of glassy aluminosilicate systems have been

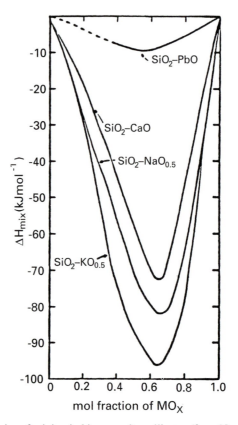

Figure 8.6. Enthalpies of mixing in binary molten silicates (from Navrotsky 1986).

studied using oxide melt solution calorimetry. Results for the systems albite–anorthite–diopside, albite–anorthite–silica, and albite–orthoclase–silica are shown in Figure 8.7 (Navrotsky 1986). The joins $NaAlSi_3O_8$–Si_4O_8 (Ab–4Q) and $KAlSi_3O_8$–Si_4O_8 (Or–4Q) show essentially zero heats of mixing. The join $NaAlSi_3O_8$–$CaMgSi_2O_6$ (Ab–Di) shows small positive heats of mixing, suggesting glass–glass immiscibility. $NaAlSi_3O_8$–$KAlSi_3O_8$ (Ab–Or) shows small negative heats of mixing and $NaAlSi_3O_8$–$CaAl_2Si_2O_8$ (Ab–An) shows substantial negative heats of mixing. The ternaries also exhibit complex behavior not predictable from their bounding binaries. Ab–An–Di shows substantial negative ternary excess heats of mixing, Ab–Ab–4Q positive ones. In Ab–Or–4Q, though all heats of mixing are small, they change sign from negative to positive (positive ternary excess terms) within the ternary.

Several factors on an atomic scale probably contribute to the overall

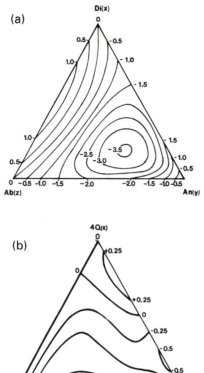

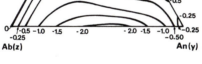

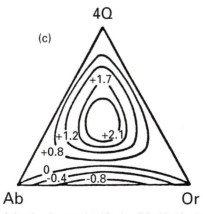

Figure 8.7. Heats of mixing in glasses: (a) Ab–An–Di, (b) Ab–An–4Q, (c) Ab–Or–4Q (from Navrotsky 1986).

enthalpy of mixing in the glass. In systems such as Ab–Di, where the degree of polymerization must change significantly with composition, a tendency toward clustering and unmixing into silica-rich (highly polymerized) and silica-poor (less polymerized) domains manifests itself in positive heats of mixing. This is analogous to, but much less pronounced than, effects in binary alkali and alkaline earth silicates already discussed. When the degree of polymerization and Al/(Al + Si) ratio remain constant but nonframework cations (e.g., Na and K in Ab–Or) mix, a small negative heat of mixing is often seen. This is analogous both to similar effects seen in molten salts (e.g., NaCl–KCl) (Hersh and Kleppa 1965) and to the "mixed alkali effect" seen in ceramic glasses (Kawamoto and Tomozawa 1982). The cause of this effect can be related to polarization of the framework anions by cations of unequal bonding power, leading to a nonrandom distribution of cation pairs. For systems where one component is an aluminosilicate glass of Al/(Al+Si) > 0.3, and the other is a low alumina glass (but not SiO_2), substantial negative heats of mixing are seen, for example, in $CaAl_2Si_2O_8$–$CaMgSi_2O_6$ (DeYoreo, Navrotsky, and Dingwell 1990; Navrotsky 1986; Navrotsky et al. 1980). The exact reasons for this are unclear, but local M-Al charge coupling, Al-avoidance, framework distortions, or depolymerization may be involved.

The enthalpies associated with charge-coupled substitution of $M_{1/n}^{n+} AlO_2$ for SiO_2 in a framework glass are shown in Figure 8.8 (Navrotsky 1986). The process of dissolving an aluminosilicate glass in molten lead borate to form a dilute (< 1 wt %) solution breaks up the aluminosilicate framework structure into isolated species, presumably silicate and aluminate tetrahedra and alkali and alkaline-earth cationic species dissolved in a borate matrix. Thus the enthalpy of solution may be considered a measure of the strength of bonding in the aluminosilicate glass, at least in a relative sense when comparing various compositions. Three points are evident from Figure 8.8. First, for Al/(Al+Si) < 0.5 the enthalpies of solution generally become more endothermic as $M_{1/n}AlO_2$ is substituted for SiO_2. This increase becomes more pronounced with decreasing field strength (or increasing basicity), that is, in the series Mg, Ca, Sr, Pb, Ba, Li, Na, K. Rb, Cs. Second, at Al/(Al + Si) > 0.5 the enthalpy of solution curves bend back toward more exothermic values, with a maximum near 0.5. This is seen most clearly in the calcium aluminate–silica system, where data exist to virtually pure $Ca_{0.5}AlO_2$, and is suggested in the sodium aluminate–silica system. This maximum reflects an exothermic enthalpy of mixing for the reaction

$$x M_{1/n}AlO_2 + (1 - x) SiO_2 = M_x Al_x Si_{1-x} O_2 \tag{8.5}$$

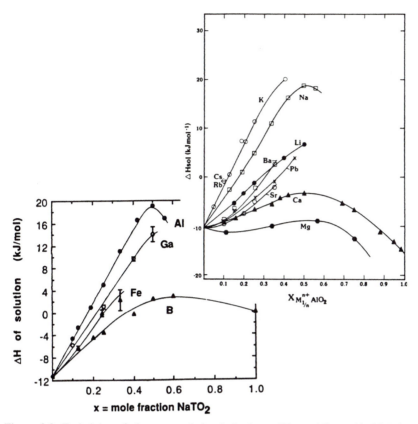

Figure 8.8. Enthalpies of charge coupled substitutions, $Si^{4+} = Al^{3+} + 1/n\ M^{n+}$ in glasses, shown as variation of enthalpy of solution in molten lead borate near 973 K; similar variation for $Si^{4+} = Na^{+} + M^{3+}$ (DeYoreo, Navrotsky, and Dingwell 1990).

which parallels, but is smaller in magnitude than, the corresponding enthalpy of formation in the crystalline state. This enthalpy of mixing appears to become more exothermic with increasing basicity of the metal oxide (e.g., Na more exothermic than Ca). Third, at high silica content, pronounced curvature occurs in the relations for the alkaline earths, whereas the relations for the alkalis are approximately linear. This curvature suggests a positive heat of mixing in this region and may presage glass–glass immiscibility.

The enthalpy of the reaction of Eq. 8.5 becomes more exothermic with increasing basicity of the oxide $M_{1/n}^{n+}O$ (decreasing field strength, z/r, of the cation M). The energetics can also be correlated in terms of the perturbation of T–O–T bond lengths and angles, both as observed in crystalline aluminosilicates and as calculated by molecular orbital methods (Geisinger, Gibbs, and Navrotsky 1985; Navrotsky, Geisinger, McMillan, and Gibbs 1985) (also see Chap. 5).

Generalizations from the preceding discussion may be drawn for multicomponent silicate glasses relevant to geology. For systems with $Al/(Al + Si) <$ 0.3, enthalpies of mixing are generally expected to be less than 10 kJ mol^{-1} in magnitude, with positive values associated with changes in degree of polymerization and negative values with nonframework cation mixing. More aluminous systems show complex behavior.

8.3.4 Models for phase equilibrium calculations

Models can be classified in two groups, with some intermediate cases. The first group consists of relatively accurate, though largely empirical, models for activities of various mineral components in a complex melt as a function of pressure (P), temperature (T), and composition (X) developed in order to decipher the high P–T adventures of a rock from the measured compositions and abundances of its mineral phases. An approach aimed at this goal leans toward experimentation on either natural rocks or their synthetic equivalents, with results applicable in detail to the compositions studied but difficult (and dangerous) to extrapolate to conditions outside the regime studied. The magnitudes of thermochemical parameters obtained in such a study form a good description of the experimental data but do not have physical significance.

The second group of models aims to develop an understanding on a fundamental level to be able to predict, at least in a general way, the variation of properties as a function of pressure, temperature, and composition. Toward this goal, one selects simple model systems with a small number of components, studies them by a variety of thermodynamic and structural means, and seeks to develop models related to interactions of molecular or ionic species inferred to be present. Because structural data are poorly known for melts at high P and T, one often relies on structural information obtained from quenched glasses. The thermodynamic models developed suffer, inherently and unavoidably, from a lack of uniqueness. Thermodynamic properties, being averages over a large number of microscopic states, are only mildly sensitive to the details of structural models used and cannot be used to discriminate reliably among various, sometimes significantly different, models. In addition, the structural models themselves, based often on interpretation of spectroscopic data, also represent average properties, and several alternate models, quite different in detail, may fit the structural data equally well. Despite this lack of uniqueness, the "structure-based" thermodynamic models of simple model systems offer a conceptual framework for extrapolation to different (P, T, X) conditions and for assembling data on binary and ternary systems to predict more complex systems. The models and hypotheses generated suggest new experiments, both thermodynamic and spectroscopic, to be tried on new model systems. Although in the short term these "microscopic" models offer a less good fit of any one set of data than do empirical models, in the long term the understanding gained from a fundamental approach will eventually pay off in predictive power. Such predictive capability is especially important if one wishes to extrapolate to extreme conditions of P and T, such as the mantle.

Several examples of purely phenomenological models may be cited. First, one can use binary oxides (e.g., Na_2O or $NaO_{1.5}$, MgO, SiO_2) as components, obtain values for the thermodynamics of formation of ternary compounds, and use polynomials to describe the binary oxide activities in the liquid state. Various series expansions describe activities or activity coefficients; the relative merits and/or equivalence of various methods have been discussed extensively recently. The most sophisticated versions of such approaches seek to optimize the thermochemical data used to be consistent with both phase equilibria and thermodynamic measurements (Berman and Brown 1987; Eriksson and Pelton 1994).

Using complex silicates rather than binary oxides as reference states allows much simpler equations to be used for mixing properties. A seventeen-component regular solution model has been used to describe metaluminous silicate liquids, with reasonable predictions of silica activity and of olivine and plagioclase crystallization (Ghiorso 1987; Ghiorso and Carmichael 1987; Ghiorso et al. 1983). However, the regular solution parameters obtained are surprisingly large in magnitude and probably have no physical significance, and such models should not be extrapolated outside their composition range (and certainly should not be applied to the bounding binary or pseudobinary systems).

The appropriate choice of components can simplify activity–composition relations in a given system to virtually Raoultian ($a_i = X_i$) behavior. This approach has been exploited by Burnham (Burnham 1975, 1981; Burnham and Nekvasil 1986). With all components except H_2O written in terms of eight-oxygen formula units, Burnham has written speciation reactions (e.g., albite → jadeite + quartz, anorthite → corundum + wollastonite + quartz), established their equilibrium constants (which vary with P, T, X), and assumed that the species so formed mix ideally. In a formal thermodynamic sense, this procedure transforms a system with a minimum number of components that mix nonideally into one with a larger number of components (not all independent because their concentrations are fixed by the speciation equilibrium constants) that mix ideally. In an intuitive physical sense, the obvious question is, Do such speciation reactions actually occur? Because most spectroscopic measurements give information mainly on nearest and next-nearest coordination in melts and glasses, and the species proposed in Burnham's reactions invoke longer range order, their existence is difficult to prove or disprove. Since a silicate melt is basically a complex ionic liquid in dynamic equilibrium at high temperature, the definitions of complex speciation are intimately related to kinetic factors and relaxation times, and not only to static configurations. Molecular dynamics simulations and NMR studies at high temperature offer promise for addressing the dynamics of silicate melts. From a statistical thermodynamic point of view, the notion of well-defined "molecules" mixing ideally (which a rigorous structural interpretation of Burnham's thermodynamic model would require to obtain the appropriate configurational entropy) appears arbitrary. Nevertheless, as a scheme for petrologic calculations, the set of empirical equilibrium constants and simple equations is attractive and useful.

Structure-based thermodynamic models for melts having compositions between 0 and 50 mol % SiO_2 can be constructed in which the degree of polymerization (and thermodynamic properties) is determined both by the bulk composition (T/O ratio) and by equilibrium constant, K, for polymerization reactions such as Eq. 8.2 (Gaskell 1973, 1977; Hess 1971) The value of K is generally assumed to be independent of chain length but K will depend on the chemical nature of the metal oxide with the equilibrium lying far to the left for basic oxides and being more closely balanced for oxides such as PbO, MgO, NiO, whose cations compete more effectively for oxygen coordination.

An additional feature of binary alkali and alkaline earth silicate melts and glasses is liquid–liquid or metastable subliquidus immiscibility on the SiO_2-rich end of the diagram. Analogous phase separation in multicomponent natural silicate melts has been discussed extensively, especially in systems high in K_2O, TiO_2, and Fe_2O_3 (Philpotts 1976; Roedder 1951).

For binary and multicomponent systems whose components are taken to be common mineral compositions (albite, diopside, etc.), the major acid–base reactions and main exothermic contributions to the heat and free energy of mixing have been accounted for in the formation of the compound (crystal, glass, or melt) from its binary oxide constituents. The changes in local atomic environments, and therefore the heats and free energies of mixing, are much smaller (see Sec. 8.3.3) for such multicomponent mixing reactions. For thermodynamic modeling these small changes are an asset since they imply that activity coefficients change rather slowly and smoothly with composition, and relatively simple empirical formulations (e.g., regular and subregular solution) might be adequate for these complex systems. Such formulations are totally inadequate to describe the strong acid–base interactions in the simple metal–silica binaries.

In aluminosilicate melts, structure-based models that consider the mixing of ionic rather than quasi-crystalline or molecular species treat the melt as a complex molten salt containing several types of ionic species. The species can be considered to be nonframework cations (Na^+, K^+, Mg^{2+}, Ca^{2+}, Fe^{2+}, etc.) and aluminosilicate anions or, alternatively, nonframework cations, framework cations (Al^{3+}, Si^{4+}), and oxide ions. In either case, the basic task is to count correctly the species being mixed and to calculate a statistical entropy of mixing. Pertinent questions are: (1) Do Al and Si mix at random and what is the distribution of bridging and nonbridging oxygens (these define the anionic species)? and (2) Do the nonframework cations mix independently of the anionic framework, or is there local charge coupling (e.g., Na^+ with Al^{3+})? The *two-lattice model* is an example of such an ionic model. In it one assumes that Al and Si mix on one set of sites, nonframework cations (network modifiers) mix on another, and all oxygens are energetically equivalent. This model has met with considerable success in describing mixing properties and calculating phase equilibria in anhydrous basaltic and granitic systems (Navrotsky 1986; Navrotsky et al. 1980). However, the intermediate degree of "aluminus avoidance" inferred from Raman and NMR studies definitely complicates the current calculation of configurational entropy. Much more work needs to be done to develop general ionic models for complex aluminosilicate melts.

8.4 The nature of the glass transition and fragility in the liquid state

The magnitude of the jump in heat capacity at the glass transition varies from system to system. It appears smallest for highly polymerized systems (silica, albite), is somewhat larger for anorthite, and is larger still for diopside and the alkali silicates. The first two cases show Arrhenius behavior of the viscosity

$$\log \eta = A + B/T \qquad (8.6)$$

the latter show more complex behavior. Thermal expansion also increases in this series. All these effects point to an increasing restructuring of the melt at and above T_g. Indeed, a number of organic and low-melting inorganic glass-forming systems show even more pronounced changes. This systematic variation has prompted Angell (1985) to classify liquids as "strong" or "fragile," depending on the amount of restructuring that occurs on heating above the glass transition temperature (see Fig. 8.9). A *strong liquid* is and remains highly polymerized, with little bond rearrangement and no major collapse of

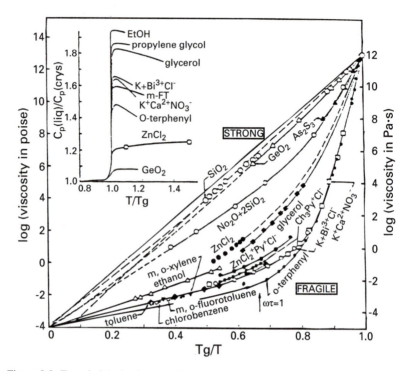

Figure 8.9. T_g-scaled Arrhenius plots for viscosities of glass-forming liquids of various types. Strong liquids fall at the top of the pattern, fragile liquids fall at the bottom. (*Insert*) Liquid-to-crystal heat capacity ratio through the glass transition temperature, showing correlation of ΔC_p with fragility (from Angell 1985).

structure as it is heated. It shows a small ΔC_p at T_g, a linear behavior in log η versus $1/T$, and a small thermal expansion. A *fragile liquid* has a large jump in heat capacity at T_g, it may continue to have an anomalously large C_p for a range of temperature interval above T_g, its viscosity decreases more strongly with increasing temperature than an Arrhenius relation would predict, and its volume changes anomalously. All this strongly suggests a major change in the structure of the melt, perhaps by the breakdown of complexes present at lower temperature, by change in coordination number, or by other mechanisms. Fragile liquids are usually less easy glass formers than strong liquids. Recently, titanosilicate liquids were found to show fragile behavior that their titanium-free analogues do not have (see Fig. 8.10) (Lange and Navrotsky 1993). This discovery may reflect change in the coordination number of titanium as well as the dissociation of alkali titanate complexes.

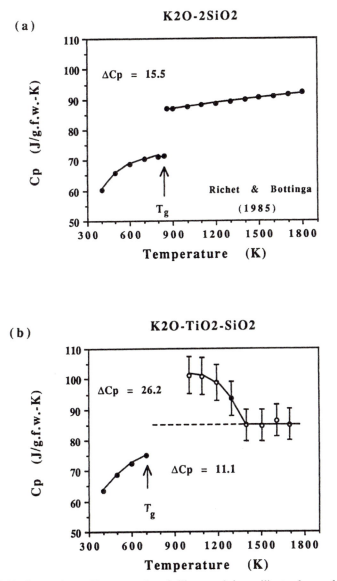

Figure 8.10. Comparison of heat capacity of silicate and titanosilicate glass and liquid (Lange and Navrotsky 1993).

The viscosity of a glass or melt can be linked to the configurations available to it and to its configurational entropy through the theory of Adam and Gibbs (1965). The physical idea follows. Motion (transport of matter) proceeds through cooperative rearrangement of the structure. This rearrangement depends on the number of configurations available to the system: If only one state could be sampled, the viscosity would be infinite. As temperature increases, the rearrangement becomes easier as the configurational entropy increases. The temperature dependence of relaxation times depends on the minimum size domain that can rearrange without requiring configurational changes elsewhere in the system. When the mathematics is carried through, the expression for viscosity, η, is of the form

$$\log \eta = A + (B/T) \, S^{\text{conf}}(T) \tag{8.7}$$

where

$$S^{\text{conf}}(T) = S^{\text{conf}}(T_g) + \int_{T_g}^{T} (\Delta C_p/T)dT \tag{8.8}$$

ΔC_p is the extra (configurational) heat capacity of the liquid compared to the glass and S^{conf} represents the configurational entropy of the melt. This configurational entropy can be derived from heat capacity measurements on crystal, glass, and melt or from viscosity data in the low- and high-temperature regions. Because the integral in Eq. 8.8 increases with temperature, the viscosity decreases faster than in an Arrhenius relation. This approach has proved very useful for relating viscosities and thermodynamic properties of mineralogically important silicate melts (Neuville and Richet 1991; Richet 1984).

8.5 Volatiles in silicate melts

In general, the solubility of a gas in a liquid increases with pressure and decreases with temperature (think of champagne). H_2O and CO_2 are the major volatiles of geological importance. Water appears to dissolve in silicate melts by a balance of two mechanism, as OH^- and as molecular H_2O (Stolper 1982). The equilibrium can be expressed as

$$H_2O \text{ (vapor)} = H_2O \text{ (molecular, melt)} \tag{8.9}$$

followed by an equilibrium

$$H_2O \text{ (molecular, melt)} + Si–O–Si = 2Si–OH \text{ (melt)} \tag{8.10}$$

Water is acting as a network modifier. However, the proton may also attack an already depolymerized site to replace a network modifier, for example, Na, with the modifier diffusing elsewhere.

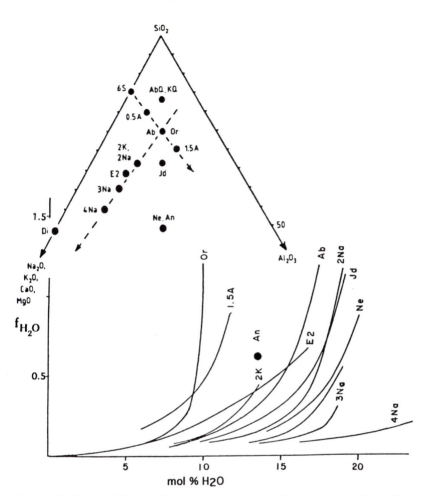

Figure 8.11. H_2O solubility in silicate glasses and melts of various compositions (from McMillan and Holloway 1987).

$$H^+ + Si–O–Na = Si–OH + Na^+ \qquad (8.11)$$

The latter mechanism may be important in the corrosion of glass by aqueous solutions.

There has been extensive study of the solubility of water in silicate melts (McMillan and Holloway 1987). Much of this work has been done by analyzing quenched glasses, particularly by infrared spectroscopy, in which vibrational bands due to molecular water and to hydroxyl can be distinguished (Silver and Stolper 1985). If all the water is present as OH^-, then the square of the mole fraction of H_2O in the melt should vary linearly with the fugacity of water in the fluid. This is approximately the case for mole fractions of water below 0.3 and pressures below 0.5 GPa for albitic melts, but definite curvature is seen at higher pressures (see Fig. 8.11). An equilibrium constant for Eq. 8.10 of about 0.2 at 1373 K fits the data reasonably well. This equilibrium constant depends on temperature, consistent with a small exothermic enthalpy of interaction of water with the melt, which has also been confirmed by calorimetry (Clemens and Navrotsky 1987). The solubility and speciation equilibria also depend on the nature of the melt. An important role of water, apparent from Eq. 8.10, is to depolymerize the melt. This lowers viscosity drastically. The dissociation into two OH^- species also explains why water lowers the dry solidus of many silicates by several hundred degrees.

CO_2 dissolves both as a molecular species and as carbonate in silicate melts (Fine and Stolper 1985). Carbonate magmas are also known, for example, at Oldonyo Lengai in Africa, and are implicated in the origin of carbonatites. Immiscibility between carbonate and silicate magmas may be important to carbonatite and kimberlite genesis.

The solubility of water and carbon dioxide decreases sharply with decreasing pressure, especially in the region between 0.2 GPa and atmospheric pressure. An ascending magma will begin to lose volatiles as pressure is released. The increased viscosity and increased water solubility of more silicic magmas explains why the Hawaiian volcanoes are relatively safe tourist attractions but the Mount St. Helens eruption was spectacularly and dangerously explosive.

8.6 Effect of pressure on glasses and melts

Components in a silicate melt generally have larger compressibility and thermal expansion than the crystalline assemblage of the same composition (compare Tables 8.3 and 6.5). This extra compression occurs largely by distorting intertetrahedral angles, changing ring size, and making internal cavities smaller (Navrotsky et al. 1985; Rigden, Ahrens, and Stolper 1988; Stolper and Ahrens 1987). Without the constraints of long-range order, many more such mechanisms are available to the melt than to the crystal. These changes make polymerized melts more fragile, and there is a marked decrease of viscosity of albite melt with pressure (Kushiro 1976).

The volume change on fusion generally decreases with increasing pressure. This tends to decrease the slope of $P-T$ melting curves (see Fig. 8.12). In some cases, the melt may actually become as dense or denser than the crystal,

Table 8.3. *Partial molar volumes, thermal expansivities, and compressibilities of oxide components in silicate melts*

$$V_{liq}(T,P,X_i) = \sum X_i[V_{i,1673\,K} \pm dV_i/dT\,(T - 1673\,K) \pm dV_i/dP(P - 1bar)]$$

Compound	$V_{i,1673\,K}$ (cc/mol)	$(dV_i/dT)1bar$ (10^{-3}cc/mol K)	$(dV_i/dP)_{1673K}$ (10^{-4}cc/mol bar)	$[(dV_i/dP)]/dT$ (10^{-7}cc/mol bar K)
SiO_2	26.90 ± .06	0.00 ± 0.50	−1.89 ± .02	1.3 ± 0.1
TiO_2	23.16 ± .26	7.24 ± 0.46	−2.31 ± .06	—
Al_2O_3	37.11 ± .18	2.62 ± 0.17	−2.26 ± .09	2.7 ± 0.5
Fe_2O_3	42.13 ± .28	9.09 ± 3.49	−2.53 ± .09	3.1 ± 0.5
FeO	13.65 ± .15	2.92 ± 1.62	−0.45 ± .03	−1.8 ± 0.3
MgO	11.45 ± .13	2.62 ± 0.61	0.27 ± .07	−1.3 ± 0.4
CaO	16.57 ± .09	2.92 ± 0.58	0.34 ± .05	−2.9 ± 0.3
Na_2O	28.78 ± .10	7.41 ± 0.58	−2.40 ± .05	−6.6 ± 0.5
K_2O	45.84 ± .17	11.91 ± 0.89	−6.75 ± .14	−14.5 ± 1.5
Li_2O	16.85 ± .15	5.25 ± 0.81	−1.02 ± .06	−4.6 ± 0.4

Source: Data from Lange and Carmichael (1990).

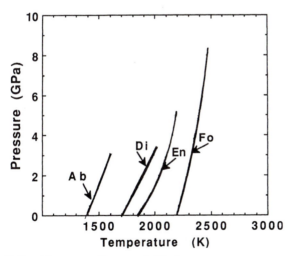

Figure 8.12. *P–T* melting curves for several silicates.

which would cause a maximum in the melting curve. A magma generated at a depth below this turnover would not be able to rise to the surface (Stolper et al. 1981).

Pressure induces coordination number changes in melts, just as it induces

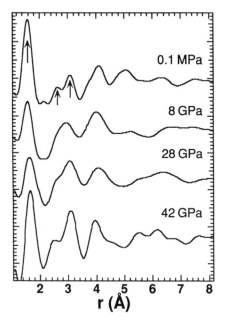

Figure 8.13. Average pair correlation functions, $G(r)$, for SiO_2 glass with increasing pressure. At ambient pressure, the Si–O, O–O, and Si–Si peaks are indicated by arrows at 1.59, 2.61, and 3.07 Å, respectively. (From Meade, Hemley, and Mao 1992.)

phase transitions involving coordination number changes in crystals. Although it was first suggested that densification at moderate pressure was due to change in Al coordination (Kushiro 1978; Waff 1975), it is now generally agreed that such coordination changes occur at higher pressure in the melt than in the crystal because, at first, densification can occur through distortional mechanisms not available to the crystal. Thus the transition from tetrahedral to octahedral aluminum, seen in crystals near 2 GPa, probably becomes significant in silicate liquids above 7 GPa (Ohtani, Taulelle, and Angell 1985). Octahedral silicon probably becomes a major constituent in geologically important melts only above 30 GPa (Williams and Jeanloz 1988), though stishovite is stable at 10 GPa and silicate perovskite near 23 GPa. The apparently flat melting curve of $MgSiO_3$ perovskite implies a close to zero volume of melting at lower mantle pressures and suggests that the melt as well as the crystal contains mainly octahedral silicon. It also places an upper limit of about 3500 K on the temperature at the core–mantle boundary; otherwise, the lower mantle would show extensive melting.

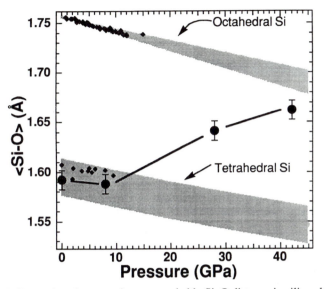

Figure 8.14. Comparison between the most probable Si–O distance in silica glass as a function of pressure (circles) and the Si–O bond length in crystalline SiO_2 polymorphs with tetrahedral and octahedral coordination (shaded areas). For tetrahedral SiO_2, the bond length at high pressures is calculated assuming an effective bulk modulus of 400 GPa and accounting for the variability due to changes in the Si–O–Si angle. The Si–O bond length for octahedral SiO_2 is derived from the equation of state of stishovite. Measured values of the shortest Si–O bond distance are plotted (diamonds) for α-quartz and stishovite at high pressure. (From Meade et al. 1992.)

The effect of pressure on glasses is even more complex. There are really two effects, an elastic compression that is reversible and can not be quenched, and a structural change that is at least partially quenchable. Thus the extent to which a glass is densified depends on its pressure–temperature–time path. Even in SiO_2 glass, compression of 10–20% can be retained. These changes seem to involve changes in intertetrahedral angles for pressures up to perhaps 25 GPa. Above that, Raman spectra in the diamond cell rapidly lose the signature of well-defined silicate tetrahedra, and severe distortion of the tetrahedra and/or coordination number change occur (Hemley et al. 1986).

Figure 8.13 shows the average pair correlation function for SiO_2 glass at various pressures, obtained by in situ study using synchrotron radiation (Meade, Hemley, and Mao 1992). Thus both quenchable and nonquenchable structural rearrangements are monitored. The change in $G(r)$ with pressure clearly shows increasing distortion of both Si–O–Si angles and of the tetrahedra themselves. This leads to the onset of silicon coordination number increase from four-fold to six-fold as confirmed by the lengthening of Si–O distance with pressure (see Fig. 8.14). At pressures near 40 GPa, most of the silicon is probably octahedrally coordinated.

A related phenomenon is pressure-induced amorphization (Hemley et al. 1988). Many high-pressure phases (including stishovite and $MgSiO_3$ perovskite), when quenched to atmospheric pressure, readily decompose to an amorphous phase when heated to even 400–600 K. Such "glass" may be structurally and energetically somewhat different from normal melt-quenched glass of the same composition (Geisinger, Navrotsky, and Arndt 1986). In nature pressure- or shock-induced amorphization occurs as primary evidence of meteorite impact, for example, at Meteor Crater, Arizona. Other high-pressure phases decompose directly to an amorphous phase on release of pressure at ambient temperature; an example is $CaSiO_3$ perovskite (Mao et al. 1989). Still others may form an amorphous phase directly at high pressure. This pressure-induced amorphization may relate to the metastable extension of a melting curve (Mishima, Calvert, and Whalley 1984), but it may also be an intrinsic manifestation of "structural frustration" (Hammack, Serghiou, and Winters 1992; Nelson 1983; Nelson and Spaepen 1989).

8.7 Other glasses and amorphous materials

Glass formation, the glass transition, and anomalous viscosity behavior are by no means limited to silicates. To show the generality of the glassy state, Table 8.4 lists a broad selection of glass-forming systems. They include silicates, hydrated salts, metals, polymers, simple organic liquids, and even water.

Furthermore, amorphous materials can be produced by processes other than rapid thermal quenching from a molten state. The pressure- or shock-induced amorphization described above is one example. Radiation damage is another. Figure 8.15 shows HRTEM electron diffraction patterns of zircons exposed to increasing radiation doses. This mimics, in a controlled fashion, the metamictization of natural minerals caused by their proximity to sources of radioactive Th and U and their bombardment by α-particles. Three stages of damage can be seen: the creation of point defects, their coalescence into tracks or heavily damaged regions, and complete amorphization (Wang and Ewing 1992). Yet this amorphization does not involve long-range diffusion, and the "metamict crystal," as the product is termed, readily anneals back to the crystalline state on heating, sometimes even preserving the morphology of the initial crystal.

Low-temperature precipitation, whether from a vapor phase or from an

Table 8.4. *Glass-forming systems*

System	Conditions for glass formation
Silicates, borates, phosphates	
mol fraction SiO_2, B_2O_3, or $P_2O_5 > 0.5$	Readily forms glass on cooling melt
mol fraction SiO_2, B_2O_3, or $P_2O_5 < 0.5$	Splat cooling, roller quenching, etc.
framework aluminosilicates	Readily forms glass on cooling melt
Other oxides	Form amorphous materials by low-temperature processing or radiation damage
Fluorides	Extensive easy glass formation in a number of systems
Oxynitrides	Easy glass formation in a number of systems, including SiAlON
Other halides	Sporadic glass formation
Chalcogenides	Extensive glass formation in (Ga, Ge) (S, Se, Te) system
Hydrated nitrates	Easy glass formation
Alcohol mixtures	Easy glass formation
Organic polymers	Frequently glassy
Metal alloys with deep eutectics	Form glasses by roller quenching

aqueous or organic solution, often produces an amorphous material. This material may reflect both small particle size and extensive disorder. These amorphous phases may be hydrous gels of silica and other oxides produced as precipitates on changing the pH of a colloidal solution, so-called sol-gel processing (Brinker and Scherer 1990; Johnson 1985; Ulrich 1990). They may also be deposited by low-temperature reactions from a gas phase, that is, chemical vapor deposited (CVD) (Huffman, Navrotsky, and Pintchovski 1986; Konnert et al. 1987). Similar amorphous materials are produced as leach layers on dissolving a mineral (aluminum and other cations dissolve more readily, leaving behind a silica-rich surface coating) (Casey et al. 1992). Amorphous iron and titanium oxides are also geologically important. Each of these materials has its own properties, is metastable with respect to crystalline products, and transforms to them at temperatures and rates that vary with its history. Thus the field of amorphous materials is very rich in diversity. At a given set of P,T,X conditions, there is only one state of lowest free energy, but an infinite number of metastable states involving continuously variable amorphous materials.

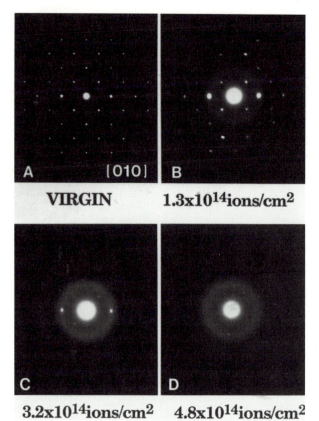

Figure 8.15. Electron diffraction patterns of zircon ($ZrSiO_4$) as a function of radiation dose (1.5 MeV Kr^+) at 25 °C. Progressive disappearance of diffraction spots and appearance of smeared-out rings of diffracted intensity document metamictization (courtesy of L. M. Wang).

8.8 Nucleation, crystal growth, and exsolution

Consider a homogeneous melt from which a crystalline phase must grow. Since the structure of the crystal is different from that of the melt, atoms must rearrange to form the initial *nucleus* from which the crystal will grow. The first few particles in the growing nucleus are all at its surface and show a positive surface energy relative to the melt, which is destabilizing and proportional to the surface area that increases as the square of the radius of the cluster. As the cluster grows, its interior becomes more like the bulk crystal and is stabilized

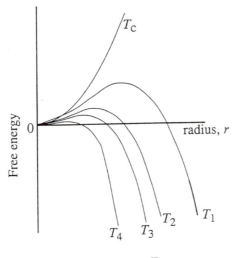

$$T_c > T_1 > T_2 > T_3 > T_4$$

Figure 8.16. Nucleation and growth kinetics: free energy versus size of nucleus for different temperatures, schematic. T_c is liquidus temperature. (From Putnis 1992.)

by the free energy of crystallization. The bulk energy is proportional to the volume, that is, to the cube of the radius. The total free energy then consists of a balance of two terms

$$\Delta G = Ar^2 - Br^3 \qquad (8.12)$$

and a barrier to nucleation (see Fig. 8.16) must be overcome before the crystal can grow. Similar arguments hold for nucleating a new crystalline phase in the solid state. The advantage in free energy for transforming the bulk phase at its equilibrium temperature is, by definition, zero. Therefore one must supercool below the equilibrium temperature to get finite transformation rates. The more supercooling, the larger the driving force for transformation. However, as temperature is lowered, it becomes harder to overcome the nucleation barrier. The result is that there is generally a maximum in nucleation and transformation rate at some temperature below the equilibrium temperature. If one cools a sample rapidly to below this maximum, it will not transform, which is the basis of quenching metastable phases, including glasses. Analogous arguments can be made for crossing a phase boundary as a function of both temperature and pressure.

8.9 References

8.9.1 General references and bibliography

Brawer, S. (1985). *Relaxation in viscous liquids and glasses, Review of phenomenology, molecular dynamics simulations, and theoretical treatment.* Columbus, OH: American Ceramic Society.

Putnis, A. (1992). *Introduction to mineral sciences.* Cambridge: Cambridge University Press.

Richet, P., and Y. Bottinga. (1986). Thermochemical properties of silicate glasses and liquids: A Review. *Rev. Geophys. 24,* 1–25.

Scarfe, C. M., ed. (1986). *Silicate melts.* Short Course Handbook, vol. 12. Toronto: Mineralogical Association of Canada.

Simmons, C. J., and O. H. El-Bayoumi, eds. (1993). *Experimental techniques of glass science.* Columbus, OH: American Ceramic Society.

Vogel, W. (1985). *Chemistry of glass.* Columbus, OH: American Ceramic Society.

Waseda, Y. (1980). *The structure of non-crystalline materials, liquids and amorphous solids.* New York: McGraw-Hill.

8.9.2 Specific references

Adam, G., and J. H. Gibbs. (1965). On the temperature dependence of cooperative relaxation properties in glass-forming liquids. *J. Chem. Phys. 43,* 139–46.

Angell, C. A. (1985). Strong and fragile liquids. In *Relaxation in complex systems,* ed. K. Ngai and G. B. Wright, 3–11. Springfield, VA: National Technical Information Service.

Berman, R. O., and T. H. Brown. (1987). *Development of models for multicomponent melts: Analysis of synthetic systems.* Reviews in Mineralogy, vol. 17, 405–42. Washington, DC: Mineralogical Society of America.

Brinker, C. J., and G. W. Scherer. (1990). *Sol-gel science.* San Diego: Academic Press.

Burnham, C. W. (1975). Thermodynamics of melting in experimental silicate-volatile systems. *Geochim. Cosmochim. Acta 39,* 1077–84.

Burnham, C. W. (1981). Nature of multicomponent aluminosilicate melts. *Phys. Earth Planets 13–14,* 191–227.

Burnham, C. W., and H. Nekvasil. (1986). Equilibrium properties of granite pegmatite magmas. *Amer. Mineral. 71,* 239–63.

Casey, W. H., C. Eggleston, P. A. Johnson, H. R. Westrich, and M. F. Hochella, Jr. (1992). *MRS Bull. 17,* 23–39.

Clemens, J. D., and A. Navrotsky. (1987). Mixing properties of $NaAlSi_3O_8$ melt-H_2O: New calorimetric data and some geological implications. *J. Geol. 95,* 173–86.

DeYoreo, J. J., A. Navrotsky, and D. B. Dingwell. (1990). Energetics of the charge-coupled substitution $Si^{4+} \rightarrow Na^{1+} + T^{3+}$ in the glasses $NaTO_2$–SiO_2 (T = Al, Fe, Ga, B). *J. Amer. Ceram. Soc. 73,* 2068–72.

Eriksson, G., and A. D. Pelton. (in press). Critical evaluation and optimization of the thermodynamic properties and phase diagrams of the CaO–Al_2O_3, Al_2O_3–SiO_2, and CaO–Al_2O_3–SiO_2 systems. *Calphad.*

Fine, G., and E. Stolper. (1985). The speciation of carbon dioxide in sodium aluminosilicate glasses. *Contrib. Mineral. Petrol. 91,* 105–21.

Gaskell, D. R. (1973). The thermodynamic properties of the masson polymerization models of liquid silicates. *Metallurgical Transactions 4,* 185–92.

Gaskell, D. R. (1977). Activities and free energies of mixing in binary silicate melts. *Metallurgical Transaction 8B,* 131–45.

Geisinger, K. L., G. V. Gibbs, and A. Navrotsky. (1985). A molecular orbital study of bond length and angle variation in framework silicates. *Phys. Chem. Minerals 11*, 266–83.

Geisinger, K. L., A. Navrotsky, and J. Arndt. (1986). Enthalpy of diaplectic labradorite glass. *Phys. Chem. Minerals 13*, 357–9.

Ghiorso, M. S. (1987). *Modeling magmatic systems: Thermodynamic relations.* Reviews in Mineralogy, vol. 17, 443–66. Washington, DC: Mineralogical Society of America.

Ghiorso, M. S., and I. S. E. Carmichael. (1987). *Modeling magmatic systems: Petrologic applications.* Reviews in Mineralogy, vol. 17, 467–99. Washington, DC: Mineralogical Society of America.

Ghiorso, M. S., I. S. E. Carmichael, M. L. Rivers, and R. O. Sack. (1983). The Gibbs free energy of mixing of natural silicate liquids; An expanded regular solution approximation for the calculation of magmatic intensive variables. *Contrib. Mineral. Petrol. 87*, 107–45.

Griscom, D. L. (1977). Borate glass structure. *Mater. Sci. Res. 12*, 11–138.

Hammack, W. S., G. C. Serghiou, and R. R. Winters. (1992). Pressure induced amorphization. In *The physics of noncrystalline solids*, ed. L. D. Pye, W. C. LaCourse, and H. J. Stevens, 208–12. London: Taylor and Francis.

Hemley, R. J., A. P. Jephcoat, H. K. Mao, L. C. Ming, and M. H. Manghnani. (1988). Pressure-induced amorphization of crystalline silica. *Nature 334*, 52–4.

Hemley, R. J., H. K. Mao, P. M. Bell, and B. O. Mysen. (1986). Raman spectroscopy of SiO_2 glass at high pressure. *Phys. Rev. Lett. 57*, 747–50.

Hersh, L. E., and O. J. Kleppa. (1965). Enthalpies of mixing in some binary liquid halide mixtures. *J. Chem. Phys. 43*, 1309–33.

Hess, P. C. (1971). Polymer model of silicate melts. *Geochim. Cosmochim. Acta 35*, 289–306.

Huffman, M., A. Navrotsky, and F. S. Pintchovski. (1986). Thermochemical and spectroscopic studies of chemically vapor-deposited amorphous silica. *J. Electrochem. Soc. 133*, 164.

JANAF thermochemical tables. (1971). Ed. D. R. Stull and H. Prophet. U.S. National Bureau of Standards. Rep. NSDRS-NBS 37. Washington, DC: GPO.

Johnson, D. W., Jr. (1985). Sol-gel processing of ceramics and glass. *Am. Ceram. Soc. Bull. 64*, 1597–602.

Kawamoto, Y., and M. Tomozawa. (1982). The mixed alkali effect in the phase separation of alkali silicate glasses. *Phys. Chem. Glasses 23*, 72–5.

Kirkpatrick, R. J. (1988). *MAS NMR spectroscopy of minerals and glasses.* Reviews in Mineralogy, vol. 18, 341–404. Washington, DC: Mineralogical Society of America.

Konnert, J., P. D'Antonio, M. Huffman, and A. Navrotsky. (1987). Diffraction studies of a highly metastable form of amorphous silica. *J. Amer. Ceram. Soc. 70*, 192–6.

Konnert, J. H., P. D'Antonio, and J. Karle. (1982). Comparison of radial distribution function for silica glass with those for various bonding topologies: Use of correlation function. *J. Non-cryst. Solids 53*, 135–41.

Kushiro, I. (1978). Viscosity and structural changes of albite melt at high pressures. *Earth Planet. Sci. Lett. 41*, 87–90.

Lange, R. A., and I. S. E. Carmichael. (1990). *Thermodynamic properties of silicate liquids with emphasis on density, thermal expansion and compressibility.* Reviews in Mineralogy, vol. 24, 25–64. Washington, DC: Mineralogical Society of America.

Lange, R. A., J. J. DeYoreo, and A. Navrotsky. (1991). Scanning calorimetric measurement of heat capacity during incongruent melting of diopside. *Amer. Mineral.* *76*, 904–12.

Lange, R. A., and A. Navrotsky. (1993). Heat capacities of TiO_2-bearing silicate liquids: Evidence for anomalous changes in configurational entropy with temperature. *Geochim. Cosmochim. Acta 57*, 3001–11.

Mao, H. K., L. C. Chen, R. J. Hemley, A. P. Jephcoat, and Y. Wu. (1989). Stability and equation of state of $CaSiO_3$-perovskite to 134 GPa. *J. Geophys. Res. 94*, 17889–94.

Masson, C. R. (1965). An approach to the problem of ionic distribution in liquid silicates, *Proc. Roy. Soc. London A287*, 201–21.

Masson, C. R. (1968). Ionic equilibria in liquid silicates. *J. Amer. Ceram. Soc. 51*, 134–43.

McMillan, P. F. (1984). Structural studies of silicate glasses and melts; Applications and limitations of Raman spectroscopy. *Amer. Mineral. 69*, 622–44.

McMillan, P. F., and J. R. Holloway. (1987). Water solubility in aluminosilicate melts. *Contrib. Mineral. Petrol. 97*, 320–32.

Meade, C., R. J. Hemley, and H. K. Mao. (1992). High pressure X-ray diffraction of SiO_2 glass. *Phys. Rev. Lett. 69*, 1387–90.

Mishima, O., L. D. Calvert, and E. Whalley. (1984). "Melting ice" I at 77 K and 10 kbar: A new method of making amorphous solids. *Nature 310*, 393–5.

Navrotsky, A. (1986). Thermodynamics of silicate glasses and melts. In *Silicate melts*, Short Course Handbook, ed. C. M. Scarfe, vol. 12, 130–53. Toronto: Mineralogical Association of Canada.

Navrotsky, A., K. L. Geisinger, P. McMillan, and G. V. Gibbs. (1985). The tetrahedral framework in glasses and melts–Inferences from molecular orbital calculations and implications for structure, thermodynamics, and physical properties. *Phys. Chem. Minerals 11*, 284–98.

Navrotsky, A., R. Hon, D. F. Weill, and D. J. Henry. (1980). Thermochemistry of glasses and liquids in the systems $CaMgSi_2O_6$–$CaAl_2Si_2O_8$–$NaAlSi_3O_8$, SiO_2–$CaAl_2Si_2O_8$–$NaAlSi_3O_8$, and SiO_2–Al_2O_2–CaO–Na_2O. *Geochim. Cosmochim. Acta 44*, 1409–23.

Navrotsky, A., D. Ziegler, R. Oestrike, and P. Maniar. (1989). Calorimetry of silicate melts at 1773 K: Measurement of enthalpies of fusion and mixing in the system diopside–anorthite–albite and anorthite–forsterite. *Contrib. Mineral. Petrol. 101*, 122–30.

Nelson, D. R. (1983). Order, frustration, and defects in liquids and glasses. *Phys. Rev. B 28*, 5515–35.

Nelson, D. R., and P. Spaepen. (1989). Polytetrahedral order in condensed matter. *Solid State Phys. 42*, 1–90.

Neuville, D. R., and P. Richet. (1991). Viscosity and mixing in molten (Ca,Mg) pyroxenes and garnets. *Geochim. Cosmochim. Acta 55*, 1011–19.

Ohtani, E., F. Taulelle, and C. A. Angell. (1985). Al^{3+} coordination changes in liquid aluminosilicates under pressure. *Nature 314*, 78–81.

Philpotts, A. R. (1976). Silicate liquid immiscibility: Its probable extent and petrogenetic significance. *Amer. J. Sci. 276*, 1147–77.

Pohl, R. O., W. F. Love, and R. B. Stephens. (1974). Lattice vibrations in noncrystalline solids. In *Amorphous and liquid semiconductors; Proceedings of the fifth international conference,* ed. J. Struke and W. Brenig, 1121–32. New York: Halsted Press.

Richet, P. (1984). Viscosity and configurational entropy of silicate melts. *Geochim. Cosmochim. Acta 48*, 471–83.

Rigden, S. M., T. J. Ahrens, and E. M. Stolper. (1988). Shock compression of molten silicate: Results for a model basaltic composition. *J. Geophys. Res. 93*, 367–82.

Roedder, E. (1951). Low temperature liquid immiscibility in the system K_2O–FeO–Al_2O_3–SiO_2. *Amer. Mineral. 36*, 282–6.

Seifert, F., B. O. Mysen, and D. Virgo. (1982). Three-dimensional network structure of quenched melts (glass) in the systems SiO_2–$NaAlO_2$, SiO_2–$CaAl_2O_4$ and SiO_2–$MgAl_2O_4$. *Amer. Mineral. 67*, 696–717.

Silver, L., and E. Stolper. (1985). A thermodynamic model for hydrous silicate melts. *J. Geol. 93*, 161–77.

Stebbins, J. F. (1988). *NMR spectroscopy and dynamic processes in mineralogy and geochemistry.* Reviews in Mineralogy, vol. 18, 405–30. Washington, DC: Mineralogical Society of America.

Stebbins, J. F. (1992). Nuclear magnetic resonance spectroscopy of geological materials. *MRS Bull. 17*, 45–52.

Stolper, E. (1982). The speciation of water in silicate melts. *Geochim. Cosmochim. Acta 46*, 2609–20.

Stolper, E. M., and T. J. Ahrens. (1987). On the nature of pressure-induced coordination changes in silicate melts and glasses. *Geophys. Rev. Lett. 14*, 1231–3.

Stolper, E. M., D. Walker, B. H. Hager, and J. F. Hays. (1981). Melt segregation from partially molten source regions: The importance of melt density and source region size. *J. Geophys. Res. 86*, 6261–71.

Taylor, M., and G. E. Brown. (1979). Structure of mineral glasses, I: The feldspar glasses $NaAlSi_3O_8$, $KAlSi_3O_8$, $CaAl_2Si_2O_8$. *Geochim. Cosmochim. Acta 43*, 61–75.

Ulrich, D. R. (1990). Prospects for sol-gel processes. *J. Non-cryst. Solids 121*, 465–79.

Waff, H. S. (1975). Pressure-induced coordination changes in magmatic liquids. *Geophys. Res. Lett. 2*, 193–6.

Wang, L. M., and R. C. Ewing. (1992). Ion beam induced amorphization of complex materials–minerals. *MRS Bull. 17*, 38–44.

Waseda, T. (1981). Current structural information on molten slags by means of a high temperature X-ray diffraction. *Canad. Metall. Quartz. 20*, 57–67.

Williams, Q., and R. Jeanloz. (1988). Spectroscopic evidence for pressure-induced coordination changes in silicate glasses and melts. *Science 239*, 902–5.

9

The interface between mineral physics and materials science

9.1 Scientific issues

Art imitates nature and nature imitates art. The Earth is a materials-processing laboratory operating with time, distance, mass, pressure, and temperature on a grand scale. From brittle fracture in the crust to convection in the core, geologic process reflects the materials properties of natural, largely inorganic materials, namely, rocks, and the fluids that permeate them. If several kilometers of the San Andreas Fault slip a few meters relative to each other during a devastating earthquake, the movement involves, on a microscopic scale, the breaking of chemical bonds in silicate minerals. The excitement of modern Earth sciences comes from this interplay of microscopic and macroscopic phenomena, the revolution in instrumentation for laboratory, computational, and field studies, and the flood of new data on all scales.

9.1.1 Rock mechanics and motion

Cold rocks near the Earth's surface deform by brittle failure, hot rocks at greater depth by plastic flow. In both cases, defect chemistry, dislocations, and other microscopic phenomena form the basis for macroscopic movement. Understanding stress–strain relationships, crack propagation, and the effect of fluids on strength are fundamental to understanding faulting and earthquakes. Plastic flow in the mantle makes that region fluid on the geologic time scale; thus geophysicists can speak of mantle convection without invoking large volumes of conventionally molten rock. Because most minerals are multicomponent solid solutions, their defect equilibria are usually extrinsic and often dominated by the oxidation–reduction reactions of iron. To understand viscous flow in the mantle, one must understand such defects to pressures of several hundred thousand atmospheres, temperatures of 2000 K, and varying oxygen fugacities. This is, both technologically and intellectually, a far greater challenge than the problem of understanding defect chemistry in chemically controllable ceramics at atmospheric pressure.

9.1.2 Phase changes at high pressure

Seismic discontinuities define the boundaries separating crust, upper mantle, transition zone, lower mantle, and core. Indicative of sharp changes in density with depth, these discontinuities have been attributed to phase changes in minerals, transformations to denser crystal structures, for example, olivine to spinel. Although we have the technology to retrieve samples from the surfaces of other planets, there is no possibility to obtain a sample from 400 km below our feet where such a phase transition might be occurring. These conditions can be created in the laboratory, although they are confined to rather small volumes, producing micrograms of material in the diamond anvil cell and milligrams of material in large multianvil presses. The mineralogy of the Earth's interior has now been charted, at least in broad outline, and the silicate phases found have the familiar crystal structures of refractory ceramics – spinel, garnet, ilmenite, perovskite. This has led to great extensions of the concepts of systematic crystal chemistry and to calculations and simulations (by lattice energy, molecular orbital, band theory, lattice dynamical models, molecular dynamics) of physical and chemical properties. Indeed, the high-pressure phases are a probe of interatomic potentials, with general interest to physics and chemistry. High pressure can also induce semiconductor-to-metal transitions, now seen in a number of oxides and sought for in the simplest of all molecules, hydrogen.

9.1.3 Melting, crystallization, and glass formation

Magmas within the Earth and lavas spilling at the surface are silicate melts, more siliceous than slags and less boron-rich than many commercial glass formulations, but similar in many respects. Crystallization occurs during cooling; the mineralogy and textures of the resulting rocks record their history. Questions of supercooling, nucleation, growth, heat transfer, and the resulting morphology are fundamentally similar in the geologic and glass-ceramic context. Rapidly cooled lavas can produce partly or totally glassy rocks; hydration of these, especially in water-rich or submarine environments, can lead to rapid degradation. These phenomena are similar to those encountered in commercial glass melting, controlled solidification, and corrosion. The use of glass to contain nuclear waste raises questions of both ceramic and geologic nature. The corrosion of such glass containment material in contact with groundwater is a special case of rock–water interaction.

9.1.4 Nature's sol-gel processing

At and near the Earth's surface, water dominates in mineral reactions. Complex fine-grained phases, previously dubbed as unintelligible mud and clay, now begin to assert their own chemical and structural identities as analytical tools become more sophisticated. Weathering involves mineral alteration and dissolution, often leaving behind amorphous hydrated layers. Sediment burial and diagenesis involve low-temperature reactions that transform, recrystallize, and densify such phases. The resulting texture and porosity are crucial to petroleum reservoir behavior. Aqueous solutions dissolve, transport, concentrate, and precipitate major and trace elements; the results are ore deposits. Catalysis, surface reactions, and kinetic rather than thermodynamic control dominate. Both the processes themselves and the need to characterize them using modern spectroscopic, chemical, and structural probes are similar, and similarly complex, whether occurring in nature or during ceramic processing. Furthermore, this low-temperature environment is the one most directly affected by human activities. Pollution and the disposal of both nuclear and chemical waste bring synthetic materials in contact with the geologic environment; their dissemination or containment are governed by the same chemical and transport processes that control aqueous geochemistry and ore deposition. Characterizing the materials properties and processes on a fundamental scale is essential to predicting long-term behavior. The detailed kinetic and mechanistic understanding needed is not yet available. These are problems involving geology, hydrology, chemistry, and materials science. Furthermore, there are scientific, engineering, economic, social, and legal dimensions to all these issues.

9.1.5 Complexity and self-organization

On the unit cell scale, minerals are often very complex, showing low-symmetry structures, a number of simultaneous coupled chemical substitutions and occasional nonstoichiometry, frequent order–disorder and phase transitions, and, almost always, the interplay of strongly bonded silicon–oxygen frameworks with much weaker cations. In some sense, several interactions compete, often on different time and distance scales, to determine a mineral's properties. Thus the resulting structure is often quite finely balanced, and small perturbations in external parameters can cause significant changes in its properties. In this sense minerals have the degree of complexity needed to be "interesting" or "smart" or "compliant," to have properties tailored for very specific applications. Natural and synthetic zeolites with controlled pore geometry offer specific ion exchange and catalytic properties. Such zeolites form the basis of multimillion dollar petroleum-based industries. Clays offer similar possibilities for catalytic applications by varying the strength of interlayer forces by compositional control and by the expansion of interlayer spacing through intercalation and pillaring. Agates, opals, banded carbonates, and many other rocks show textures and repeating geometric patterns that suggest either cyclic variations in depositional conditions (the classical view) or self-organized structures governed by specific growth mechanisms or the irreversible thermodynamics of processes far from equilibrium. These mesoscale structures may be inorganic analogues of simple prebiological organization, and the idea of directed organic synthesis at clay surfaces, including the origin of chirality and eventually of life itself, is not new. Geological systems operating on long time scales may be models from which simpler synthetic systems can be abstracted.

9.2 Interaction between Earth science and materials science communities

It is obvious that Earth materials research has benefited immensely from the revolution in technology available to materials research. Electron microscopy at both high and mid-resolution, EXAFS, surface analytical techniques, solid-state NMR, molecular dynamics simulations – the list of techniques applicable and already applied to Earth materials goes on and on. Being a small discipline, Earth science has also benefited from the larger scale on which materials science can operate in terms of equipment availability and joint use. At the same time, Earth science is becoming more quantitative, rigorous, and mechanism-oriented in terms of concepts taken from chemistry, physics, and materials science.

Less obvious, perhaps, are the ways in which Earth science contributes to materials science. The first is in terms of people. Students in the United States well trained in crystal chemistry, crystallography, and solid-state chemistry come largely from the Earth science community. It is no accident that mineralogical crystallographers first solved the crystal structure of the oxide superconductors; they were comfortable with the level of complexity of those structures. Such modern mineralogists, or, as many prefer to be called, mineral physicists, have found employment in industrial laboratories of firms like Corning, DuPont, Exxon, Motorola, and Mobil, and in federal labs doing materials research. The second contribution is that of a broad view. When an Earth scientist looks at a material, the dimension of pressure as well as temperature is obvious, and questions of how this material will behave on long time scales are natural. The Earth scientist is usually well versed in phase equilibria, and materials compatibility is a familiar geologic problem. The Earth materials community leads the materials science community in high-pressure research and in the control of thermodynamic intrinsic parameters (e.g., the fugacity of water, oxygen, or carbon dioxide) at atmospheric and high pressure. Electron microprobe analysis and the more classical, but still very useful, optical microscopy techniques are natural to the geologist. Coming to materials science from a different starting point, the Earth scientist speaks with a different and unique voice.

9.3 Mineral physics, applied mineralogy, and the environment

Earth science is undergoing a period of rapid change at present in the way the field defines its relation to society at large. For over two hundred years, geology was obviously useful in exploring for and exploiting resources – iron, copper, gold, uranium, coal, oil. Geologists and mining companies worked hand in hand. Obtaining the minerals at the lowest possible cost was the main concern, with environmental issues being far less important. From that point of view, mineralogy was a useful industrial trade; to identify minerals and characterize ore deposits was societally useful, and mineralogists found ready employment. Over the last decade, societal emphasis has shifted significantly from exploration and exploitation to management of resources, with growing concern about issues of pollution, safety, and remediation.

The whole sequence of resource use from mining to manufacturing to product use to waste disposal is receiving increasing scrutiny. We worry about natural and anthropogenic changes in environment and climate. The movement of trace pollutants in groundwater, soil, and air is an increasingly critical concern. Thus processes on and in the Earth, not over millions of years, but over days, months, years, and centuries, are of major importance. Environmental consulting firms are probably hiring more Earth scientists than are the traditional mining and petroleum interests. The environmental firms are also hiring chemists, engineers, computer scientists, and, of course, lawyers. They provide an interdisciplinary workplace.

What is the role of mineral physics in relation to these changing societal priorities? First, our students must and will look to where the jobs are. To attract students, the field of Earth materials research must lead to jobs and a feeling that one is working on important problems. Second, many fundamental unsolved problems related to how minerals react with groundwater, how pollutants are transported and adsorbed, and how complex geochemical cycling takes place are essentially applications of mineral physics. The characterization of fine-grained materials (clays, soils, amorphous phases), surface chemistry, adsorption, catalysis – these are all environmentally relevant aspects of mineral physics. These processes are governed by the fundamental physics and chemistry of the materials involved, and the microscopic to macroscopic approach can be expected to be as useful and valid when applied to the Earth's surface as when applied to the lower mantle. Because of the lower temperatures, more complex phases, and greater role of kinetic factors, mineral physics of surface and near-surface processes is probably more complex than that of the Earth's deep interior. But the field is mature enough to tackle new challenges, and solution of the near-surface problems, in addition to being intellectually exciting, may indeed hold the key to the quality of life, if not the very survival of life, on this planet.

9.4 References

Ewing, R. C., and Navrotsky, A., eds. (1992). Earth materials [Special issue]. *MRS Bull. 17*, no. 5.
The following materials journals frequently contain articles concerning complex oxides and silicates, thus are bridging to mineral physics:
Journal of Solid State Chemistry
Journal of Materials Research
Journal of the American Ceramic Society
Materials Research Society Bulletin
Ceramic Bulletin
Chemistry of Materials
Progress in Solid State Chemistry
Advances in Materials Science
Proceedings of Materials Research Society Meetings
Physics and Chemistry of Glasses
Journal of Non-crystalline Solids
Zeolites

Index

acid–base reactions, 261–2, 299, 383, 304
Al_2SiO_5, 58–9, 105, 326
amorphous materials, 5, 107–8, 126, 131, 409–10
  enthalpy of mixing, 383–8
  fragility, 391–4
  glass transition, 391–4
  at high pressure, 396–400
  non-oxide glasses, 400–1
  Q-nomenclature, 132–4
  radiation damage, 400, 402
  structure, 132–4, 253, 371–7
  thermodynamics, 377–91
  viscosity, 391, 392, 394
  volatiles, 394–6
  volumes, 397
amphibole, 66–9, 80, 82–3, 102, 111, 119, 140
Avrami equation 340

Boltzmann factor, 274–5
bonding, 4, 172–266
  bond length–bond strength correlations, 259–61
  electron density map, 100, 244–6
  ionic model, 228–51
  ligand field theory, 221–8
  optical basicity, 261–3
  silicates, 211–21
  tetrahedrites, 190–2
  types of, 174
bond lengths, 100
  in melts and glasses, 373
  in rare earth compounds, 227
  in silicates, 213–21, 399
  and solid solution formation, 341–5
  in transition metal compounds, 227
  variation with temperature and pressure, 263–6

Born-Haber cycle, 231–4
Bragg Williams model, 359–60

calcite, 47, 120, 144, 357–60
calorimetry, 148–56
carbonate, 47, 120, 144, 299, 357–60
cesium chloride, 38, 258, 321–5
clay, 70–2, 108
cluster variation method (CVM), 359–60
color, 117
corundum, 50
crystal chemistry, 4
  ABX_3 compounds, 46–52, 309–12
  AB_2X_4 compounds, 52–6, 304–9
  AX compounds, 34–9, 238, 255–9, 321–5, 343
  AX_2 compounds, 39–46
  chair silicates, 61–9
  close packing, 25–30, 34
  framework silicates, 73–6
  metals, 30–2
  nonmetallic elements, 32–3
  orthosilicates, 56–61
  pyrosilicates, 60–1
  sheet silicates, 69–73
  silicate classification, 56–7, 60
  sorting diagrams, 255–9
  systematics, 247–8, 254–66
crystallography (*see also* diffraction, crystal chemistry), 1, 4
  crystal system, 12–14
  d-spacing, 14–15, 98
  lattice, 7, 98
  Miller indices, 14, 98
  point groups, 10
  space groups, 13, 98
  superstructure, 16
  symmetry, 7–10
  unit cell, 10, 13, 15, 98, 101, 103–5, 177